Optoelectronics Applications Manual

Prepared by The Applications Engineering Staff of the
HEWLETT-PACKARD OPTOELECTRONICS DIVISION

Stan Gage, *Applications Engineering Manager*
Dave Evans, *Applications Engineer*

Mark Hodapp, *Applications Engineer*
Hans Sorensen, *Applications Engineer*

McGraw-Hill Book Company

New York   St. Louis   San Francisco   Auckland   Bogotá
Düsseldorf   Johannesburg   London   Madrid   Mexico
Montreal   New Delhi   Panama   Paris   São Paulo
Singapore   Sydney   Tokyo   Toronto

Library of Congress Cataloging in Publication Data
Hewlett-Packard Company.   Optoelectronics Division.
  Applications Engineering Staff.
  Optoelectronics applications manual.
  Includes index.
  1. Light emitting diodes.   I. Gage, Stan.
II. Title.
TK7871.89.L53H48 1977   621.3815'42   77-5529
ISBN 0-07-028605-1

34567890 HDHD 78654321098

Hewlett-Packard assumes no responsibility for the use of any circuits described herein and makes no representations or warranties, express or implied, that such circuits are free from patent infringement.

TABLE OF CONTENTS

TABLE OF CONTENTS (Continued)

TABLE OF CONTENTS (Continued)

PREFACE

Over the last decade the commercial availability of the Light Emitting Diode has provided electronic system designers with a revolutionary component for application in the areas of information display and photocouplers.

As the use of the LED and associated derivative products has become more common, many electronic engineers have encountered the need for a resource of information about the application of and designing with LED products. This book is intended to serve as an engineering guide to the use of a wide range of solid state optoelectronic products. Wherever appropriate, throughout the book, use has been made of approximations to simplify the formulas and design equations. As such, some formulas will differ substantially in form from those found in a physics book or the subject; however, the end result will be an excellent approximation of the behavior of real devices.

To treat the various major aspects of LED products the book is divided into chapters covering each of the generalized LED product types. Additional chapters treat such peripheral information as contrast enhancement techniques, photometry and radiometry, LED reliability, mechanical considerations of LED devices, photodiodes and LED theory.

The authors are grateful to Bill Jensen and Mike Abbey for their preparation of artwork and Barbara Lee for her accuracy and patience in typing of the manuscript. Thanks is also due to Al Petrucello and George Liu for their effort in breadboarding and performance checking of the circuits detailed in the book.

Section I
LED Theory

1.0   LED THEORY

Electroluminescence in solids is a phenomenon which has been well known and intensively studied for many years. Perhaps the most commonly utilized application of electroluminescence is in the screen of a television set. Better known as cathodluminescence, this form of electroluminescence is caused by the collision of high energy electrons with the phosphor coating which lines the inside surface of a cathode ray tube. Though numerous other less common types of electroluminescence have been observed, one particular type has given rise to an entirely new field of technology. This is the phenomenon of p-n junction injection electroluminescence. The emission of light (photons) from a p-n junction was first noted in naturally occurring junctions by Lossew in 1923. More recently, (circa 1962) studies of GaAs revealed that it was quite feasible to achieve relatively high levels of electroluminescent emission from p-n junctions. Starting in 1962, an intensive development program was undertaken at Hewlett-Packard and elsewhere to produce a useful, manufacturable, visible-light emitting p-n junction device. An electroluminescent device of this type was thought to be attractive because of:

1)   The low currents and voltages required to produce useful light output.

2)   The precision with which the light emitting area could be defined through the use of semiconductor photo lithographic processes.

3)   The high speed at which the device could be switched.

1.1   The Theory of P-N Junction Electroluminescence

As a background to the theory of electroluminescent diodes, it is useful to review the basic concepts of current flow in p-n junctions.

1.1.1   Semiconductor Energy Gap

In semiconductor materials, as in all crystalline solids, electrons can assume only certain levels of energy. The Valence band and the Conduction band are the terms assigned to the two bands having the highest energy levels for electrons in normal semiconductor materials. The separation between the top of the valence band and the bottom of the conduction band is called the energy gap. In a pure semiconductor material, electrons cannot exist in (i.e., assume energy levels within the range of) this forbidden gap.

1.1.2   Semiconductor Doping

It is possible, however, to introduce impurities into the semiconductor which will produce electronic states in the forbidden gap. Impurities which add electrons to the conduction band are called donors and give raise to n-type conductivity. Impurities which produce electron vacancies, holes, in the valence band are called acceptors and give raise to p-type conducitivity. The energy bands and the relative donor and acceptor levels are shown in Figure 1.1.3-1a.

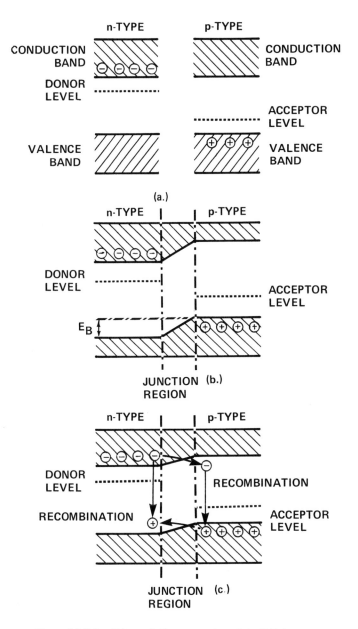

Figure 1.1.3-1   Schematic Representation of the P-N Junction

1.1.3   The P-N Junction

A p-n junction can be formed in a semiconductor material by doping one region with donor atoms and an adjacent region with acceptors. In this situation, electrons and holes will flow in opposite directions across the junction (without any applied bias) until an equilibrium is reached. This will give rise to a built-in potentional barrier, E_B, which is slightly less than the energy gap. Figure 1.1.3-1a and -1b represent schematically the material before and after this equilibrium is reached. Note that the natural potential difference across the junction makes it difficult for an electron, for instance, to move from the n to the p region. If, however, an external electric bias is applied across the junction in a manner such as to counteract the built-in potential, additional electrons and holes will flow (be injected) across the junction boundary. See Figure 1.1.3-1c. These carriers will then recombine via one of the mechanisms described in Section 1.1.4. This recombination, may give rise to photon emission as well as current flow. Within a few diffusion lengths of the junction, virtually all of the injected carriers will have recombined and the bulk current flow will occur by means of majority carrier diffusion.

1.1.4   Recombination

The injected electrons or holes are annihilated by a carrier of opposite type through one of several possible recombination processes. These recombination processes are either "radiative" or "non-radiative". The energy release in a radiative process is in the form of photons (light); whereas, the energy release in a non-radiative process is in the form of phonons (heat). The two important radiative processes in LEDs are shown schematically in Figure 1.1.4-1. Band-to-band recombination, 1, is the direct recombination of an electron near the bottom of the conduction band with a hole near the top of the valence band. The photon energy is then approximately equal to the band-gap energy of the crystal. This mechanism of radiative recombination is predominant in direct band-gap materials such as GaAs.

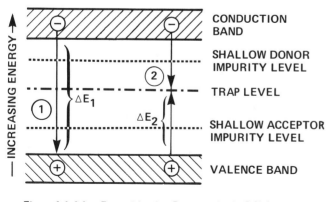

Figure 1.1.4-1   Recombination Processes in the P-N Junction

The second type, 2, of radiative recombination occurs by the formation and annihilation of a bound exciton at an "iso-electronic" center. Iso-electronic centers (associated with specific impurities in the crystal) are normally neutral, but introduce a local potential which is attractive to electrons. In p-type material, an injected electron is first trapped at the center. The negatively charged center then captures a hole from the valence band to form the bound exciton. The subsequent annihilation of this hole-electron pair yields a photon with an energy equal to the band gap minus an energy approximately equal to the binding energy of the center. This mechanism of radiative recombination is predominant in indirect band-gap materials such as GaP. The photon energy can be converted to wavelength by equation 1.1.4-1:

$$\lambda = \frac{1240}{\Delta E} \text{ (nm)} \qquad (1.1.4\text{-}1)$$

where ΔE is the energy transition in electron volts.

1.1.5   Materials Available for LED Devices

Equation 1.1.4-1 relates the expected wavelength of photon emission to energy differential experienced by an electron undergoing recombination. The maximum possible energy of the emitted photons is determined by the band-gap energy of the solid in which the p-n junction is formed. There are numerous elements and elemental compounds which have band-gap energies that lie in the region which could produce emission ranging from ultraviolet to infrared. However, very few of these materials are viable candidates for practical LED devices. Table 1.1.5-1 lists a few of the available materials and the associated band gap and emission wavelength. Some of these materials are not useful because they cannot be doped to form a p-n junction, some do not have emission at a useful wavelength, some have too low a conversion efficiency to be useful. At the present time, the only commercially available LED devices are manufactured using GaAs, GaP or the ternary compound Ga(As,P).

MATERIAL	BAND GAP ENERGY	EMISSION λ nm	TRANSITION TYPE
	eV		
Ge	0.66	1880	INDIRECT
Si	1.09	1140	INDIRECT
GaAs	1.43	910	DIRECT
GaP	2.24	560	INDIRECT
GaAs$_{60}$P$_{40}$	1.91	650	DIRECT
Al Sb	1.60	775	INDIRECT
In Sb	0.18	6900	DIRECT
Si C	2.2-3.0	563-413	INDIRECT

Table 1.1.5-1 Some of the Materials Available for LED Devices

1.1.6 Direct and Indirect Band-Gap Materials

In Section 1.1.4, the concepts of recombination were presented. In this section, the concepts of direct and indirect band-gap transitions are considered. The conduction band edge and the valence band edge as a function of momentum for allowed electronic states in the $GaAs_{1-x}P_x$ compound system (where x = mole fraction) are plotted in Figure 1.1.6-1 for different values of x. As indicated, there are two wells or minima; one designated as direct and the other as indirect. Electrons in the conduction band will generally occupy states in the lowest energy minimum, while holes will occupy states near the valence band maximum. Electrons in the direct minimum and holes at the top of the valence band have equal momentum;

whereas, electrons in the indirect minimum have different momementum. Since momentum is conserved, band-to-band transitions may occur with high probability for electrons in the direct minimum. The probability of a band-to-band transition for an electron in an indirect minimum is nearly zero, since a third component (phonon) must participate in the process in order to conserve momentum. Note that GaAs and $GaAs_{1-x}P_x$ up to $x \cong .4$ are primarily direct gap materials in contrast to $GaAs_{1-x}P_x$ with $x > .4$ and GaP which are primarily indirect gap materials. (Also note that the band gap energy increases with increasing x.) Thus, without the incorporation of special recombination centers, indirect gap materials, like GaP, are very inefficient light emitters, since the dominant recombination processes are non-radiative.

1.1.7 Enhanced Photon Emission in Indirect Gap Materials

More recent developments in LED technology have led to significantly enhanced radiative recombination in indirect gap materials, such as GaP. These have been achieved by the incorporation of appropriate impurities to form iso-electronic trapping centers, as discussed in Section 1.1.4. Because the trapped electron is highly localized at the center, its momentum is diffused. Thus, momentum can be conserved and the probability of direct recombination is greatly enhanced.

In GaP, two types of iso-electronic centers are utilized. One is formed by replacing a Phosphorous atom with an equivalent Nitrogen atom in the lattice. Another type is formed by replacing an adjacent Gallium and Phosphorous pair of atoms with a Zinc-Oxygen pair having the same total valence electrons. The resulting trap states will lie in the band gap at energies somewhat below the edge of the conduction band. This structure is illustrated schematically in Figure 1.1.7-1.

For N doped GaP, bound exciton recombination gives rise to emission at 565 nm (green). For Zn-O doping, the corresponding emission wavelength has a peak at about 700 nm (red). Nitrogen doping can be utilized in indirect gap $GaAs_{1-x}P_x$ to similarly produce yellow and red emission with relatively high quantum efficiencies.

1.2 Quantum Efficiency of LED Devices

The efficiency with which an electroluminescent material can convert current flow into detectable photon emission is of paramount importance in determining whether usable devices can be manufactured from that material. The percentage of the current flow which results in recombinations which give rise to photons of the desired

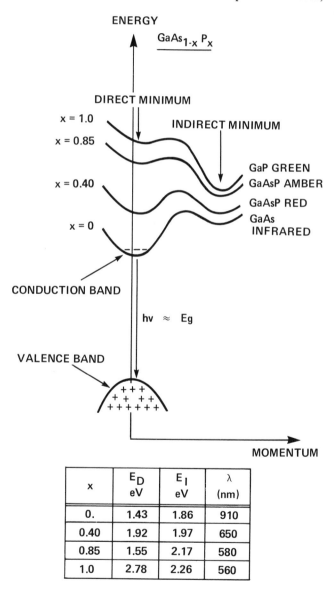

x	E_D eV	E_I eV	λ (nm)
0.	1.43	1.86	910
0.40	1.92	1.97	650
0.85	1.55	2.17	580
1.0	2.78	2.26	560

Figure 1.1.6-1 Plot of Momentum vs. Bandgap Energy for Various Compounds of the GaAs/GaP System

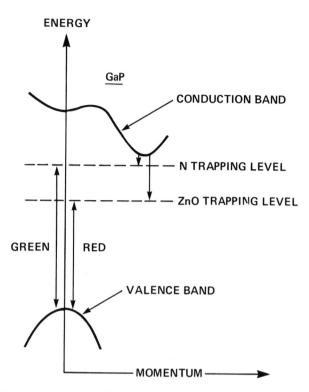

Figure 1.1.7-1 **Plot of Momentum vs. Bandgap Energy for Indirect GaP Materials Showing Special Trapping Levels**

wavelength is a measure of the internal conversion efficiency, η_{int} of the diode. Obviously, a material which has a very low η_{int} would be of little interest as a practical electroluminescent device. However, even a material which has an η_{int} of 100% may not be useful if the emitted photons cannot be efficiently coupled from the devices to a detector. Two major factors control the internal to external coupling coefficient. One factor is direct reabsorbtion of the emitted photon in the bulk material, basically a measure of the opacity of the material. The other factor is internal reflection at the crystal/air interface which causes the photon to be reflected back into the crystal and subsequently reabsorbed.

No matter how efficient a device is in terms of η_{ext} and η_{int}, the output power cannot be detected unless the wavelength is matched to available detectors. In the vast majority of LED applications, the detector of interest is either the human eye or a silicon photodetector. The nominal range of spectral sensitivity of the silicon detector is from about 300 nm to 1100 nm. The human eye has a much narrower range of sensitivity with useful responsivity only in the range from 400 nm to 700 nm. It is generally desirable to optimize the total coupling efficiency between the emitting device input signal and detector output response. For the case of matching a standard, non-Nitrogen doped $GaAs_{1-x}P_x$ emitter to the human eye, this is accomplished by picking the wavelength at which the product of the relative response of the eye and the relative

efficiency of the diode is largest. This maximum occurs at 655 nm and x = .4. This relationship is illustrated in Figure 1.2-1. The plots of $\overline{y}_{REL}$ and η_{extREL} are normalized to the peak values. $\eta_{LUMINOUS}$ represents the product of the two values.

For diodes having shorter peak emitting wavelength, such as N doped Ga(As,P) emitters at 565 nm (green) and 585 nm (yellow), η_{ext} is substantially lower but $\overline{y}_{REL}$ is higher. This has resulted in a spectrum of red through green emitters which all produce about the same $\eta_{LUMINOUS}$ (within an order of magnitude). The relative emission bands and responsivities of several different emitters and detectors are depicted in Figure 1.2-2. When trying to achieve matching with a silicon detector, factors other than achieving maximum responsivity may come into play. These factors are covered in Section 3.1.1.

1.3 Relative Efficiency

The mechanism of current flow in an LED as discussed in Section 1.1.3 represents the function of an ideal diode. Just as in silicon semiconductor technology, a p-n junction in Ga(As,P) material exhibits numerous discrepancies between the ideal and the actual diode. Surface recombination, tunneling phenomena, space charge recombination, current crowding and bulk recombination due to anomalous impurities tend to reduce the efficiency of photon emission of an LED. The relative effects of these non-radiative phenomena tend to be dependent on the current density (amps per square centimeter of junction area) and the perimeter to area ratio of the diode. At low forward currents (<1 A/cm^2), nearly all of the current flowing in a practical diode may result from one of these non-ideal mechanisms.

As the current density in the junction is increased to the value of 10 amps/cm^2, the non-ideal mechanisms will tend to have saturated, so that non-radiative current becomes a progressively smaller function of total current. As the junction current density becomes very large (>500 amps/cm^2) current crowding in the junction will tend to reduce emitting efficiency. The net effect of these mechanisms is to cause an LED to have a peak operating efficiency at a current which is dependent on the area and geometry of the junction and the size of the electrical contact. Figure 1.3-1 is a plot of normalized operating efficiency vs. current density for an .011" square diode. Note that at low current densities, a doubling of current may lead to a factor of 5 increase in luminous flux whereas at high current densities, doubling the current may result in slightly less than a doubling of light output.

The practical benefits of this aspect of LED devices are discussed in Sections 2 and 5.

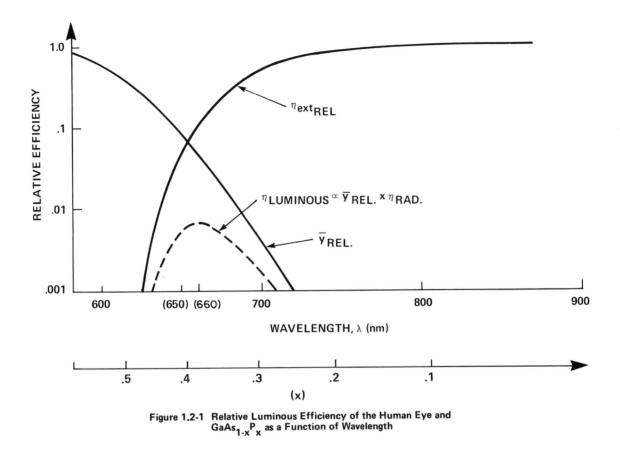

Figure 1.2-1 Relative Luminous Efficiency of the Human Eye and
$GaAs_{1-x}P_x$ as a Function of Wavelength

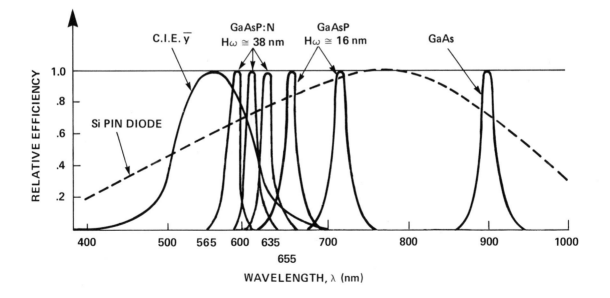

Figure 1.2-2 Normalized Responsivities of Different Emitters and
Detectors

1.5

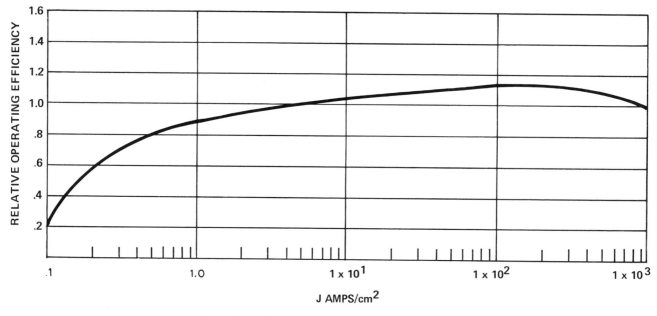

Figure 1.3-1 Normalized Operating Efficiency vs. Current Density for an LED

1.4 Material Processing

The processes used to manufacture LED devices are basically an outgrowth of the techniques perfected for the manufacture of silicon semiconductor devices. Crystal growing, epitaxial deposition, controlled impurity diffusions, photolithography, and vapor deposition of thin films all play an important role in the production of LEDs.

1.4.1 LED Structure

Since Ga(As,P) is a non-congruently melting material, a single crystal having proper GaP concentration is extremely difficult, if not impossible, to produce. The Ga(As,P) is, therefore, produced as an epitaxial layer grown on a substrate of either GaAs or GaP. The substrates are wafers sliced from ingots grown by the Czochralski method. These wafers are mechanically and chemically polished to provide a crystal substrate which is nearly free of lattice imperfections and unwanted chemical impurities. During the first stage of the epitaxial growth, the crystal layer has the same composition (GaP or GaAs) as the substrate. As the growth of the layer proceeds, the composition is changed gradually so as to maintain monocrystallinity in the epi layer. During the final mil (25 μm) of growth, the P/As ratio is held at the desired level so that later when the p-type material is diffused into the epitaxial layer, the resulting p-n junction will exist in homogeneous GaAs$_{1-x}$P$_x$ of the desired composition (typically x = .4 for direct-gap 655 nm RED LEDs). The n-type doping and any special impurities required to generate the bound exciton states are also introduced in the gas stream of the vapor-phase epitaxy system used to grow the layer.

Following growth of the epi layer, the wafer is coated with silicon nitride which acts as a barrier to the p-type diffusant and passivates the resulting p-n junction. To produce shaped p-n junctions, photolithography is used as in silicon wafer processing to shape the openings in the silicon nitride. Through the openings in the silicon nitride, p-type diffusant (generally Zinc enters the Ga(As,P) to form the p-n junction. It is not possible to use a layer of silicon dioxide because the p-type diffusant used in LED processing penetrates silicon dioxide too rapidly. Figure 1.4.1-1 depicts the crossectional and top view of a typical LED device.

1.4.2 Transparent vs. Opaque Substrate

The photons generated at the junction of a p-n electroluminescent diode are emitted in all directions. If the diode substrate is opaque, as in the case with GaAs, only those photons which are emitted upward within a critical angle defined by Snell's Law will be emitted as useful light. All other photons emitted into or reflected into the bulk crystal will be absorbed. This phenomena is illustrated in Figure 1.4.2-1a. GaP is nearly transparent by comparison with GaAs. Diodes formed in an epitaxial layer grown on a GaP substrate will exhibit improved efficiency due to the emission of photons which would be absorbed in the GaAs substrate structure. The resulting structure is depicted in Figure 1.4.2-1b. In actual manufacturing practice, indirect gap devices are generally fabricated on GaP substrates while direct-gap devices are fabricated on GaAs substrates.

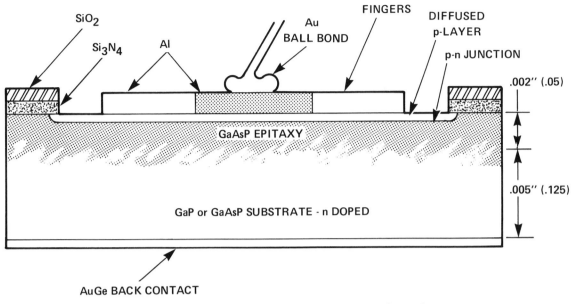

Figure A. CROSS SECTION OF AN LED

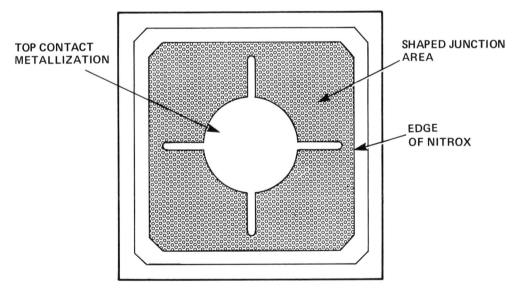

Figure B. PLAN VIEW OF AN LED

Figure 1.4.1-1   Crossectional and Top View of a Typical LED

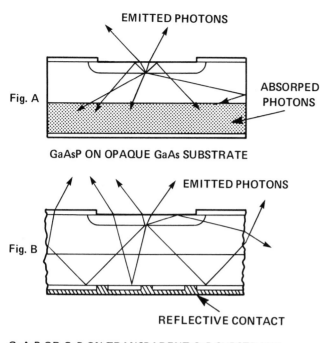

EMITTED PHOTONS

Fig. A

ABSORBED PHOTONS

GaAsP ON OPAQUE GaAs SUBSTRATE

EMITTED PHOTONS

Fig. B

REFLECTIVE CONTACT

GaAsP OR GaP ON TRANSPARENT GaP SUBSTRATE

Figure 1.4.2-1   Effects of Transparent and Opaque Substrates on Photons Emitted at the Junction

1.5   The Effect of Temperature Variation on LED Parameters

The physical parameters of light emitting diodes, as in all semiconductor devices, exhibit a dependence on absolute temperature. The forward voltage/forward current relationship, quantum efficiency, and emitted wavelength are the temperature variant parameters of greatest interest to the LED user.

1.5.1   Forward Voltage as a Function of Temperature

The forward voltage/forward current relationship for an LED can be expressed by equation 1.5.1-1

$$I_F = I_o \exp \left(\frac{q \, V_F}{n \, kT} \right) \qquad (1.5.1\text{-}1)$$

where n is a function of temperature, I_F, and the nature of the recombination mechanism. Empirical results for both direct and indirect gap LEDs exhibit temperature coefficients of -1.3 mV/°C to -2.3 mV/°C depending on forward current. Figure 1.5.1-1 depicts this relation.

1.5.2   Change in Peak Wavelength as a Function of Temperature

The effective energy gap in both direct and indirect gap semiconductors tends to become slightly smaller with increasing temperature. This will result in slight increases in

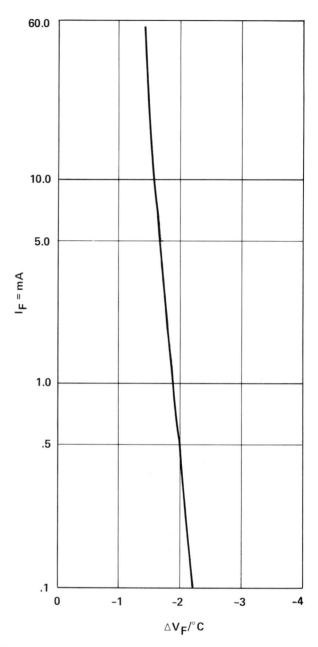

Figure 1.5.1-1   Temperature Coefficient of Forward Voltage as a Function of Forward Current

the emitted wavelength. For direct gap emitters wavelength will increase by 0.2 nm/°C. The wavelength of Nitrogen doped indirect gap emitters shows a somewhat lower dependency with typical positive variations of about 0.09 nm/°C.

1.5.3   Change in Output Power vs. Temperature

The radiant power of an LED decreases as a function of increasing temperature. Variations on the order of -1% per °C are typical for both direct and indirect gap materials. For LED devices in applications using the eye as a detector, the variation in responsivity of the eye as a function of wavelength must be added to the temperature

1.8

dependent variation of the LED radiant emission. In the 650 nm Red region, the eye response is changing at about -4.3%/nm. In the 565 nm Green region, the responsivity is changing by about -.86%/nm. If the change in wavelength as a function of temperature is .2 nm/°C for a 655 nm direct gap device, then apparent optical intensity will be decreasing by about:

$$(1.5.3\text{-}1)$$

$$\frac{\Delta I_V}{\Delta T}\left(\frac{\%}{°C}\right) = \left(\frac{-4.3\%}{nm}\right)\left(\frac{.2\ nm}{°C}\right) - \frac{1\%}{°C} = -1.86\ \%/°C$$

Conversely, an indirect gap device in the 565 nm green region will exhibit:

$$(1.5.3\text{-}2)$$

$$\frac{\Delta I_V}{\Delta T}\left(\frac{\%}{°C}\right) = \left(\frac{-.86\%}{nm}\right)\left(\frac{.09\ nm}{°C}\right) - \frac{1\%}{°C} = -1.08\ \%/°C$$

The change in luminous intensity of an LED exhibits a logarithmic relationship over large changes in temperature. Calculation of expected changes should be done using the following relationship:

$$I_{V\ TEMP\ 1} = I_{V\ TEMP\ 0}\ exp\ k\ \Delta T \qquad (1.5.3\text{-}3)$$

where $I_{V\ TEMP\ 1}$ = luminous intensity at temperature 1

$I_{V\ TEMP\ 0}$ = luminous intensity at reference temperature

$\Delta T = T_0 - T_1$

$k = \ln(1 - \text{Temperature Coefficient})$

NOTES

NOTES

Section 2

Lamps

2.0 LED LAMPS

2.1 Physical Properties of an LED Lamp Device

Incandescent, flourescent and neon lamps have been used for a long period of time in a wide variety of applications. As a result, their physical properties are well understood.

In recent years, the semiconductor LED lamp has been replacing these earlier devices in many applications. Also, new applications, designed specifically for the LED lamp, are being developed almost every day. Therefore, in order to effectively utilize an LED lamp in an application, a designer should be familiar with the physical properties of an LED device.

2.1.1 Plastic Encapsulated LED Lamp

Most commercial LED lamps are manufactured by encapsulating an LED chip inside a plastic package with a lens surface directly above the LED junction. If the plastic is undiffused, this configuration forms an immersion lens. The effect of this immersion lens construction is to enlarge the apparent size of the emitter. The magnification is a direct function of the index of refraction of the encapsulating material. The immersion construction increases the light output from the LED chip by reducing fresnel loss and increasing the critical angle.

Not every photon generated within the LED's p-n junction emerges from the surface of the junction to reach the eye of an observer. Three separate loss mechanisms contribute to reduce the quantity of emitted photons: 1) loss due to absorbtion within the LED chip material, 2) fresnel loss and 3) critical angle loss.

GaAsP/GaAs LED devices are opaque to light and absorb approximately 85% of the photons emitted at the junction, allowing a material efficiency factor of $\eta \approx .15$. A significant improvement is observed with GaAsP/GaP devices where $\eta \approx .76$.

2.1.2 Fresnel Loss

When light passes from a medium whose index of refration is n_1 to a medium whose index of refraction is n_2, a portion of the light is reflected back at the medium interface. This loss of light is called fresnel loss. The reflection coefficient is:

$$R = \left(\frac{n_2 - n_1}{n_2 + n_1}\right)^2 \qquad (2.1.2\text{-}1)$$

Since the numerator $(n_2 - n_1)$ is squared, the same reflection loss occurs whether the light is passing from a low-index to a high-index medium or from a high-index to a low-index

medium. The transmission coefficient across the medium interface is:

$$T = 1 + R = 1 - \left(\frac{n_2 - n_1}{n_2 + n_1}\right)^2 \qquad (2.1.2\text{-}2)$$

$$T = \frac{4 n_2 n_1}{n_2^2 + 2 n_2 n_1 + n_1^2}$$

The Fresnel Loss Efficiency Factor, η_{Fr}, is obtained by dividing numerator and denominator of equation 2.1.2-2 by $n_2 n_1$:

$$\eta_{Fr} = \frac{4}{n_2^2 + n_2/n_1 + n_1/n_2} \qquad (2.1.2\text{-}3)$$

For an unencapsulated GaAsP LED chip with an index of refraction of 3.4 emitting directly into air with index of refraction of 1.0, the fresnel loss efficiency factor is 0.702:

$$\eta_{Fr} = \frac{4}{2 + 1/3.4 + 3.4/1} = .702$$

Thus, only 70.2% of the light reaching the chip surface is transmitted across the chip/air interface.

If the LED chip is coated with an intermediate material having a suitable index of refraction, η_{Fr} can be improved. Ideally, this material should have an index of refraction $n_x = \sqrt{n_1 n_2} = \sqrt{(3.4)(1)} = 1.84$. Then $\eta_{Fr} = T_1 T_2$, where:

$$T_1 = \frac{4}{2 + n_1/n_x + n_x/n_1}$$

$$T_2 = \frac{4}{2 + n_2/n_x + n_x/n_2}$$

With an ideal coating having $n_x = 1.84$,

$$T_1 = T_2 = \frac{4}{2 + \sqrt{\frac{n_1}{n_2}} + \sqrt{\frac{n_2}{n_1}}} = .912$$

and the overall transmission is: $\eta_{Fr} = T_1 T_2 = (.912)^2 = .832$. This is an improvement of 18.5%.

Substantial improvement is obtained when a LED is encapsulated in plastic having an index of refraction of 1.5. The fresnel efficiency factor is calculated to be 0.816:

$$\eta_{Fr} = T_1 T_2$$
$$= \left[\frac{4}{2 + 3.4/1.5 + 1.5/3.4}\right] \left[\frac{4}{2 + 1/1.5 + 1.5/1}\right]$$
$$= [.850][.960] = .816$$

This is an improvement of 16.2% over that of an unencapsulated LED and just 2.3% less than ideal.

2.1.3 Critical Angle Loss

The third efficiency loss is due to total internal reflection of photons incident to the chip surface at angles greater than the critical angle. This effect is shown diagramatically in Figure 2.1.3-1. As is depicted in Figure 2.1.3-1a, a ray of light passing from the interior of the crystal to the outer surface is refracted according to Snell's Law:

$$n_1 \sin \theta_1 = n_2 \sin \theta_2 \qquad (2.1.3-1)$$

where: θ_1 = The angle of incidence inside, at the surface of the crystal.

θ_2 = The angle of refraction outside, at the surface of the crystal.

n_1 = The index of refraction of the crystal.

n_2 = The index of refraction of the medium outside of the crystal.

The angle of incidence, θ_1, at which the angle of refraction θ_2 is equal to 90°, is called the critical angle, θ_c. The critical angle is calculated for an LED chip as follows:

$$n_1 \sin \theta_c = n_2 \sin 90° \qquad (2.1.3-2)$$
$$\sin \theta_c = n_2/n_1$$
$$\theta_c = \sin^{-1} (n_2/n_1) = \sin^{-1} (1/3.4)$$
$$\theta_c = 17.1°$$

Light rays from within the crystal reaching the surface at an angle greater than 17.1° are totally reflected back into the crystal.

The Critical Angle Efficiency Factor for an LED emitting into air is 0.0865:

$$\eta_{Cr} = \left(\frac{n_2}{n_1}\right)^2 = \left(\frac{1}{3.4}\right)^2 = .0865$$

Encapsulating an LED in a medium with a high index of refraction, n_x, increases the amount of flux that can escape from the crystal. However, if a flat surface is used, this increase in flux is lost because the refraction angle, θ_x, at the crystal-to-encapsulant interface becomes the incidence angle at the encapsulant-to-air interface, see Figure 2.1.3-1b. The value of the critical angle is not improved.

By shaping the encapsulant into a dome lens, the incidence angle at the encapsulant-to-air interface is less than the refraction angle at the crystal to encapsulant interface. As shown in Figure 2.1.3-1c, flux that would have been trapped by a flat surface is permitted to pass through the dome lens. If no flux is trapped within the encapsulant, the critical angle efficiency for a plastic dome lens become 0.195:

$$\eta_{Cr} = \left(\frac{n_x}{n_1}\right)^2 = \left(\frac{1.5}{3.4}\right)^2 = .195$$

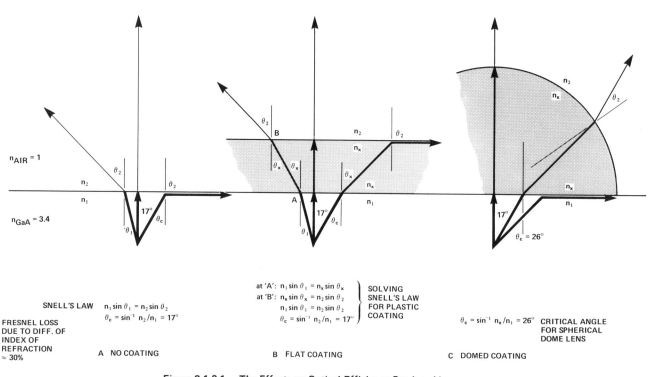

Figure 2.1.3-1   The Effects on Optical Efficiency Produced by an Optical Coating.

This improvement of 2.25:1 over an unencapsulated LED chip. The critical angle at the LED chip to dome interface is increased from $\theta_c = 17°$ to $\theta_c = 26°$.

$$\theta_c = \sin^{-1}(n_x/n_1) = \sin^{-1}(1.5/3.4) = 26.2°$$

2.1.4 Optical Efficiency

The optical efficiency of an LED lamp is the product of the absorbtion efficiency, fresnel efficiency and critical angle efficiency:

$$\eta_{optical} = \eta_A \cdot \eta_{Fr} \cdot \eta_{Cr} \qquad (2.1.4\text{-}1)$$

The optical efficiency for a T-1 3/4 clear high-efficiency red lamp is 0.121:

$$\eta_{optical} = (.76)(.816)(.195) = .121$$

2.1.5 External Quantum Efficiency

External quantum efficiency, η_{qext}, is defined by the following expression:

$$\eta_{q_{ext}} = \frac{\phi_e \,(W)}{I_F \,(A)\left[\dfrac{1240}{\lambda(nm)} \cdot \dfrac{eV}{photon}\right]} = \frac{photon}{electron} \qquad (2.1.5\text{-}1)$$

The value 1240 is the product of Plank's Constant and the speed of light:

$$\lambda = \frac{hc}{Eg}$$

$$= \frac{(6.626 \times 10^{-34} \,joules \cdot sec)(2.998 \times 10^{8} \,meters/sec)}{Eg\,(1.602 \times 10^{-19} \,joules/eV)}$$

$$= \frac{1.240 \times 10^{-6}\,m - eV}{Eg\,(eV)}$$

$$\lambda\,(nm) = \frac{1240}{Eg(eV)}$$

A T-1 3/4 high-efficiency red lamp with $2\theta_{\frac{1}{2}} = 35°$ and produces 12 mcd at 10 mA radiates 96.6 μW. The external quantum efficiency for this lamp is 4.95×10^{-3} photons/electron.

$$\eta_{q_{ext}} = \frac{96.6 \times 10^{-6}\,W}{(.010A)\left(\dfrac{1240}{635\,nm}\right)} = 4.95 \times 10^{-3} \frac{photons}{electron}$$

2.1.6 Internal Quantum Efficiency

The current flowing through an LED is composed of two components, a radiative component and a non radiative component. The internal quantum efficiency, $\theta_{q_{int}}$, is the ratio of the radiative component to the total current. The internal quantum efficiency may be calculated by dividing the external quantum efficiency by the optical efficiency:

$$\eta_{q_{int}} = \frac{\eta_{q_{ext}}}{\eta_{optical}} = photons/electron \qquad (2.1.6\text{-}1)$$

2.1.7 Calculating Radiated Flux

2.1.7.1 Luminous Efficacy and Power Per Unit Solid Angle

The ratio of luminous flux (lumens) to radiant flux (watts) is called luminous efficacy, η_v (lm/w). For a complete discussion of luminous efficacy, see the section on photometry. The value of luminous efficacy is given on each lamp data sheet. These values for luminous efficacy are:

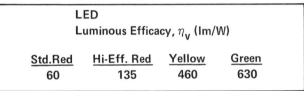

LED Luminous Efficacy, η_v (lm/W)			
Std.Red	Hi-Eff. Red	Yellow	Green
60	135	460	630

In some applications, such as providing a light source for a lens focusing system which subtends a specified solid angle, the amount of radiated power in microwatts per steradian, I_e (μW/sr), is important. The value of radiated power per unit solid angle may quickly be calculated from the following equation:

$$I_e \,(\mu W/sr) = \frac{I_v \,(mcd = m\,lm/sr)}{\left(\dfrac{\eta_v \,(lm/W)}{1000}\right)} \qquad (2.1.7.1\text{-}1)$$

For a T-1 3/4 clear high-efficiency red lamp with $2\theta_{\frac{1}{2}} = 35°$ and $I_v = 12$ mcd, the radiated power per sterdian is 89 μW/sr.

$$I_e = \frac{(12\,m\,lm/sr)(1000)}{135\,lm/W} = 89\,\mu W/sr$$

2.1.7.2 Calculating Total Power

In applications where the total radiated flux is to be utilized, the lamp radiation pattern must be known. The section on photometry provides a detailed discussion on calculating the total radiated flux using the method of zonal integration. In this section, an illustrative example of this method is presented which calculates the total flux for a T-1 3/4 undiffused high-efficiency red lamp with $2\theta_{\frac{1}{2}} = 35°$ and producing 12 mcd.

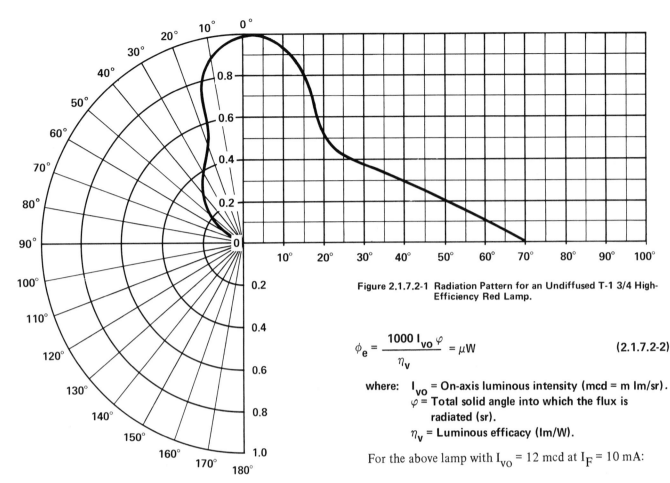

Figure 2.1.7.2-1 Radiation Pattern for an Undiffused T-1 3/4 High-Efficiency Red Lamp.

$$\phi_e = \frac{1000 \, I_{vo} \, \varphi}{\eta_v} = \mu W \qquad (2.1.7.2\text{-}2)$$

where: I_{vo} = On-axis luminous intensity (mcd = m lm/sr).
φ = Total solid angle into which the flux is radiated (sr).
η_v = Luminous efficacy (lm/W).

For the above lamp with I_{vo} = 12 mcd at I_F = 10 mA:

$$\phi_e = \frac{1000 \, (12) \, (1.0869)}{135} = 96.6 \, \mu W$$

Figure 2.1.7.2-1 is the polar/linear plot of the radiation pattern for the above lamp. The linear graph is used to determine the relative luminous intensity points for use in the calculation. Since the grid is marked off in 5° increments, we shall take our summation over 5° intervals.

$$\Delta = 5° = \frac{180}{N} \text{ , then N = 36}$$

$$\varphi\left(\frac{mA}{I_o}\right) = \frac{1}{2} I_r (M\Delta) \, C_Z (M\Delta) + \sum_{m=1}^{m-1} I_r (m\Delta) \, C_Z (m\Delta)$$

$$(2.1.7.2\text{-}1)$$

where: $C_Z (M\Delta) = \frac{2\pi^2}{N} \sin (M\Delta) = .5483 \sin (M\Delta)$

$I_r (M\Delta)$ = Relative Luminous Intensity from linear graph at the angle $m\Delta$.

The calculation is set up in a tabular format. The calculation needs to include angles only up to $m\Delta = 70°$, M = 14.

m	mΔ(°)	I_r (mΔ)	C_Z (mΔ)	I_r (mΔ) C_Z (mΔ)
1	5	.981	.0478	.0469
2	10	.938	.0952	.0893
3	15	.801	.1419	.1137
4	20	.516	.1875	.0968
5	25	.435	.2317	.1008
6	30	.379	.2742	.1039
7	35	.342	.3145	.1076
8	40	.298	.3524	.1050
9	45	.248	.3877	.0961
10	50	.199	.4200	.0836
11	55	.149	.4492	.0669
12	60	.099	.4749	.0470
M−1 ► 13	65	.056	.4969	$\sum_{m=1}^{m-1}$ .0278 / 1.0854
M ► 14	70	.006	.5152	.0031

The total radiated power is now calculated from the following equation:

$$\varphi\left(\frac{M\Delta}{I_o}\right) = \frac{1}{2} (.0031) + 1.0854 = 1.0869 \text{ sr}$$

2.4

2.1.8 Magnification and Luminous Intensity

The luminous intensity of an LED lamp is a function of the magnification provided by the immersion lens. Using the parameters of Figure 2.1.8-1 and assuming paraxial light rays, the magnification may then be defined as the ratio of the image size to the object size and approximated by the following formula:

$$\text{Magnification} = m \overset{\triangle}{=} \frac{y_2}{y_1} \approx \frac{1}{1 - \frac{x_1}{r}\left(1 - \frac{n_2}{n_1}\right)} \qquad (2.1.8\text{-}1)$$

For a T-1 3/4 lamp, the focal length is: $X_1 =$ 4.70 mm, r = 2.44 mm and $n_1 = 1.53$. The magnification is approximately 3:

$$m \approx \frac{1}{1 - \frac{4.70 \text{ mm}}{2.44 \text{ mm}}\left(1 - \frac{1}{1.53}\right)} = 3.01$$

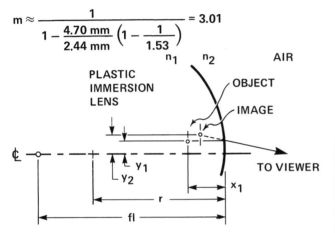

Figure 2.1.8-1   Parameters of a Spherical Immersion Lens.

The magnification may also be calculated from the focal length of the immersion lens. If the surface of the dome lens is regarded as the principal plane, a focal point may be defined as that point on the lens axis from which paraxial light rays emerge from the dome lens parallel to the lens axis. The principal plane is the locus of equivalent refracting points for a lens where extended entering and emerging rays intersect. The focal length of the lamp's spherical dome lens may be determined from the following formula:

$$fl = \frac{r}{1 - \left(\frac{n_2}{n_1}\right)} \qquad (2.1.8\text{-}2)$$

$$fl = \frac{2.44 \text{ mm}}{1 - \left(\frac{1}{1.53}\right)} = 7.044 \text{ mm}$$

The magnification may then be calculated from this focal length:

$$m = \frac{1}{1 - \left(\frac{X_1}{fl}\right)} \qquad (2.1.8\text{-}3)$$

The effect of magnification on luminous intensity and radiation pattern for an undiffused package is illustrated in Figure 2.1.8-2. If the LED is encapsulated very close to the dome surface, curve A, the dome surface is effectively flat. There is no magnification and the radiation pattern remains the lambertian pattern of the unencapsulated LED chip. In the lambertian radiation pattern, the luminous intensity varies as the cosine of the off-axis angle, θ.

$$I(\theta) = I_o \cos\theta \text{ (Lambertian radiator)} \qquad (2.1.8\text{-}4)$$

where: I_o is the on-axis luminous intensity

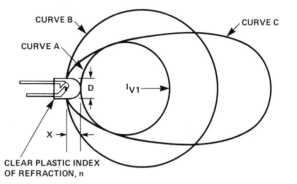

CURVE	DISTANCE X	ON-AXIS LUMINOUS INTENSITY, I_{VO}	LUMINOUS FLUX, ϕ_V	RADIATION PATTERN
A	≈0	$I_V: \approx I_{VO}$ of unencapsulated LED	πI_{V1}	Lambertian
B	D/2=r	$n^2 I_{V1}$	$\pi n^2 I_{V1}$	Lambertian
C	>D/2	$>n^2 I_{V1}$	$<n^2 \pi I_{V1}$	Non Lamberian

Figure 2.1.8-2   Effect of Magnification on Luminous Intensity and Radiation Pattern for an Undiffused Plastic Lamp.

The luminous flux is equal to π times the luminous intensity.

At distances from the dome surface of $o < x < r$, the LED chip is magnified and the luminous intensity is directly proportional to the square of the magnification. When x = r, the magnification is equal to the index of refraction of the encapsulant and the luminous intensity is then equal to that of the unencapsulated chip multiplied by the index of refraction squared. For a spherically shaped dome lens, the radiation pattern is still lambertian, curve B. The total luminous flux increased by the factor of index of refraction squared.

At distances greater than the radius of the dome lens further increases the magnification above that value equal to the index of refraction. The luminous intensity increases but the radiation pattern narrows reducing the viewing angle, curve C. The maximum practical magnification is that at which the image of the LED's emitting area equals the diameter of the dome lens.

An aspheric dome lens can be used to achieve a high value of on-axis luminous intensity along with a wider radiation pattern than is obtainable with a spherical dome lens. An aspheric dome lens is used as the package of T-1 3/4 low profile lamps. The comparison of the radiation pattern of an aspheric dome lens with a spherical dome lens of equal magnification is illustrated in Figure 2.1.8-3.

At this point, the reader is reminded that when evaluating the specifications for an LED lamp, it is necessary to take into account both the on-axis luminous intensity and the lamps radiation pattern. Even though two lamps may have the same luminous intensity at a specified forward current, one may have a wider radiation pattern. The lamp with the wider radiation pattern may have a more efficient dome lens or a more efficient LED chip.

2.1.9 Diffused and Undiffused LED Lamps

The immersion lens concept applies to a lamp which has the LED encapsulated in undiffused plastic. The result is a beam of light with a high value of luminous flux which is concentrated in a narrow radiation pattern. This lamp configuration is especially useful for backlighting applications and for applications requiring a concentrated light source.

A front panel indicator lamp requires a very wide off-axis viewing angle. To achieve this wide viewing angle, diffusant is added to the lamp to disperse the light rays emitting from the LED. The result is a lamp with a wide radiation pattern and a reduced value of on-axis luminous intensity. Figure 2.1.9-1 pictorially illustrates the differences between an undiffused and diffused lamp. Dye coloring is added to tint the diffused lamp to enhance on/off contrast.

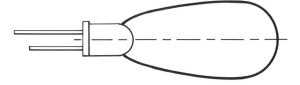

T-1 3/4 LAMP WITH SPHERICAL DOME LENS

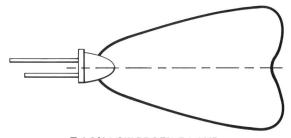

**T-1 3/4 LOW PROFILE LAMP
WITH AN ASPHERIC DOME LENS**

Figure 2.1.8-3   Radiation Patterns for Undiffused Lamps with Spherical and Aspheric Dome Lenses.

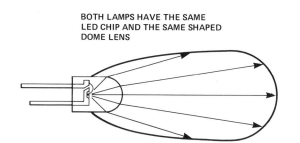

BOTH LAMPS HAVE THE SAME
LED CHIP AND THE SAME SHAPED
DOME LENS

UNDIFFUSED PLASTIC LAMP: HIGH VALUE OF ON-AXIS LUMINOUS INTENSITY
WITH A NARROW RADIATION PATTERN

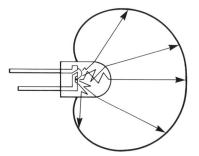

DIFFUSED PLASTIC LAMP: REDUCED VALUE OF ON-AXIS LUMINOUS
INTENSITY WITH A WIDE RADIATION PATTERN

Figure 2.1.9-1   Comparison Between Undiffused and Diffused Plastic LED Lamps.

2.2   LED Lamp Packaging

The function of a lamp package is to utilize the various physical properties described in Section 2.1 to effect the best coupling of the emitted light from an LED to an observer or electronic detector. The desired package configuration depends upon the specific requirements of each application. These requirements will determine such parameters as the size of the device, if the lamp is to be tinted or untinted and whether the encapsulating epoxy is to be diffused or undiffused. Some requirements may dictate a lamp device where epoxy encapsulating is not desired, such as in a high-reliability application.

It is obvious that there could well be as many lamp package configurations as there are applications. However, certain lamp packages have been developed by the optoelectronics industry which cover a wide range of the more common applications. Prior to discussing the commonly available package configurations, it is a benefit to the reader to first understand the basic lamp packaging process.

2.2.1   Lead Frame Packaging

Immensely popular because of its low cost, the lead frame technique is used for the packaging of LED lamp devices. Basically, in lead frame packaging, LED dice are die attached to one element of a metal frame and a wire bond is made from the top of each die to another element of the frame. During the die-attach and bonding operations, the elements are joined for mechanical support by metal straps called "dam bars". Plastic encapsulation applied around the devices surrounds also the elements of the frame while the dam bars remain outside of the plastic. When cured, the plastic provides mechanical support to the elements and the dam bars may then be sheared away. Shearing of the dam bars leaves the opposite ends of the lead frame protruding from the plastic; these then become the external leads of the finished device. Only after the dam bars are removed can the devices be directly energized for testing.

The material of the lead frame is thick enough that the dice can be attached on the edge (rather than the face) of the frame material. This places the die on the end of an element, called a "die-attach post", which is canted, allowing the die to be centered with respect to those portions of the lead frame which later become the external leads, as shown in Figure 2.2.1-1. The bonding wire is then connected from the top contact of the die to the end of the adjacent element, called a "bonding post". After the bonding wires are in place, the lead frame, bearing a number of devices is clamped into a fixture which controls precisely the distance by which the posts of the frame are lowered into cavities that have been previously filled to a

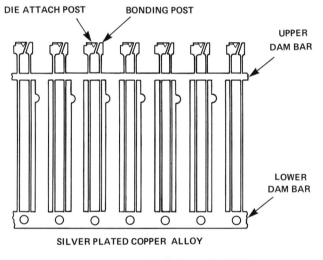

Figure 2.2.1-1   A Typical Lead Frame for LED Lamps.

precise level with the uncured liquid epoxy. The shape of these cavities, called "mold cups", determines the shape of the lenses of the lamp packages, and the distance to which the posts are inserted in the mold cups determines the magnification and hence the radiation pattern of the finished product. Dye is used to tint the epoxy to absorb ambient light and make the lamp appear darker when it is off; since the dye does not absorb appreciably the light from the LED, this raises the ON/OFF contrast ratio. In some devices, a diffusant material is added to the epoxy to cause light scattering, thereby increasing the effective viewing angle.

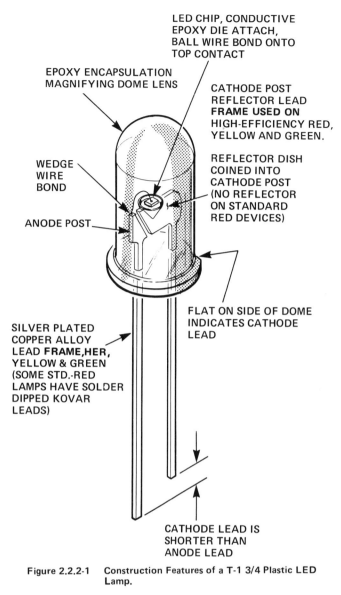

Figure 2.2.2-1   Construction Features of a T-1 3/4 Plastic LED Lamp.

Curing of the epoxy is done in two stages. A pre-cure hardens the plastic to a point which permits removal from the mold cups and shearing of the lead frame and dam bars. The post-cure is then applied, which further develops and stabilizes the mechanical properties of the epoxy, enabling the lamps to withstand the abuse they may later receive when being installed and connected into circuits. Except for testing, manufacture is now complete.

2.2.2   The Industry Standard T-1 3/4 & T-1 LED Lamps

When LED lamp devices were first introduced, two established miniature sized packages were borrowed from the incandescent lamp. These are the T-1 and T-1 3/4 sizes. In the T-X designation, the number indicates the lamp diameter in 1/8ths of an inch. These two sizes have now become standard in the optoelectronic industry.

The construction features of a T-1 3/4 lamp are illustrated in Figure 2.2.2-1. The lead frame for a GaP transparent substrate LED has a dish shaped reflector coined into the top of the cathode post. A GaP transparent substrate LED emits light from the sides of the chip as well as from the top surface. The reflector directs this side emitted light towards the dome lens in an analogous fashion to the common hand held flashlight. The result is a lamp with a front emitter equal to the larger area of the reflector dish. A standard red lamp does not require a reflector, since a GaAsP/GaAs LED emits light only from the top surface of the junction.

A variation of the standard T-1 3/4 lamp is the low profile T-1 3/4 lamp. The construction is identical to the standard lamp except for the height and shape of the dome lens. The dome is 2/3rds the height of the standard lamp and is an aspheric lens which provides high magnification coupled with a wide viewing angle. For a visual comparison, see Figure 2.1.8-3.

The construction of the T-1 lamp is essentially the same as for T-1 3/4 lamp. However, the reflector dish is too large to be successfully encapsulated inside this small lamp package. It is for this reason, that a more compact lead frame, without a reflector, is used in all of the T-1 lamp products.

Initially, the two standard lamp sizes were used only as indicators on front panels and printed circuit boards. However, as the variety of uses expanded into other areas, such as backlighting and various array configurations, the T-1 and T-1 3/4 packages could not be easily adapted to meet the shape and size requirements for all of these more sophisticated applications. As a result, new lamp packages have been developed to conform to many of these needs.

2.2.3 The Subminiature LED Lamp

Many applications require a very small, low cost LED lamp to be used in a location where space is at a premium. Such might be the case for the colon in a small desk clock, for an indicator on a small hand held instrument or for use in an array where a high packing density is required. To meet this need, the subminiature lamp package has been developed.

The subminiature device offers the customer a lamp that is low in cost, smaller in size than a T-1 lamp and has superior optical consistency device-to-device. The small size allows for a very high packing density in an array, see Section 2.4.4 on Arrays. The optical consistency of each lamp is derived from the transfer molding process used to encapsulate the device.

Figure 2.2.3-1 illustrates the construction features of the subminiature lamp package. The subminiature package uses a radial, flexible, rectangular lead frame suitable for

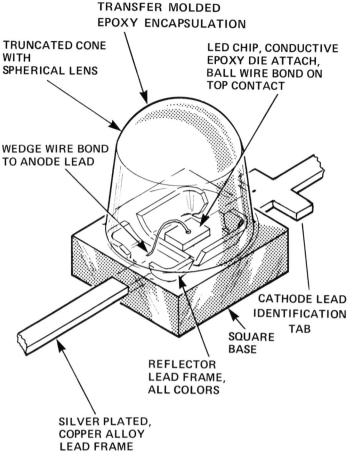

Figure 2.2.3-1   Construction Features of a Subminiature Plastic LED Lamp.

handling in the transfer molding process. From a user's point of view, this lead frame offers two important benefits. First, unlike the T-1 lead frame which does not have a reflector, a small reflector cup is formed by bending up the sides of the die attach and bonding pads. The optical effect is similar to the T-1 3/4 lamp, producing a small lamp with very high sterance. The second advantage is a choice of installation procedures; the leads may be left in the radial position for soldering to pads on the face of a printed circuit board, or the leads may be bent 90° into the axial position for insertion into plated through holes and subsequent flow soldering or for insertion into a socket.

The package dome is constructed as a truncated cone with a spherical lens to obtain high magnification for high intensity combined with a very wide viewing angle. The package base is square shaped to provide firm support for the lead frame, resulting in a high degree of mechanical reliability, and to aid in positive alignment in high density array applications.

2.2.4 The Rectangular LED Lamp

The rectangular package is designed for use in those specific applications where a cylindrically shaped device is not effective. For instance, the rectangular LED lamp is most effective for illuminating a legend by directly backlighting a transparent character so that it will stand out more vividly. As another example, a bar graph is easily implemented with rectangular LED lamps, as they may be either end stacked or side stacked to form a continuous bar of light.

The construction of the rectangular lamp is an extention of the T-1 3/4 package, as both use the same reflector lead frame. As shown in Figure 2.2.4-1, the difference is in the encapsulation. The encapsulating epoxy is formed into a rectangular light pipe with an integral diffusing layer at the top. The light pipe is the optical path through which the LED light travels to diffusing layer. The diffusing layer spreads the light to form an evenly lighted rectangular source.

2.2.5 The Hermetic LED Lamp

Although plastic devices are used in the majority of LED lamp applications, they become vulnerable when placed into adverse environments. The problem is the limit to which the encapsulating epoxy can withstand temperature extremes, moisture or other detrimental environmental conditions without losing its optical properties or exerting failure causing stresses on the LED die attach and wire bonds. Therefore, a hermetically sealed device is used for many military lamp applications and applications where the lamp may be exposed to an adverse industrial environment.

The hermetic lamp is assembled in a TO-18 package as illustrated in Figure 2.2.5-1. After die attach and wire bond, a coating of silicone jell is applied. At the top of the metal cap is an optical window with a hermetic seal at the glass to metal interface. An epoxy dome is cast on top of the optical window to increase viewing angle and to provide good on/off contrast. The metal cap is welded to the TO-18 header to complete the assembly. The exterior of the package is gold plated to resist corrosion.

2.2.6 LED Lamps that Include Other Components

The preceding sections have described the LED lamp package as a device containing only one component. Electrically, an LED lamp is a two terminal device and is capable of utilizing a limited number of other components, in addition to the LED chip, to perform certain functions. Such devices are termed "integrated LED Lamps".

A commonly used integrated lamp is the resistor LED lamp. In this device, the package contains an integral current limiting resistor, that is die attached to the anode post and wire bonded to the LED top contact. This integral resistor is a nominal 215Ω, allowing the lamp to be driven directly from 5.0 volt supply when controlled by a TTL gate.

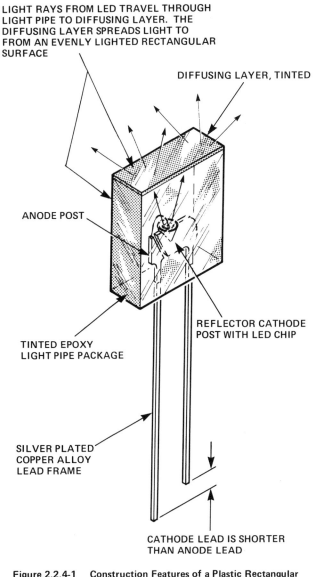

LIGHT RAYS FROM LED TRAVEL THROUGH LIGHT PIPE TO DIFFUSING LAYER. THE DIFFUSING LAYER SPREADS LIGHT TO FROM AN EVENLY LIGHTED RECTANGULAR SURFACE

DIFFUSING LAYER, TINTED

ANODE POST

TINTED EPOXY LIGHT PIPE PACKAGE

REFLECTOR CATHODE POST WITH LED CHIP

SILVER PLATED COPPER ALLOY LEAD FRAME

CATHODE LEAD IS SHORTER THAN ANODE LEAD

Figure 2.2.4-1   Construction Features of a Plastic Rectangular LED Lamp.

Another integrated lamp device is the voltage sensing LED lamp. In this device, an integrated circuit turns the lamp on or off depending upon the level of the applied voltage. A reference threshold level, V_{TH} is built into the integrated circuit. If the applied voltage is above V_{TH} the lamp is ON, and below V_{TH} the lamp is OFF. This device is commonly used as a battery test indicator.

Figure 2.2.6-1 shows schematic representations for these two integrated lamp devices.

2.3   LED Lamp Characterization Information

A lamp data sheet contains specific characterization information to aid the designer in selecting the correct lamp for his application, determining the maximum worst case operating limits and establishing nominal operating

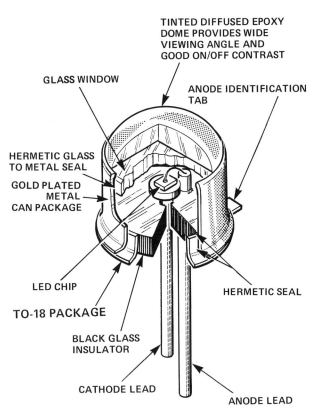

GLASS WINDOW

TINTED DIFFUSED EPOXY DOME PROVIDES WIDE VIEWING ANGLE AND GOOD ON/OFF CONTRAST

ANODE IDENTIFICATION TAB

HERMETIC GLASS TO METAL SEAL

GOLD PLATED METAL CAN PACKAGE

LED CHIP

HERMETIC SEAL

TO-18 PACKAGE

BLACK GLASS INSULATOR

CATHODE LEAD

ANODE LEAD

Figure 2.2.5-1   Construction Features of a Hermetic LED Lamp.

conditions. The data sheet contains this information in the following basic sections:

Product Description and Package Dimensions

Absolute Maximum Ratings

Electrical/Optical Characteristics

Operating Curves

Once the lamp package configuration has been selected, the designer is ready to consider those characteristics of most importance:

1.   Light output and color matching

2.   Maximum temperature derated operating limits

3.   Pulsed operating conditions

4.   Time average luminous intensity

Other considerations may also be of importance such as reverse breakdown voltage, capacitance or speed of response, depending upon the application.

For a designer to be able to effectively utilize the information contained in a data sheet, he needs an appreciation of what the numbers mean and on what basis they have been derived.

2.3.1   Light Output and Color Matching

Of initial concern to a designer is the selection of that lamp which is of the proper color and has the most light output at a specified current. The electrical/optical characteristics contain four parameters that quantitatively aid the designer in making this selection. These are peak wavelength, dominant wavelength, luminous intensity and the included angle between half intensity points.

The color of the lamp, as perceived by the eye, is dependent upon the radiated spectrum of the LED. Not only is the radiated spectrum of importance for determining color in visual applications, but it is very important in determining the coupling to a detector in non-visual applications. Peak wavelength and dominant wavelength are the two quantitative parameters which describe the radiated spectrum.

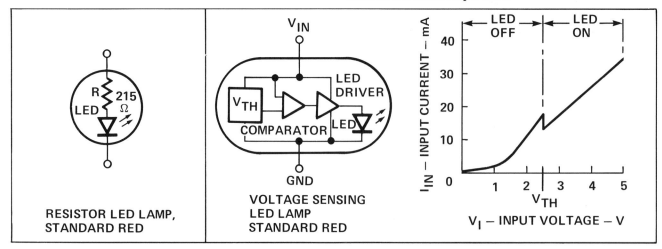

RESISTOR LED LAMP, STANDARD RED

VOLTAGE SENSING LED LAMP STANDARD RED

Figure 2.2.6-1   Schematic Representation of Two Common Integrated LED Lamps.

Peak wavelength, λ_p, is that wavelength at the peak of the radiated spectrum. It is maintained within narrow limits during the growth of the epitaxial layer, with each wafer being inspected by a photo-luminescence measurement on a production basis. To the designer of a non-visual application, λ_p becomes important when determining the coupling efficiency of the LED light to a photodetector, signal loss through a fiber optic conductor or photographic film sensitivity. In visual applications, the total amount of LED emitted light passing through an optical filter, used for contrast enhancement, is approximately equal to the filter's relative transmission at λ_p. Therefore, the relative transmission at λ_p is a quantitative measure of the optical density of a contrast filter.

The color of an LED device is especially important to the designer if lamps from various manufacturers might possibly be installed in the same array. Not only does the eye detect intensity differences, it also detects color differences. The dominant wavelength, λ_d, is a quantitative measure of the color of an LED device as perceived by the eye. Two devices of somewhat different radiated spectra will appear as the same color if they both have the same λ_d. The dominant wavelength is not necessarily dependent upon peak wavelength. Conceptually, λ_d may be envisioned as that wavelength near the centroid of the radiated spectrum. Specifically, λ_d is that wavelength when mixed with an equal amount of the light from a 6500°K lamp will be perceived by the eye as the same color as is produced by the radiated spectrum of the LED.

It has been common practice of many designers to use the data sheet luminous intensity as the only parameter necessary to evaluate light output. This is not the only important consideration in the determination of the total available luminous flux. A designer needs to consider both the axial luminous intensity and the radiation pattern in making a determination of total light output. It is the radiation pattern that has the most influence.

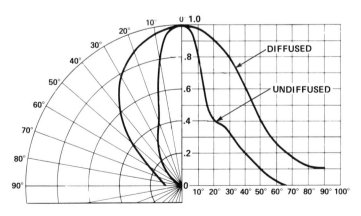

Figure 2.3.1-1   Radiation Pattern for a T-1 3/4 High-Efficiency Red LED Lamp.

Luminous Intensity, I_v, is a measure of axial light output. The unit of measure is the candela, which is luminous flux per unit solid angle. Thus, the measure of luminous intensity gives no information concerning the radiation pattern. The axial luminous intensity of each device is measured on a production basis, using a calibrated photometer, to insure that it meets a minimum value at a specified drive current.

The importance of the radiation pattern to a designer is that it defines the apparent luminous intensity of the device when viewed at some off-axis angle. The radiation pattern is quantitatively described at a single point, which is the included off-axis angle where the luminous intensity is equal to one half of the axial intensity. This included angle is referred to as $2\theta_{1/2}$. The value of $\theta_{1/2}$ is controlled by the distance of the LED from the focal point of the dome lens. Larger values of $2\theta_{1/2}$ result in a lamp which may be viewed at greater off axis angles. Also, a lamp which has a greater $2\theta_{1/2}$, but the same axial intensity will actually have a much greater total flux. A good example of this is the comparison of two undiffused T-1 3/4 LED lamps that have the same axial luminous intensity but differ in the value of $2\theta_{1/2}$. The first lamp has $2\theta_{1/2} = 35^\circ$ and the second lamp has $2\theta_{1/2} = 25^\circ$. Depending upon the exact shape of the two radiation patterns with respect to each other, the first lamp could produce up to three times the luminous flux that is produced by the second lamp. An examination of the data sheet curve of the radiation pattern reveals the intensity at any specific off-axis angle, as illustrated in Figure 2.3.1-1.

Relying only on a minimum or a typical intensity value for lamps used in an array does not guarantee intensity matching. It is best to install lamps that have been categorized for light output. Each light output category should have an $I_{v\ MAX}/I_{v\ MIN}$ ratio of 2:1 or less. A single array is then assembled with all lamps from the same category. To an observer, the lamps will appear evenly illuminated throughout the array.

In summary, peak wavelength, dominant wavelength, axial luminous intensity and radiation pattern are necessary to completely specify the optical characteristics of an LED device.

2.3.2   Maximum Temperature Derated Operating Limits

Once the optical characteristics of an LED device have been established, the next step is to determine the maximum temperature derated limits at which the device may be operated. Forward current, power dissipation, thermal resistance and LED junction temperature are all interrelated in establishing absolute maximum ratings.

The absolute maximum ratings of an LED device have been determined theoretically and by extensive reliability testing and are those limits beyond which reliable operation cannot be assured. These maximum ratings cannot be taken as limits in concert, as operation at one maximum limit may preclude the operation at another maximum limit. The best example of this exclusion rule is that a device cannot be reliably operated at maximum power dissipation in the maximum allowed ambient temperature. Operation at the maximum ambient temperature is only allowed with proper power derating.

Maximum ratings are based on package temperature limitations and the current density limit within the LED junction. The package temperature limitation may be based on the glass transition temperature, T_G, of the encapsulating epoxy. Above T_G, the cross linkages between the epoxy molecules change allowing the molecules to move with respect to each other. The epoxy changes from an ordered structure to an amorphous structure, analogous to that of glass, and the coefficient of thermal expansion drastically increases. The result is that thermal stresses above T_G may be sufficient to cause a catastrophic failure. The maximum current rating is based on (1) the level of current density within the LED junction that produces no more than an acceptable amount of light output degradation, and (2) the amount of power dissipation generated by the forward current that maintains the LED junction at an acceptable temperature level below T_G.

The limiting factor for operation in any ambient temperature is the LED junction temperature, T_J. For a hermetic device, T_J should be less than 125°C to realize an acceptable rate of light output degradation, and for a plastic device, T_J should be at least 10°C below the T_G of the encapsulating epoxy to prevent catastrophic failure. The encapsulating epoxies currently being used in the manufacture of LED lamps have a nominal $T_G = 120°C$.

The two parameters that effect T_J, which a designer has control of, are forward current and thermal resistance. The maximum forward current is derived from the maximum allowed temperature derated power dissipation and the thermal resistance is dependent upon the method used to install the lamp in the circuit.

The maximum allowed temperature derated power dissipation is obtained by derating the absolute maximum power rating at the rate of -1.6 mW/°C above an ambient of 50°C. This is a linear derating from maximum power at an ambient of 50°C to zero power at an ambient of 125°C, as shown in Figure 2.3.2-1.

Average power dissipation is the product of average forward current and peak forward voltage. An equivalent circuit for an LED, consisting of a dc voltage source in series with a dynamic resistance, may be derived from the diode's forward characteristics. Also, a lamp data sheet lists a value

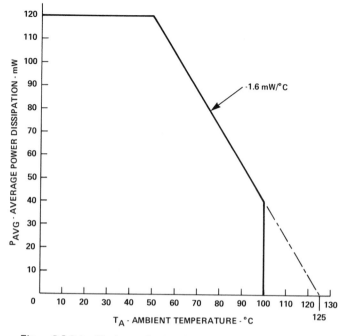

Figure 2.3.2-1   Maximum Average Power Derating for a Plastic LED Lamp.

of forward voltage at a specific current. From these two observations, the following equation may be derived to calculate the power dissipation in an LED lamp:

$$P_{AVG}(W) = I_{AVG}[V_F(DS) + R_s(I_{PEAK} - I_F(DS)] \quad (2.3.2\text{-}1)$$

where: I_{AVG} = Average forward current, amperes.

I_P = Peak forward current, amperes.

R_s = LED dynamic resistance, ohms.

$V_F(DS)$ = Data sheet forward voltage, volts.

$I_F(DS)$ = Data sheet forward current where $V_F(DS)$ is specified, amperes.

For standard red devices, $R_s = 1.6\Omega$ typical and 5Ω maximum; for GaP transparent substrate devices $R_s = 21\Omega$ typical and 35Ω maximum.

Once the maximum tolerable operating conditions have been established for pulsed operation, as described in Section 2.3.3, equation 2.3.2-1 may be used as a check to insure that the average power dissipation does not exceed the maximum derated limit. For dc operation, the dc power dissipation may be calculated by setting $I_{DC} = I_{AVG} = I_{PEAK}$.

LED junction temperature is the sum of the operating ambient temperature and the temperature rise above this ambient.

$$T_J (°C) = T_A + \Delta T_J \quad (2.3.2\text{-}2)$$
$$T_J (°C) = T_A + \theta_{JA} P_{AVG}$$

2.12

where: T_A = Ambient temperature immediately surrounding the LED lamp, °C.

P_{AVG} = Average power dissipation, Watts.

θ_{JA} = Thermal resistance LED junction-to-ambient of the lamp installed into the circuit, °C/W.

The value of θ_{JA} is the sum of the device thermal resistance, LED junction-to-lead, θ_{JC}, and the thermal resistance to ambient of the supporting structure, θ_{CA}:

$$\theta_{JA}(^\circ C/W) = \theta_{JC} + \theta_{CA} \qquad (2.3.2\text{-}3)$$

For most printed circuit boards, θ_{CA} ranged between 35°C/W and 50°C/W, depending upon metallization pattern. Figure 2.3.2-2 illustrates this concept.

A numerical example utilzing these equations is presented in Section 2.4.3.

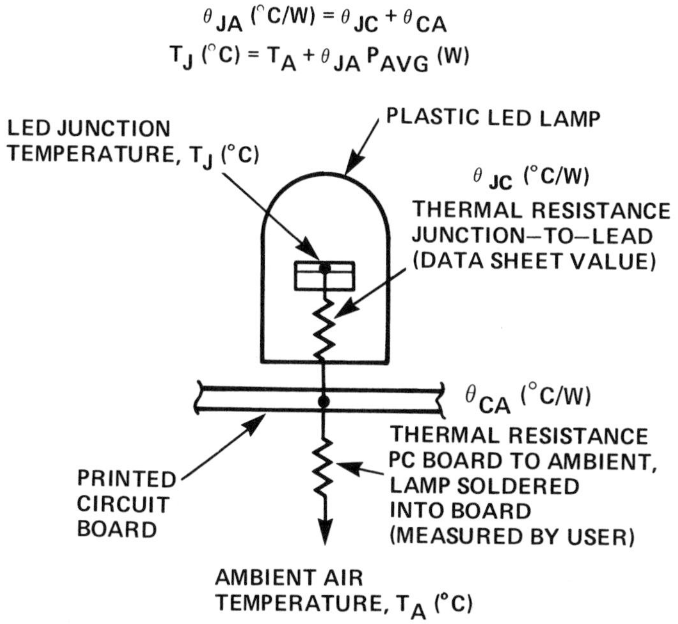

$$\theta_{JA}(^\circ C/W) = \theta_{JC} + \theta_{CA}$$
$$T_J(^\circ C) = T_A + \theta_{JA} P_{AVG}(W)$$

LED JUNCTION TEMPERATURE, T_J (°C)

PLASTIC LED LAMP

θ_{JC} (°C/W)

THERMAL RESISTANCE JUNCTION—TO—LEAD (DATA SHEET VALUE)

θ_{CA} (°C/W)

THERMAL RESISTANCE PC BOARD TO AMBIENT, LAMP SOLDERED INTO BOARD (MEASURED BY USER)

PRINTED CIRCUIT BOARD

AMBIENT AIR TEMPERATURE, T_A (°C)

Figure 2.3.2-2   A Schematic Representation of the Thermal Resistance Paths for a Plastic LED Lamp.

2.3.3   Pulsed Operating Conditions

When a design requires an LED lamp to be operated in the pulsed mode, maximum tolerable operating limits need to be established. These maximum tolerable limits should not raise the LED junction temperature above that which would be obtained by operating the lamp at the maximum dc current. This limitation on T_J imposes a definite interrelationship between peak current, pulse duration, and refresh rate. This interrelationship is most easily obtained by establishing combinations of peak current and pulse duration for various refresh rates, maintaining the maximum T_J at that value obtained by operating at the maximum dc current. These data points are plotted on a

log-log scale to form the family of curves shown in Figure 2.3.3-1.

The curve for any specific refresh rate is the locus of maximum tolerable operating conditions which maintain the limitation of T_J. Any combination of operating

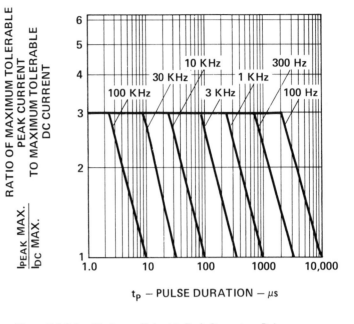

Figure 2.3.3-1   Maximum Tolerable Peak Current vs. Pulse Duration for a T-1 3/4 High-Efficiency Red LED Lamp.

conditions at, or below, a line of constant refresh rate is permissible. Operation above a line of constant refresh rate violates the limitation on junction temperature.

One procedure for determining a set of operating conditions from Figure 2.3.3-1 consists of the following five steps:

Step 1. Determine the desired duty factor, DF. Example: DF = 30%.

Step 2. Determine the desired refresh rate, f. Use DF to calculated pulse duration, tp. Example: f = 1 KHz; tp = DF/f = .30/1 KHz = 300 μsec.

Step 3. Enter Figure 2.3.3-1 at the calculated value of tp. Move vertically to refresh rate line and record the corresponding value of $I_{PEAK\ MAX}/I_{DC\ MAX}$. Example: At tp = 300 μsec and f = 1 KHz, $I_{PEAK\ MAX}/I_{DC\ MAX}$ = 2.4.

Step 4. Calculate $I_{PEAK\ MAX}$ using the data sheet maximum average forward current for $I_{DC\ MAX}$. Calculate I_{AVG} from I_{PEAK} and DF. Example: from data sheet, the maximum average current = 20 mA.

2.13

$$I_{PEAK\,MAX} = (2.4)(20\ mA) = 48\ mA$$
$$I_{AVG} = (.3)(48\ mA) = 14.4\ mA$$

Step 5. Refer to Equation 2.3.2-1 and calculate P_{AVG} as a check to insure that the above operating conditions are within the required power derating corresponding to the operating ambient temperature. Should P_{AVG} fall above the required power derating, decrease tp to reduce I_{AVG} (or reduce I_{PEAK}) in order to lower P_{AVG} to an acceptable level.

An alternate procedure is presented in the form of a numerical example in Section 2.4.3.5.

2.3.4 Time Average Luminous Intensity

The axial luminous intensity value listed on the data sheet is a dc measurement. The measurement is made by driving the lamp at the specified current for a time duration of 20 msec to 50 msec and measuring the intensity with a calibrated photometer. The luminous intensity of each lamp will be at least equal to the specified minimun, but on the average will be close to the typical value listed on the data sheet. This measurement does not give a designer any insight as to what the intensity will be at some other drive condition.

Of interest to a designer is the light output, as perceived by an observer, when the lamp is driven at some other dc current or is operated in the pulsed mode. Since the eye is a time averaging detector, it is the time average luminous intensity that is of specific interest.

For dc operation, the time average intensity is a supralinear function of forward current.

$$\frac{I_v}{I_{vo}} = \left(\frac{I_F}{I_{FO}}\right)^n \qquad (2.3.4\text{-}1)$$

where: I_{vo} = An initial luminous intensity at a reference dc current, I_{FO}.

I_v = The luminous intensity at the operating dc current, I_F.

n = A supralinear exponent that ranges between 1.1 and 1.4, depending upon the LED product and the forward current.

Equation 2.3.4 is derived in the following manner. A measurement of luminous intensity at various drive currents is obtained from a large sample of devices selected from many different lots. A curve of relative luminous intensity vs. forward current is derived from this data, normalized to 1.0 at the current where the luminous intensity is specified, and presented as a graph on the data sheet. Equation 2.3.4

is the mathematical representation of this curve, with the value of n having been derived directly from the curve.

The determination of time average intensity, observed with pulsed operation, involves the concept of relative efficiency. Once the concept of relative efficiency is understood, a designer is able to calculate the time average intensity for any given set of pulsed operating conditions.

The efficiency of an LED device may be stated as intensity per unit current, such as millicandelas/milliampere. Relative efficiency is the ratio of the efficiency at one peak current to the efficiency at another peak current for the same average current.

A graph of relative efficiency is obtained in the following manner. An LED device is measured for luminous intensity at the test current specified in the data sheet. Then the device is strobed at different peak currents, but with the same average current as specified in the data sheet, and measured for luminous intensity. The resulting curve of luminous intensity vs. peak current is normalized to 1.0 at the data sheet test current. This is the curve of relative efficiency for that device. The data sheet relative efficiency curve represents the average of a large sample of devices from many different lots.

To obtain the time average intensity of an LED lamp being operated in the pulsed mode, a designer need only multiply the ratio of the operating average current to the data sheet average current by the product of the relative efficiency at the operating peak current and the luminous intensity specified on the data sheet.

Section 2.4.2 on relative efficiency illustrates the use of the two data sheet graphs, Relative Luminous Intensity vs. Forward Current and Relative Efficiency vs. Peak Current, in determining time average intensity.

2.4 Visual Applications of LED Lamps

2.4.1 Introduction

The largest usage of LED lamps is in visual applications. LED lamps have long been used as panel mounted indicators, printed circuit status indicators, and both x-y addressable and linear arrays. With the introduction of the High Efficiency LED lamps, applications that previously could only use neon and incandescent lamps, now in many instances, can also use LED lamps. These high intensity applications include backlighting a legend or illuminating a push button. LED lamps offer many advantages to the designer. They are small, light weight, and mechanically rugged. Since they operate at low voltages and currents, they can interface directly to most digital logic families. Because LEDs are solid state devices, they have a projected operating life of over 100,000 hours. These features benefit the end user by substantially reducing field maintenance costs due to lamp replacement.

Figure 2.4.1-1 shows some of the traditional uses for LED lamps, such as panel indicators and printed circuit status indicators. The designer has the flexibility of soldering the lamps directly into a printed circuit board and positioning the board behind the front panel as shown, or by using a clip and ring to attach the lamp to the front panel and soldering or wirewrapping leads to the LED lamp. When several lamps are driven by a common LED driver, they are commonly called an array. While it is not important that the lamps be positioned together, special purpose displays can be formed by specific patterns of lighted lamps in a specific grouping. An x-y addressable array is a group of LED lamps that are connected so that one particular lamp is illuminated through the application of proper signals to an x and a y coordinate. Numeric and alphanumeric information can be displayed with 35 LED lamps arranged in a 5x7 matrix. Fewer LEDs are required if only numeric information is to be displayed (an example of a modified 4x7 matrix to display hexidecimal information is illustrated in Figure 5.1.1.1-1). Analog information can be displayed

with a linear display. Two types of linear displays are the bar graph display and the position indicator display In the bar graph display, all LED lamps that represent a value smaller than the input quantity are turned on. Only the single LED lamp that represents a value closest to the input quantity is turned on in the position indicator display. Another application of LED lamps is to highlight or backlight a printed legend. Traditionally, LED lamps have been used to attract the attention of a viewer to a message printed near the LED lamp. The introduction of the High Efficiency LED lamps, now allows LEDs to backlight a legend printed on translucent or diffused film.

Electrically, LED lamps behave similarly to silicon or germanium diodes. LED lamps emit light only when they are forward biased. Because the LED lamp has a very small dynamic resistance above the turn-on voltage of the device, LEDs are normally driven by a current source. For most applications, a battery in series with a resistor can be considered as a current source if $V_{BAT} \gg V_F$ and $R \gg R_S$. Figure 2.4.1-2 shows how the current flowing through an LED lamp can be solved graphically by superimposing a load line over the forward characteristics of the LED device. The forward current through the LED should be

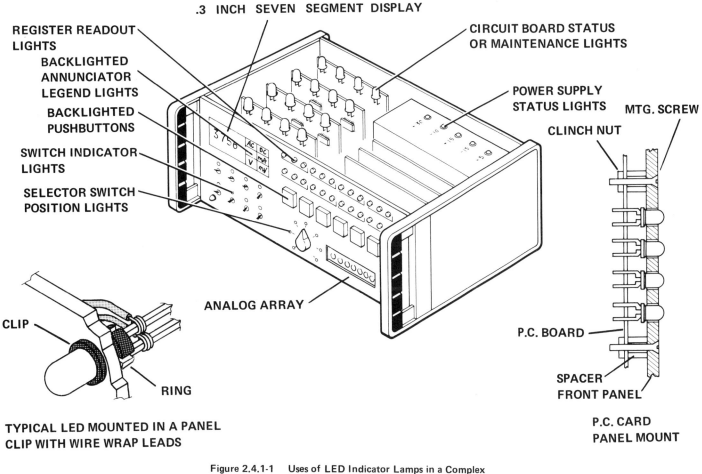

Figure 2.4.1-1   Uses of LED Indicator Lamps in a Complex Instrument.

selected to give the desired time averaged luminous intensity for worst cased values of power supply voltages and circuit tolerances. Maximum forward current is also constrained by the maximum allowable average power dissipation based on ambient temperature and the maximum tolerable peak current for a specified pulse duration and repetition rate. Section 2.4.3 outlines some of the techniques to operate an LED lamp in dc or pulsed mode and shows several worst cased circuit designs using these techniques.

Since an LED lamp is an optoelectronic device, the luminous intensity, radiation pattern, and visual appearance of the lamp should dictate the choice of LED packages and whether the lamp should be diffused, non diffused, tinted or non tinted. The luminous intensity of the lamp should be large enough to achieve a desired contrast ratio between the lamp and the background around the lamp. A lamp mounted on a reflective surface will require a higher luminous intensity than the same lamp mounted on a dull, non reflective surface. If the ambient luminous incidence is increased, the luminous intensity of the lamp will have to increase proportionately. Once the desired luminous intensity is known, the required drive current to obtain that luminous intensity can be calculated using either the figures of "Relative Luminous Intensity vs. I_F" or "Relative

Efficiency vs. I_{PEAK}" that are shown in the data sheet for that particular device. Section 2.4.2 describes these calculations in more detail.

2.4.2 Relative Efficiency

Traditionally, LED lamps have been characterized for light output (I_V) vs. dc forward current (I_F). An example of this characterization is shown in Figure 2.4.2-1. The abcissa is dc current in milliamperes and the ordinate is luminous intensity in millicandelas or luminous intensity normalized to one at a particular value of input current. The use of such a figure is relatively simple if one variable is known. For example, if an LED lamp emits 4.0 mcd at 10 mA I_F, then typically it will emit 9.9 mcd at 20 mA I_F.

Two other representations of Figure 2.4.2-1 are useful in strobed applications. An LED lamp will emit a certain number of photons per milliamp of input current. This is known as the efficiency of the LED. This efficiency varies according to the peak current through the LED. The relative efficiency curve such as Figure 2.4.2-2 is used to show this relationship. The abcissa is peak current in milliamps and the ordinate is millicandelas per milliamp normalized to one at a particular value of input current.

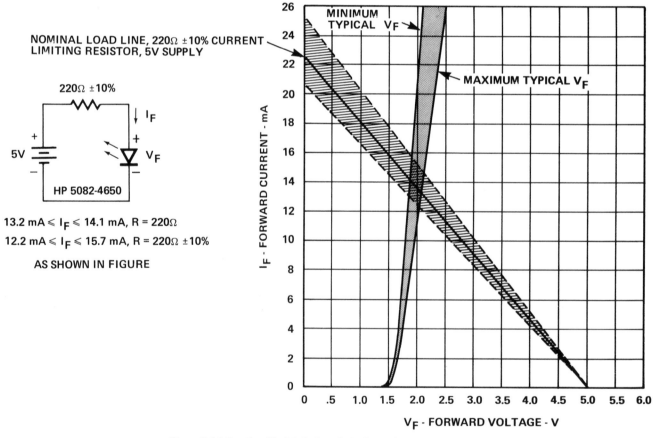

Figure 2.4.1-2   Graphical Solution of the Current Flowing through a High Efficiency Red LED.

Figure 2.4.2-2 can be used to calculate the time averaged luminous intensity for any particular peak current and duty cycle with equation 2.4.2-1.

$$(2.4.2\text{-}1)$$

$$I_{V\ TIME\ AVG} = \frac{[I_{PEAK}]\,[DUTY\ CYCLE]\,[\eta(I_{PEAK})]\,[I_{V\ SPEC}]}{[I_{SPEC}]\,[\eta(I_{PEAK})]}$$

where I_{PEAK} is the desired peak current; DUTY CYCLE is the ratio of time the LED is "ON" to total time; η is the relative efficiency of the LED at I_{PEAK} or at I_{SPEC}; I_{SPEC} is the dc current at which $I_{V\ SPEC}$ is tested; and $I_{V\ SPEC}$ is the luminous intensity of the LED according to the test conditions.

For example, using Figure 2.5.2-2, if the desired I_{PEAK} is 60 mA, the desired duty cycle is 1/8, and the LED has a luminous intensity of 4.0 mcd at 10 mA dc, then the time averaged I_V can be calculated as follows:

$I_{V\ TIME\ AVG}$ (mcd)

$$= \frac{(60\ mA)\,(.125)\,(1.54)\,(4.0\ mcd)}{(10\ mA)\,(1.00)}$$

$$= 4.6\ mcd\ at\ 60\ mA\ I_{PEAK},\ 1/8\ DUTY\ CYCLE$$

One final representation of Figure 2.4.2-1 is a time averaged luminous intensity curve. An example of this representation is shown in Figure 2.4.2-3. The abcissa is average current in milliamps and the ordinate is time averaged luminous intensity. Time averaged luminous intensity is then plotted for different values of peak current or duty factor. With this representation, time averaged luminous intensity can be read directly from the curve. For example, if the desired I_{PEAK} is 60 mA, the desired duty cycle is 1/8 and the LED has a luminous intensity of 4.0 mcd at 10 mA dc, then the time averaged I_V can be read directly off Figure 2.4.2-3 as 4.6 mcd (I_{AVE} = 7.5 mA). For lamps with a different luminous intensity than 4.0 mcd at 10 mA dc, the final result will need to be linearly scaled to reflect the difference. Figure 2.4.2-3 shows the advantages of strobing the LED as a means for achieving a higher luminous intensity at the same average current or by reducing the required average current and still maintaining the desired luminous intensity. For example, a typical device will emit 1.55 mcd at 5 mA as shown by Figure 2.4.2-3. The same device will emit 3.1 mcd at 60 mA peak, 5 mA average current or 1.55 mcd at 60 mA peak, 2.5 mA average current.

Ignoring the effect of junction heating due to the average power dissipation in the LED (the temperature coefficient of I_V is about -1%/°C), Figures 2.4.2-1, 2.4.2-2, and 2.4.2-3 are equivalent. The designer can derive one figure from either of the other two if he prefers. The relative efficiency

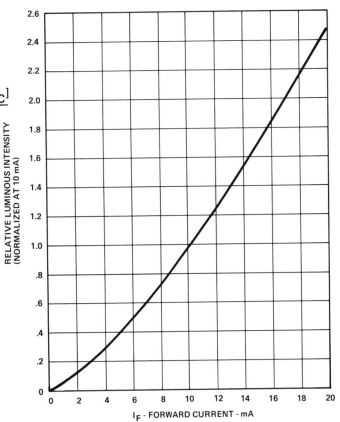

Figure 2.4.2-1 Relative Luminous Intensity vs. Current for High Efficiency Red LED.

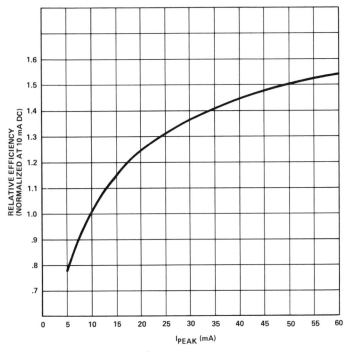

Figure 2.4.2-2 Relative Efficiency vs. Peak Current for High Efficiency Red LED.

curve is normally given on all data sheets. As a final example assume a lamp is strobed at 20 mA peak current at a 10% duty cycle and the lamp has a luminous intensity of 3.0 mcd at 10 mA dc:

Figure 2.4.2-1: $I_V = (2.48)(3.0)(.10) = .74$ mcd

Figure 2.4.2-2: $I_V = \dfrac{(20)(.10)(1.24)(3.0)}{(10)(1.00)(1.00)} = .74$ mcd

Figure 2.4.2-3: $I_V = \dfrac{(1.0)(3.0)}{4.0} = .75$ mcd

2.4.3 Driving an LED Lamp

2.4.3.1 LED Electrical Characteristics

Figure 2.4.3.1-1 shows typical electrical characteristics for standard red, high efficiency red, yellow and green lamps. Above 1.5 volts V_F, the current flowing through an LED increases very rapidly. The dynamic resistance can be considered to be the slope of the diode characteristic $(\Delta V_F/\Delta I_F)$ in the forward region. The standard red lamp has a very low dynamic resistance, while the high efficiency red, yellow and green lamps have a somewhat higher dynamic resistance. Since the dynamic resistance is so small, LED lamps should not be connected in parallel. Small variations in V_F or dynamic resistance can cause current hogging by the LED with the lowest V_F. This current hogging can cause variations in luminous intensity and excessive power dissipation in the lamp. However, LED

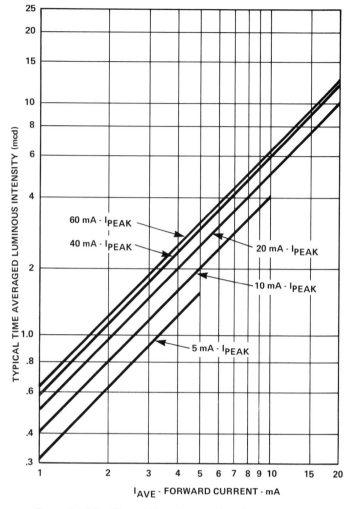

Figure 2.4.2-3   Typical Time Averaged Luminous Intensity vs. Average Current for a High Efficiency Red LED.

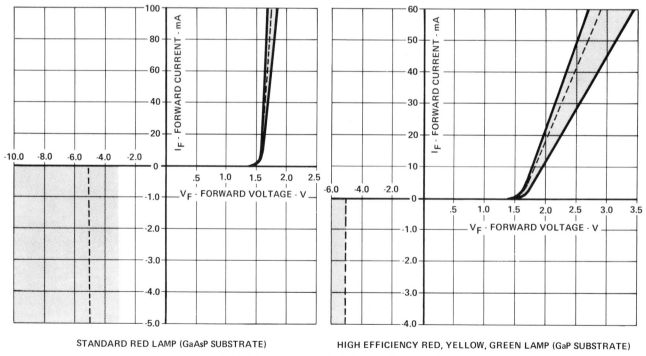

STANDARD RED LAMP (GaAsP SUBSTRATE)   HIGH EFFICIENCY RED, YELLOW, GREEN LAMP (GaP SUBSTRATE)

Figure 2.4.3.1-1 Typical Electrical Characteristics of LED Lamps.

2.18

lamps can be connected in series as long as the combined V_F doesn't exceed the power supply potential.

Negligable current flows through an LED in the reverse direction until the breakdown voltage is exceeded. Above the breakdown voltage, BV_R, the reverse current increases very rapidly, such as shown in Figure 2.4.3.1-1. Exceeding the BV_R will not harm the LED lamp as long as the reverse current is externally limited to prevent excessive power dissipation in the LED. When several LED lamps are connected in an array, reverse leakage current can cause unwanted ghosting of normally off LED's. This can be prevented by using LED drivers with a high impedance off state.

2.4.3.2 Resistive Current Limiting

When LED lamps are driven from a regulated power supply, a resistor can be used to limit the current flowing through the LED. The LED current, I_F, is determined by the following equation:

$$I_F = \frac{V_{CC} - V_F - V_{CE\ SAT}}{R} \qquad (2.4.3.2\text{-}1)$$

where V_{CC} is the power supply potential, $V_{CE\ SAT}$ is the "on" voltage of the LED driver, I_F and V_F are the forward characteristics of the LED, and R is the current limiting resistor.

If V_{CC} is considerably larger than $V_F + V_{CE\ SAT}$, then small variations of V_F or $V_{CE\ SAT}$ will have only negligable effects on I_F. However, variations of V_{CC} or R will cause a corresponding variation of I_F. For example, suppose that a standard red LED lamp is to be driven from a 5.0V supply by a transistor switch with a $V_{CE\ SAT}$ of .2V at 18 mA I_F. Then the nominal value of resistance is equal to:

$$R = \frac{5.0 - .2 - 1.65}{18} = 175\Omega$$

Figure 2.4.3.2-1 shows both numerical and graphical solutions to this example assuming tolerances of V_{CC}, R, $V_{CE\ SAT}$, and V_F.

Resistor-LED lamps are also available that are designed to operate off of a 5 volt supply.

2.4.3.3 Constant Current Limiting

In some applications, it may be desirable to drive LED lamps with a current source. The current source can be used to regulate the current through the LED regardless of power supply variations or variations in V_F between LED lamps. Figure 2.4.3.3-1 shows some examples of simple current sources constructed of npn transistors. For both circuits, the current through the LED string remains constant as long as V_{CC} is greater than $V_{CC\ (min)}$. The

NUMERICAL SOLUTION:

$$I_{F(max)} = \frac{V_{CC(max)} - V_{CE\ SAT\ (min)} - V_{F(min)}}{R(min)}$$

$$I_{F(min)} = \frac{V_{CC(min)} - V_{CE\ SAT(max)} - V_{F(max)}}{R(max)}$$

EXAMPLE:

ASSUME $V_{F(min)}$ = 1.62 @ 20 mA
$\quad\quad V_{F(max)}$ = 1.67 @ 20 mA
$\quad\quad V_{CC}$ = 5.0V ±10%
$\quad\quad R$ = 180Ω ±10%
$\quad\quad V_{CE\ SAT}$ = .2V @ 20 mA

1. V_{CC} = 4.5V, $I_{F(max)} = \dfrac{4.50 - .20 - 1.62}{(180)(.9)}$ = 16.5 mA

$\quad\quad I_{F(min)} = \dfrac{4.50 - .20 - 1.67}{(180)(1.1)}$ = 13.3 mA

2. V_{CC} = 5.0V ∴ 15.8 mA ⩽ I_F ⩽ 19.6 mA

3. V_{CC} = 5.5V ∴ 18.3 mA ⩽ I_F ⩽ 22.7 mA

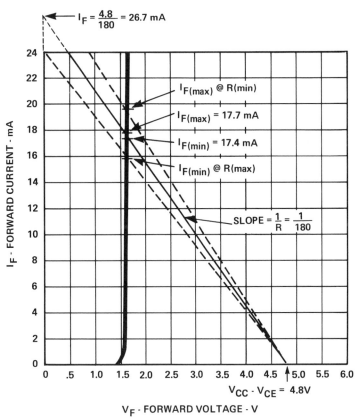

Figure 2.4.3.2-1 Numerical and Graphical Solution to Resistive Current Limiting.

first circuit uses an LED lamp as a voltage reference. Since $\Delta V_{F(T)} \cong \Delta V_{BE(T)} \cong -2$ mV/°C, the current source will remain stable over temperature. In the second circuit, since V_{BE} varies with temperature, the current source will also vary with temperature. However, this change is typically about:

$$\frac{\Delta V_{BE}/\Delta T}{V_{BE}} \cong \frac{-2}{650} \cong -.3\%/°C$$

Commercially available current regulator IC's can also used.

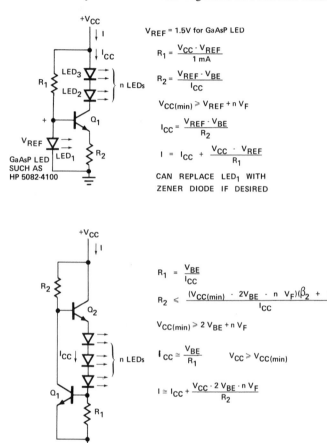

V_{REF} = 1.5V for GaAsP LED

$$R_1 = \frac{V_{CC} \cdot V_{REF}}{1 \text{ mA}}$$

$$R_2 = \frac{V_{REF} \cdot V_{BE}}{I_{CC}}$$

$$V_{CC(min)} \geqslant V_{REF} + n V_F$$

$$I_{CC} = \frac{V_{REF} \cdot V_{BE}}{R_2}$$

$$I = I_{CC} + \frac{V_{CC} \cdot V_{REF}}{R_1}$$

CAN REPLACE LED$_1$ WITH ZENER DIODE IF DESIRED

$$R_1 = \frac{V_{BE}}{I_{CC}}$$

$$R_2 \leqslant \frac{(V_{CC(min)} \cdot 2V_{BE} \cdot n V_F)(\beta_2 + 1)}{I_{CC}}$$

$$V_{CC(min)} \geqslant 2 V_{BE} + n V_F$$

$$I_{CC} \cong \frac{V_{BE}}{R_1} \qquad V_{CC} \geqslant V_{CC(min)}$$

$$I \cong I_{CC} + \frac{V_{CC} \cdot 2 V_{BE} \cdot n V_F}{R_2}$$

Figure 2.4.3.3-1 Some Examples of Constant Current LED Drivers that Regulate I_{CC} Regardless of V_{CC}.

2.4.3.4 LED-Logic Interface

Since LED lamps operate at low voltages and currents, they can interface to most digital logic families directly. Figure 2.4.3.4-1 shows some of the ways that an LED lamp can be used to interface to digital logic. TTL logic families are guaranteed to sink a minimum amount of current (I_{OL}) which can drive most LED lamps without additional buffering. Table 2.4.3.4-1 lists the guaranteed I_{OL} of most common TTL families:

74S: $I_{OL} \leqslant 20$ mA	74LS: $I_{OL} \leqslant 8$ mA
74H: $I_{OL} \leqslant 20$ mA	74L: $I_{OL} \leqslant 3.6$ mA
74: $I_{OL} \leqslant 16$ mA	

TABLE 2.4.3.4-1 TTL Interface

If additional current is required, such as in a strobed application, TTL buffers are also available. CMOS buffers can also drive many LED lamps directly. Table 2.4.3.4-2 lists some of the commonly available CMOS buffers:

RCA 4049, 4050	$I_{OL} \geqslant 3$mA, V_{OL}=.4V, V_{CC}=5V
4009, 4010	$I_{OL} \geqslant 8$mA, V_{OL}=.5V, V_{CC}=10V
National 74C906	$I_{OL} \sim 8$mA, V_{OL}=.5V, V_{CC}=4.75V
	$I_{OL} \sim 20$mA, V_{OL}=.5V, V_{CC}=10V
National 74C901 74C902	$I_{OL} \geqslant 3.8$mA, V_{OL}=.4V,

TABLE 2.4.3.4-2 CMOS Interface

OPEN COLLECTOR GATES

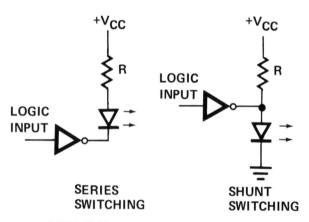

SERIES SWITCHING

SHUNT SWITCHING

ACTIVE PULLUP - TOTEM POLE GATES

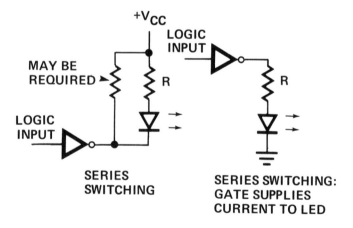

SERIES SWITCHING

SERIES SWITCHING: GATE SUPPLIES CURRENT TO LED

Figure 2.4.3.4-1 Digital Logic Can Interface Directly to LED Lamps.

Low cost transistors can be used when larger currents are required than a logic gate can supply directly. Figure 2.4.4.2-3 shows some of the commonly used LED drive schemes implemented with NPN, PNP, and FET transistors.

2.4.3.5 Worst Case Design

Regardless of whether the LED lamp design uses resistive or constant current limiting, is DC driven or strobed, the designer should consider worst case design techniques. The purpose of worst case design is two-fold. First, the design should guarantee that the LED lamps will operate within their recommended operating conditions. This will insure a long operating lifetime for the LED devices -- probably longer than the expected lifetime of the instrument. Secondly, worst case design can determine whether power supply, resistor and device tolerances will cause a noticeable variation in luminous intensity from lamp to lamp. In many applications, this requirement for luminous intensity matching is not important, but when several LED lamps are used in a closely packed array, wide variations in luminous intensity can be very objectionable to the viewer.

When a lamp is dc driven, a worst case design analysis can determine whether the lamp operates within its recommended operating conditions. The maximum operating conditions for an LED lamp under dc operation are determined by two factors. Normally each LED device is derated over temperature to keep the junction temperature below a specified maximum temperature. Typically, this maximum junction temperature is 110°C for a plastic encapsulated lamp. Secondly, the current density through the LED junction should be limited to prevent rapid degradation of light output. The maximum peak current for a standard red lamp is typically 1000 mA and 60 mA for high efficiency red, yellow, and green lamps.

Depending on the maximum expected ambient temperature for the particular application, the designer can determine the maximum average power dissipation, P_{AVG}, from the power derating curve for the device. An example of a power derating curve is shown in Figure 2.3.2-1. The average power dissipation in a lamp is the product of the average forward current times the peak forward voltage. Equation 2.3.2-1 can be used to calculate the average power dissipation in an LED lamp:

$$P_{AVG(W)} = I_{AVG} [V_F + R_S (I_{PEAK} - I_F)] \qquad (2.3.2\text{-}1)$$

where I_{AVG} is the average forward current, I_{PEAK} is the peak forward current, I_F and V_F are the LED lamp test conditions, and R_S is the LED dynamic resistance.

In a dc application, I_{AVG} is equal to I_{PEAK}. Equation 2.3.2-1 and Figure 2.3.2-1 can be used to calculate the

maximum dc current through the LED lamp as shown below:

$$\text{(2.4.3.5-1)}$$
$$P_{AVG(mW)} = I_{AVG(mA)} [V_F + R_s (I_{AVG(mA)} - I_F)/1000]$$

For a standard red lamp, $V_{F\,(max)}$ = 2.0V @ 20 mA I_F, $R_{S\,(max)}$ = 5Ω

$$\therefore P_{AVG} = I_{AVG} [2.0 + 5 (I_{AVG} - 20)/1000] \qquad (2.4.3.5\text{-}2)$$

$$\therefore I_{AVG(mA)} = \frac{\sqrt{(1.9)^2 + (.02) P_{AVG(mW)}} - 1.9}{.01}$$

So at 50°C since P_{AVG} = 120 mW, then I_{AVG} = 55 mA and at 70°C where P_{AVG} = 87 mW, then I_{AVG} = 41 mA. Since the maximum average current is specified in the data sheet as 50 mA, then I_{AVG} should be restricted to less than 50 mA regardless of the results of equation 2.4.3.5-3. For additional reliability, the designer can operate at currents less than $I_{AVG\,(max)}$.

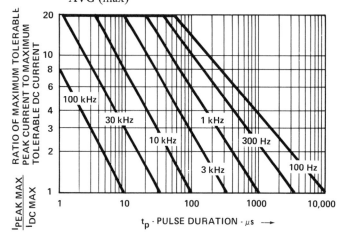

Figure 2.4.3.5-1 Maximum Tolerable Peak Current vs. Pulse Duration for a T-1 3/4 Red Lamp ($I_{DC\ MAX}$ per Figure 2.3.2-1 and Eqn. 2.4.3.5-3).

For strobed applications, a curve of maximum peak current, pulse width, and repetition rate can be used to determine the maximum recommended operating conditions for an LED lamp. The curve is determined by comparing the peak junction temperature of a lamp under strobed conditions to the average junction temperature under maximum allowable dc conditions. At any operating point, the peak junction temperature should not exceed the average junction temperature under maximum allowable dc conditions. An example of such a curve is shown in Figure 2.4.3.5-1. $I_{DC\ MAX}$ is the maximum average current calculated by equation 2.4.3.5-1 or as restricted by the data sheet. The ordinate of Figure 2.4.3.5-1 is the ratio of maximum allowable peak current to maximum allowable dc current. At any specified repetition rate, the relationship between maximum peak current and pulse width is shown. For reliable operation, the device should be operated at or

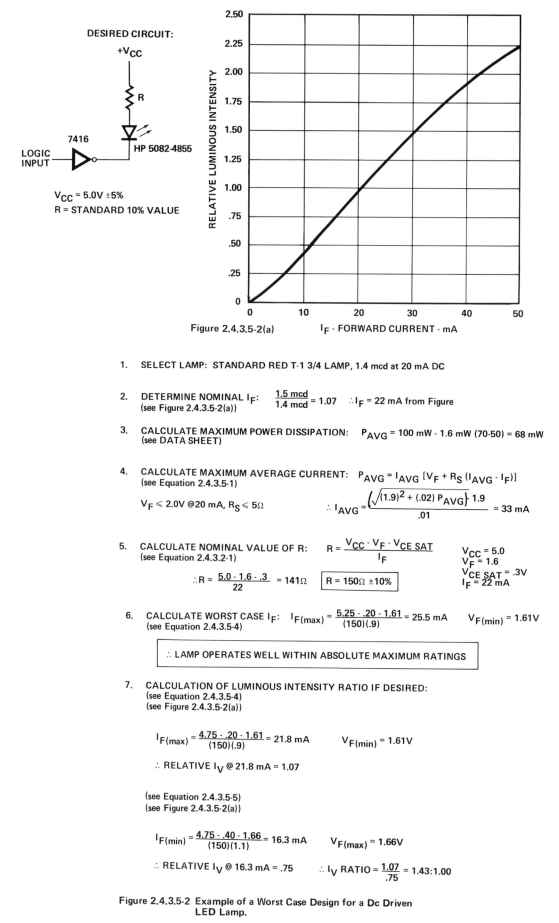

DESIRED CIRCUIT:

$+V_{CC}$

R

7416

LOGIC
INPUT

HP 5082-4855

V_{CC} = 5.0V ±5%
R = STANDARD 10% VALUE

Figure 2.4.3.5-2(a) I_F - FORWARD CURRENT - mA

1. SELECT LAMP: STANDARD RED T-1 3/4 LAMP, 1.4 mcd at 20 mA DC

2. DETERMINE NOMINAL I_F: $\frac{1.5 \text{ mcd}}{1.4 \text{ mcd}} = 1.07$ $\therefore I_F$ = 22 mA from Figure
    (see Figure 2.4.3.5-2(a))

3. CALCULATE MAXIMUM POWER DISSIPATION: P_{AVG} = 100 mW - 1.6 mW (70-50) = 68 mW
    (see DATA SHEET)

4. CALCULATE MAXIMUM AVERAGE CURRENT: $P_{AVG} = I_{AVG} [V_F + R_S (I_{AVG} - I_F)]$
    (see Equation 2.4.3.5-1)

    $V_F \leqslant$ 2.0V @20 mA, $R_S \leqslant$ 5Ω $\therefore I_{AVG} = \frac{\left(\sqrt{(1.9)^2 + (.02) P_{AVG}}\right) 1.9}{.01}$ = 33 mA

5. CALCULATE NOMINAL VALUE OF R: $R = \frac{V_{CC} - V_F - V_{CE\ SAT}}{I_F}$ V_{CC} = 5.0
    (see Equation 2.4.3.2-1) V_F = 1.6
                                                                          $V_{CE\ SAT}$ = .3V
    $\therefore R = \frac{5.0 - 1.6 - .3}{22}$ = 141Ω $\boxed{R = 150Ω \pm 10\%}$ I_F = 22 mA

6. CALCULATE WORST CASE I_F: $I_{F(max)} = \frac{5.25 - .20 - 1.61}{(150)(.9)}$ = 25.5 mA $V_{F(min)}$ = 1.61V
    (see Equation 2.4.3.5-4)

    $\boxed{\therefore \text{LAMP OPERATES WELL WITHIN ABSOLUTE MAXIMUM RATINGS}}$

7. CALCULATION OF LUMINOUS INTENSITY RATIO IF DESIRED:
    (see Equation 2.4.3.5-4)
    (see Figure 2.4.3.5-2(a))

    $I_{F(max)} = \frac{4.75 - .20 - 1.61}{(150)(.9)}$ = 21.8 mA $V_{F(min)}$ = 1.61V

    $\therefore$ RELATIVE I_V @ 21.8 mA = 1.07

    (see Equation 2.4.3.5-5)
    (see Figure 2.4.3.5-2(a))

    $I_{F(min)} = \frac{4.75 - .40 - 1.66}{(150)(1.1)}$ = 16.3 mA $V_{F(max)}$ = 1.66V

    $\therefore$ RELATIVE I_V @ 16.3 mA = .75 $\therefore I_V$ RATIO = $\frac{1.07}{.75}$ = 1.43:1.00

Figure 2.4.3.5-2 Example of a Worst Case Design for a Dc Driven
LED Lamp.

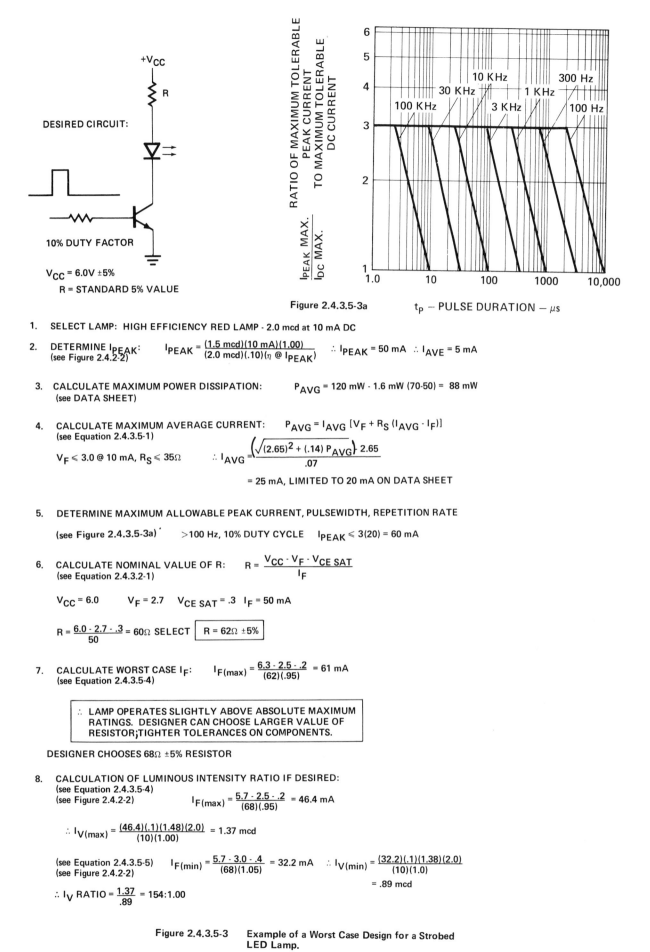

DESIRED CIRCUIT:

+V$_{CC}$

R

10% DUTY FACTOR

V$_{CC}$ = 6.0V ±5%

    R = STANDARD 5% VALUE

Figure 2.4.3.5-3a

t$_P$ – PULSE DURATION – μs

RATIO OF MAXIMUM TOLERABLE PEAK CURRENT TO MAXIMUM TOLERABLE DC CURRENT

$\dfrac{I_{PEAK\ MAX.}}{I_{DC\ MAX.}}$

100 KHz   30 KHz   10 KHz   3 KHz   1 KHz   300 Hz   100 Hz

1. SELECT LAMP: HIGH EFFICIENCY RED LAMP - 2.0 mcd at 10 mA DC

2. DETERMINE I$_{PEAK}$: $I_{PEAK} = \dfrac{(1.5\ mcd)(10\ mA)(1.00)}{(2.0\ mcd)(.10)(\eta\ @\ I_{PEAK})}$ ∴ I$_{PEAK}$ = 50 mA ∴ I$_{AVE}$ = 5 mA
   (see Figure 2.4.2-2)

3. CALCULATE MAXIMUM POWER DISSIPATION: P$_{AVG}$ = 120 mW - 1.6 mW (70-50) = 88 mW
   (see DATA SHEET)

4. CALCULATE MAXIMUM AVERAGE CURRENT: $P_{AVG} = I_{AVG}\ [V_F + R_S\ (I_{AVG} \cdot I_F)]$
   (see Equation 2.4.3.5-1)

   V$_F$ ≤ 3.0 @ 10 mA, R$_S$ ≤ 35Ω   ∴ $I_{AVG} = \dfrac{\left(\sqrt{(2.65)^2 + (.14)\,P_{AVG}}\right) - 2.65}{.07}$

           = 25 mA, LIMITED TO 20 mA ON DATA SHEET

5. DETERMINE MAXIMUM ALLOWABLE PEAK CURRENT, PULSEWIDTH, REPETITION RATE

   (see Figure 2.4.3.5-3a)   >100 Hz, 10% DUTY CYCLE   I$_{PEAK}$ ≤ 3(20) = 60 mA

6. CALCULATE NOMINAL VALUE OF R: $R = \dfrac{V_{CC} - V_F - V_{CE\ SAT}}{I_F}$
   (see Equation 2.4.3.2-1)

   V$_{CC}$ = 6.0   V$_F$ = 2.7   V$_{CE\ SAT}$ = .3 I$_F$ = 50 mA

   $R = \dfrac{6.0 - 2.7 - .3}{50} = 60Ω$ SELECT $\boxed{R = 62Ω\ ±5\%}$

7. CALCULATE WORST CASE I$_F$: $I_{F(max)} = \dfrac{6.3 - 2.5 - .2}{(62)(.95)} = 61\ mA$
   (see Equation 2.4.3.5-4)

       $\boxed{\begin{array}{l}∴\ \text{LAMP OPERATES SLIGHTLY ABOVE ABSOLUTE MAXIMUM} \\ \text{RATINGS. DESIGNER CAN CHOOSE LARGER VALUE OF} \\ \text{RESISTOR;TIGHTER TOLERANCES ON COMPONENTS.}\end{array}}$

   DESIGNER CHOOSES 68Ω ±5% RESISTOR

8. CALCULATION OF LUMINOUS INTENSITY RATIO IF DESIRED:
   (see Equation 2.4.3.5-4)
   (see Figure 2.4.2-2) $I_{F(max)} = \dfrac{5.7 - 2.5 - .2}{(68)(.95)} = 46.4\ mA$

   ∴ $I_{V(max)} = \dfrac{(46.4)(.1)(1.48)(2.0)}{(10)(1.00)} = 1.37\ mcd$

   (see Equation 2.4.3.5-5) $I_{F(min)} = \dfrac{5.7 - 3.0 - .4}{(68)(1.05)} = 32.2\ mA$ ∴ $I_{V(min)} = \dfrac{(32.2)(.1)(1.38)(2.0)}{(10)(1.0)}$
   (see Figure 2.4.2-2)
                                                      = .89 mcd

   ∴ I$_V$ RATIO = $\dfrac{1.37}{.89}$ = 154:1.00

Figure 2.4.3.5-3   Example of a Worst Case Design for a Strobed
                                   LED Lamp.

below these conditions. For example, suppose the design requires a standard red lamp to be driven at 200 mA peak on a 10% duty cycle at $25°C$. At 100 Hz repetition rate, the maximum allowable peak current for a 1000 μS pulse width is 3.7 x 50 = 185 mA. However at a 300 Hz repetition rate, the maximum allowable peak current is 4.6 x 50 = 230 mA on a 10% duty cycle. Thus, by operating the device at 200 mA peak current, 10% duty factor, at a 300 Hz repetition rate, the maximum junction temperature will not exceed the junction temperature obtained by operating the device at 50 mA dc. For additional reliability, the designer can derate $I_{DC\ MAX}$ below the maximum temperature derated $I_{AVG(MAX)}$ as calculated by equation 2.4.3.5-1 or as restricted by the data sheet.

When LED lamps are to be grouped together in an array, the designer should also consider the requirement for luminous intensity matching of adjacent LEDs. Variations in power supply voltages, resistor tolerances, driver tolerances, and device tolerances all contribute to variations in I_F and hence luminous intensity variations. The usual procedure to determine whether variations in I_F will cause noticeable luminous intensity variations is to calculate the worst case minimum and maximum values of I_F. Then the relative luminous intensity or relative luminous efficiency curves can be used to determine the worst case variation in luminous intensity. In general, the maximum luminous intensity ratio between LED lamps should be less than 2.0:1.0, and ratios greater than 2.3:1.0 will be objectionable to an observer. When LED lamps are driven by a common power supply, variations in the power supply voltage will cause only a small change in the maximum ratio of LED forward currents since the forward currents will all change proportionally due to power supply variations. However, the component tolerances will have the greatest effect on I_F at the minimum power supply voltage. When resistive current limiting is used, the minimum and maximum I_F can be calculated as shown below:

$$(2.4.3.5\text{-}4)$$

$$I_{F(MAX)} = \frac{V_{CC} - V_{CE\ SAT(MIN)} - V_{F(MIN)}}{R\ (MIN)}$$

$$(2.4.3.5\text{-}5)$$

$$I_{F(MIN)} = \frac{V_{CC} - V_{CE\ SAT(MAX)} - V_{F(MAX)}}{R\ (MAX)}$$

1.   Design Example

An array of standard red LED lamps is to be dc driven with a desired typical luminous intensity of 1.5 mcd at $25°C$ with a maximum ambient temperature of $70°C$. The array will be driven by 7416 TTL hex inverters from the 5.0V TTL supply. The desired circuit and calculations are shown in Figure 2.4.3.5-2.

2.   Design Example

An array of high efficiency red LED lamps is to be strobed on a 10% duty factor with an npn transistor. The desired luminous intensity is 1.5 mcd at $25°C$ with a maximum ambient temperature of $70°C$. The desired circuit and calculations are shown in Figure 2.4.3.5-3.

2.4.4   LED Arrays

2.4.4.1 Introduction

When several LED lamps are used in an application, the cost of the associated LED drive circuitry can often be reduced by connecting the LEDs in a multiplexed array. For example, suppose 16 LED lamps are used as status indicators on a panel. Each LED can be driven by an individual transistor or logic gate as shown in Section 2.4.3. This configuration requires 16 LED drivers, 16 current limiting resistors, and 17 address lines. The LED lamps can also be connected as a 4x4 multiplexed array. In this configuration, only 8 LED drivers, 4 current limiting resistors, and 8 address lines are required. In general, $p \cdot q$ LED lamps can be driven by $p \cdot q$ transistors or gates and $p \cdot q$ current limiting resistors on a DC basis. Each LED lamp can be considered to be an element, a_{ij}, where i = 1, 2, ... p and j = 1, 2, ... q, of an x-y addressable array of $p \cdot q$ individual LEDs, such as shown in Figure 2.4.4.1-1. If the LED lamps are connected as a multiplexed x-y array with $p \cdot q$ elements, then only p+q transistors or gates will be required to drive the $p \cdot q$ lamps and p or q current limiting resistors will be needed. In many applications, such as with a microprocessor, the information is available on a multiplexed basis and minimal logic is required for proper decoding. If dc signals are available, they can be multiplexed with inexpensive multiplexers or a bus configuration. One final advantage of the multiplexed array is that the number of wires required to connect the LED lamps to the drive circuitry is reduced. The dc driven array requires (p·q+1) wires to connect the lamps but the multiplexed x-y array requires only p+q wires.

Figures 2.4.4.1-2 and 2.4.4.1-3 show the two basic types of LED arrays. The array shown in Figure 2.4.4.1-2 is dc driven because only one LED lamp is turned on at any one time. By selecting one x axis and one y axis, the single LED specified by the coordinates (x_j, y_i) is turned on. Multiple LED lamps can be driven on a dc basis if they can all be addressed by one common x or y axis. Figure 2.4.4.1-3 shows the more general type of x-y addressable array where any combination of LED lamps can be turned on. One axis is sequentially selected on a 1/p or 1/q duty factor. While the proper axis is selected, any combination of address lines of the opposite axis can be enabled, turning on the corresponding LED lamps. If the display is refreshed at a

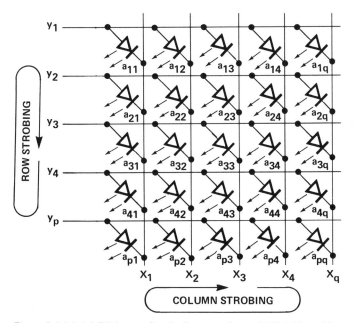

Figure 2.4.4.1-1 LED Lamps Can Be Connected as an X-Y Addressable Array of PQ Elements with P Rows and Q Columns.

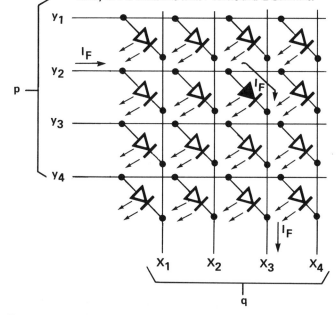

Figure 2.4.4.1-2 One LED Will Be Turned On By Applying the Proper Signal to One X Axis and One Y Axis.

fast enough refresh rate (>100Hz) then all LED lamps will appear to be dc driven. The present nomenclature used for x-y addressable arrays is that the y axis addresses (y_i; i = 1, 2 ... p) are called rows and the x axis addresses (x_j; j = 1, 2 ... q) are called columns. If the x axis addresses are sequentially selected, then the array is column strobed and if the y axis addresses are sequentially selected, then the array is row strobed.

Two basic applications exist for LED arrays. One application connects the LED lamps in an x-y addressable array as a means of simplifying the LED drive circuitry. However, each LED is individually mounted on a large surface. The second application of an x-y addressable array

is for character generation, such as an alphanumeric display, where the LEDs are mounted in close proximity to each other. In applications of this type, besides designing the drive circuitry to obtain the proper time averaged luminous intensity as described in Section 2.4.2, each LED lamp used in the array should be preselected to have less than a specified maximum luminous intensity ratio. Luminous intensity ratios of 2.3:1.00 between adjacent LEDs will be noticable by an observer. A luminous intensity ratio of 2.0:1.0 or less is recommended for this application.

2.4.4.2 Designing An X-Y Addressable LED Array

The first step in the design of an x-y addressable array is the selection of the LED lamp package that is to be used. The designer can choose between hermetic lamps, T-1 3/4 lamps, T-1 3/4 low dome lamps, T-1 lamps, subminiature lamps, or rectangular lamps. The designer can mix package styles or LED colors in the same array, although LEDs in a single row for a column strobed application or single column for a row strobed application should have similar electrical and luminous intensity characteristics. As a guide, standard red GaAsP substrate LED lamps; high-efficiency red or yellow GaP substrate LED lamps; and green GaP substrate LED lamps have substantial electrical or luminous intensity differences. To compensate for these differences, these three lamp categories should be driven in separate rows of a column strobed circuit (or separate columns of a row strobed circuit) so that a different current limiting resistor can be used for each lamp category. For character generation applications, such as an alphanumeric display, the lamp package should be chosen depending on the desired array size. Figure 2.4.4.2-1 shows the minimum LED lamp spacing for T-1 3/4, T-1, subminiature, and rectangular lamps.

Now the desired luminous intensity of the array should be specified. The desired luminous intensity sets a limit on the minimum duty factor that can be used to strobe the x-y array because of the peak current and pulse width limitations of the LED lamps. The minimum duty factor by which the x-y array can be driven is given by Equation 2.4.2-1, reproduced below:

(2.4.4.2-1)

$$\text{DUTY FACTOR} \geqslant \frac{[I_{V\ TIME\ AVG}]\ [I_{SPEC}]\ [\eta(I_{SPEC})]}{[I_{PEAK}]\ [\eta(I_{PEAK})]\ [I_{V\ SPEC}]}$$

where: all variables are defined in Section 2.4.2 and I_{PEAK} is the maximum tolerable peak current.

Table 2.4.4.2-1 shows four representative lamps and gives the maximum peak current and duty factor limitations to obtain a desired luminous intensity. The duty factors given

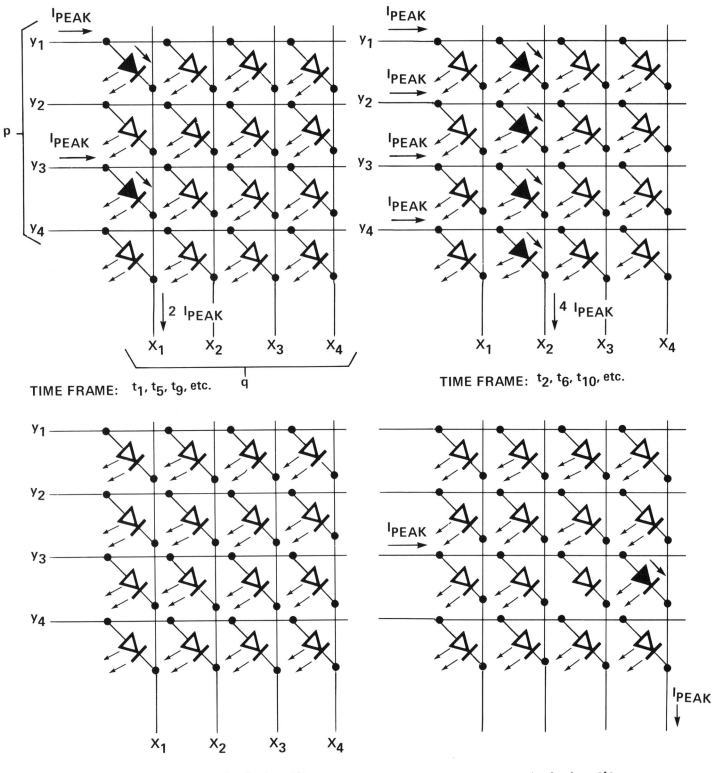

TIME FRAME: t_1, t_5, t_9, etc.

TIME FRAME: t_2, t_6, t_{10}, etc.

TIME FRAME: t_3, t_7, t_{11}, etc.

TIME FRAME: t_4, t_8, t_{12}, etc.

Figure 2.4.4.1-3 In a Multiplexed Array Any Combination of LEDs
Can Be Turned On by Sequentially Strobing One
Axis.

2.26

DUTY FACTOR	8-bit Latch	16-bit Latch	24-bit Latch	32-bit Latch
DC	8x 1 = 8 LEDs	16x 1 = 16 LEDs	24x 1 = 24 LEDs	32x 1 = 32 LEDs
1/2	7x 2 = 14 LEDs	15x 2 = 30 LEDs	23x 2 = 46 LEDs	31x 2 = 62 LEDs
1/4	6x 4 = 24 LEDs	14x 4 = 56 LEDs	22x 4 = 88 LEDs	30x 4 = 120 LEDs
1/8	5x 8 = 40 LEDs	13x 8 = 104 LEDs	21x 8 = 168 LEDs	29x 8 = 232 LEDs
1/16	4x16 = 64 LEDs	12x16 = 192 LEDs	20x16 = 320 LEDs	28x16 = 448 LEDs
1/32	3x32 = 96 LEDs	11x32 = 352 LEDs	19x32 = 608 LEDs	27x32 = 864 LEDs

TABLE 2.4.4.3-1 Maximum Size of X-Y Addressable Arrays that Can Be Driven by a Microprocessor.

Lamp Type	Typical Light Output	Maximum Peak Current	Minimum Duty Factor to Obtain Desired Time Averaged I_v			
			.5 mcd	1.0 mcd	1.5 mcd	2.0 mcd
GaAsP Red (655nm)	.8 mcd — 20 mA dc $\eta(1A)/\eta(20mA)=.8$	1A	1/64	1/32	------	------
GaP HER (635nm)	2.0 mcd — 10 mA dc $\eta(60mA)/\eta(10mA)=1.54$	60 mA	1/40	1/18	1/12	1/9
GaP Yellow (583nm)	1.8 mcd — 10 mA dc $\eta(60mA)/\eta(10mA)=1.54$	60 mA	1/32	1/16	1/11	1/8
GaP Green (565nm)	1.8 mcd — 20 mA dc $\eta(60mA)/\eta(20mA)=1.44$	60 mA	1/16	1/8	1/5	1/4

TABLE 2.4.4.2-1 Calculations of Minimum Duty Factor for Some Typical LED Lamps.

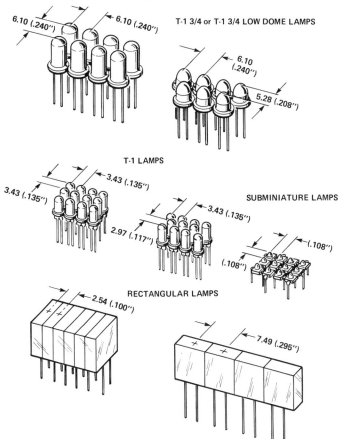

Figure 2.4.4.2-1 Minimum Dimensions of LED Arrays.

in the table can be decreased only by specifying brighter lamps or by exceeding the maximum tolerable peak current ratings of the LED lamps. When several different types of LED lamps are used in a single array, the lamp with the largest minimum duty factor determines the minimum duty factor for the entire array. The minimum duty factor as specified above sets an upper limit on one dimension of the x-y array. For example, suppose that an array is constructed of the standard red, high-efficiency red, yellow and green lamps specified in Table 2.4.4.2-1 and the desired time averaged luminous intensity of the display is 1.0 mcd for standard red lamp and 1.5 mcd for the other three lamps. The minimum duty factor of the array is 1/5, but a 1/4 duty factor is selected to simplify logic decoding. The other dimension of the x-y addressable array is determined by the total number of LED lamps in the array and the number of each type of LED lamp with different electrical or luminous intensity characteristics. Suppose that in the previous example, the array consists of 3 red lamps, 2 high-efficiency red lamps, 6 yellow lamps, and 1 green lamp. The x-y addressable array would be dimensioned as 4x4 even though only 12 lamps are used. Since the duty factor and desired time averaged luminous intensity for each lamp is known, the peak current can be determined for each type of LED lamp. Figure 2.4.4.2-2 shows the completely specified array described in the text.

RED: I_{PEAK} = 97 mA
HIGH EFFICIENCY RED: I_{PEAK} = 24 mA
YELLOW: I_{PEAK} = 26 mA
GREEN: I_{PEAK} = 50 mA

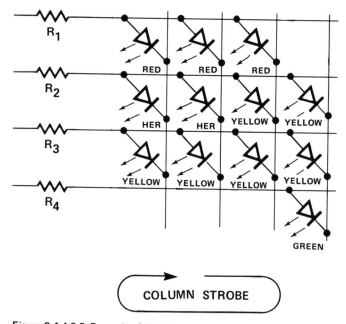

COLUMN STROBE

Figure 2.4.4.2-2 Example of X-Y Addressable Array Using 3 Red LEDs, 2 High Efficiency Red LEDs, 6 Yellow LEDs and 1 Green LED.

The final step in designing the LED x-y addressable array is the design and specification of the associated LED drive circuitry. The LED anode drivers must be able to source current to the LED array and the LED cathode drivers must be able to sink current from the array. The anode and cathode drivers can be implemented with NPN, PNP, FET transistors or commercially available LED drivers. Figure 2.4.4.2-3 shows some of the commonly used LED drive schemes implemented with NPN, PNP or FET transistors. The transistors should be selected so that the average power supply dissipation within the device or multiple transistor array is less than the manufacturers specifications. R_1 and R_2 are selected so that the transistor drivers remain in saturation under worst case conditions of V_{CC}, V_{OL}, V_{OH}, transistor H_{FE}, and resistor tolerances at the maximum peak currents specified for the LED array. Finally, the LED current limiting resistors are calculated as shown in Section 2.4.3. Figures 2.4.4.2-4 and 2.4.4.2-5 show examples of row and column strobed x-y addressable arrays using PNP anode drivers and NPN cathode drivers. The calculations of the maximum average and peak currents of each transistor assume that all LED lamps in the array are on. Active pullup drivers can also be used to drive the array provided that the B_{VR} restrictions of the LED lamps are not exceeded. If an LED lamp is connected to a low impedance driver with a potential greater than BV_R, current can flow

through the lamp in the reverse direction. This reverse leakage current will not harm the LED unless the maximum power dissipation of the lamp is exceeded, however, this reverse leakage current can flow through adjacent LED lamps in the forward direction and thus cause unwanted LED ghosting.

2.4.4.3 Design of a Microprocessor Controlled LED Array

A microprocessor can be used to control an x-y addressable array of LED lamps. The external circuitry that is required for the array is minimal and only a small portion of microprocessor time is used to refresh the LED array. The technique that is used is to periodically strobe data into a latch from the microprocessor and use the outputs of the latch to address the array by row or by column. A specified time later, new information for another row or column is strobed into the latch. If the x-y addressable array is row or comun strobed at a refresh rate greater than 100 Hz, then the entire array will appear to be dc driven. Refresh timing for the array is handled by generating an interrupt after a specific time interval. The latch is partitioned so that n bits specify one of 2^n rows (row strobe) or 2^n columns (column strobe) and the remaining bits specify the coordinates of the LED lamps of the opposite axis which should be turned on. The total number of LED lamps that can be addressed is determined by the minimum duty factor of the array and the size of the latch. Table 2.4.4.3-1 shows the maximum number of LED lamps that can be driven with latches ranging from 8 to 32 bits and for duty factors ranging from 1/32 to dc. Figure 2.4.4.3-1 shows an example of a 5x8 LED array that is designed to interface to an Intel 8080A microprocessor. A low to high transition of output Q of the monostable multivibrator requests an interrupt from the microprocessor. The interrupt circuitry (not shown) forces a RST7 instruction into the microprocessor. Following the RST7 instruction, the microprocessor executes the program shown in Figure 2.4.4.3-2. The microprocessor then updates output port (n) which is the 8 bit latch used by the x-y addressable array. If the microprocessor uses a 2 MHz clock, then the percentage of total time required to refresh the LED array at a 100 Hz repetition rate is as follows:

(2.4.4.3-1)

$$\text{REFRESH TIME} = \frac{(143C-1)\ R}{\text{MICROPROCESSOR CLOCK RATE}}$$
$$= 5.7\% \text{ FOR } C = 8,\ R = 100\ Hz$$

where: C is the number of columns in the display and R is the refresh rate.

For the remaining 94% of the time, the microprocessor can be used to update the contents of the LED array in RAM and perform countless other tasks required by the system.

2.28

To address larger LED arrays, several 8 bit latches can be used. Each 8 bit latch is addressed by a separate output address. For N eight bit latches, each latch would be addressed as output (n+i), where i = 0, 1, 2, ... N-1. The program is modified by inserting INX HL, MOV A, M and OUT (n+i) instructions into the program as shown in Figure 2.4.4.3-2 and changing the CPI $(17)_{16}$ instruction to CPI $(OF+8N)_{16}$. The monostable should be triggered by the high to low transition of the strobe input of the final 8 bit latch. To prevent a small amount of ghosting while data is being strobed into the latches, the output of the monostable can be used to disable the 1 of N decoder and thus turn all columns off from the time that the interrupt is requested until all the data has been strobed into the latches. The extra time required to address several 8 bit latches as compared to a single 8 bit latch is negligible. For example, suppose four 8 bit latches are used to address a 29x8 LED array. If the microprocessor uses a 2 MHz clock, the total percentage of time required to refresh the LED array at a 100 Hz repetition rate will increase to:

$$(2.4.4.3\text{-}2)$$

$$\text{REFRESH TIME} = \frac{[C(121+22N)-1]\,R}{\text{MICROPROCESSOR CLOCK RATE}}$$
$$= 8.4\% \text{ FOR } C = 8, N = 4, R = 100 \text{ Hz}$$

where: C is the number of columns in the display, N is the number of eight bit latches, and R is the refresh rate.

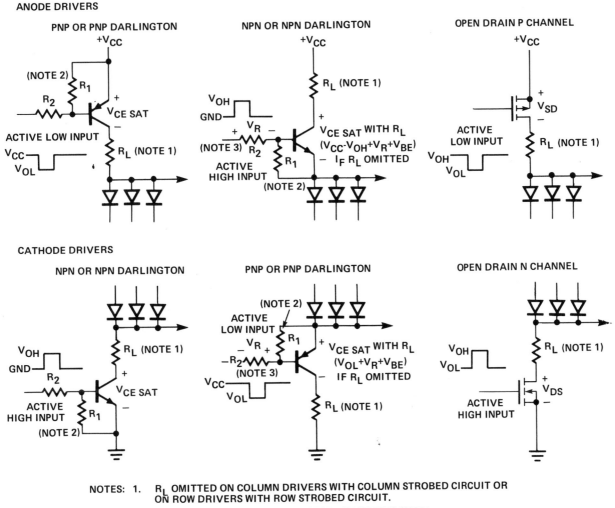

NOTES: 1. R_L OMITTED ON COLUMN DRIVERS WITH COLUMN STROBED CIRCUIT OR ON ROW DRIVERS WITH ROW STROBED CIRCUIT.
2. R_1 CAN BE OMITTED IF TRANSISTOR LEAKAGE IS SMALL.
3. R_2 CAN BE OMITTED IF R_L OMITTED AND TRANSISTOR REMAINS ACTIVE.

Figure 2.4.4.2-3 Common Transistor Drive Schemes.

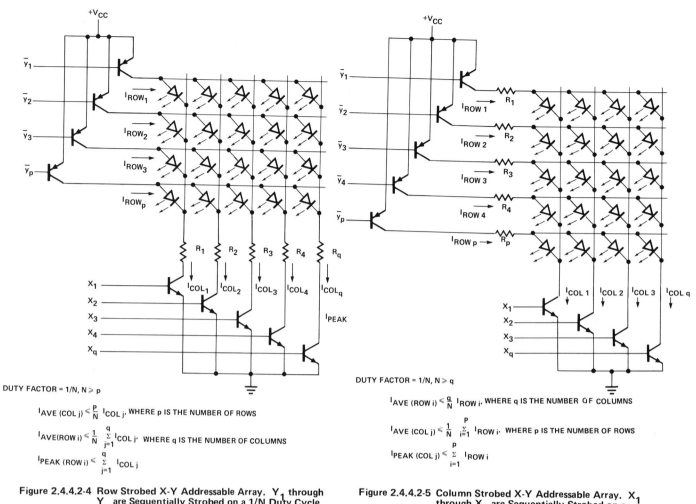

DUTY FACTOR = 1/N, N ≥ p

$$I_{AVE\,(COL\,j)} \leq \frac{P}{N}\, I_{COL\,j},\ \text{WHERE p IS THE NUMBER OF ROWS}$$

$$I_{AVE(ROW\,i)} \leq \frac{1}{N} \sum_{j=1}^{q} I_{COL\,j},\ \text{WHERE q IS THE NUMBER OF COLUMNS}$$

$$I_{PEAK\,(ROW\,i)} \leq \sum_{j=1}^{q} I_{COL\,j}$$

Figure 2.4.4.2-4 Row Strobed X-Y Addressable Array. Y_1 through Y_p are Sequentially Strobed on a 1/N Duty Cycle, N ≥ P.

DUTY FACTOR = 1/N, N ≥ q

$$I_{AVE\,(ROW\,i)} \leq \frac{q}{N}\, I_{ROW\,i},\ \text{WHERE q IS THE NUMBER OF COLUMNS}$$

$$I_{AVE\,(COL\,j)} \leq \frac{1}{N} \sum_{i=1}^{P} I_{ROW\,i},\ \text{WHERE p IS THE NUMBER OF ROWS}$$

$$I_{PEAK\,(COL\,j)} \leq \sum_{i=1}^{p} I_{ROW\,i}$$

Figure 2.4.4.2-5 Column Strobed X-Y Addressable Array. X_1 through X_q are Sequentially Strobed on a 1/N Duty Cycle, N ≥ P.

2.4.4.4 Analog Bar Graph Arrays

In many applications, analog information must be converted into a visual display. Traditionally analog panel meters have been used for low cost applications requiring only moderate accuracy and resolution. LED linear displays can often be substituted for these applications. LED linear displays consist of a linear or circular array of LED lamps that are driven by a device that decodes the varying analog or digital signal into a bar graph or position indicator display. The input signal can be decoded so that all LEDs with thresholds below the input are turned on (bar graph) or that only the LED with its threshold closest to the input is turned on (position indicator). Some examples of these type of displays are shown in Figure 2.4.4.4-1. LED linear arrays have many advantages over analog panel meters. These advantages include higher reliability, higher resistance to mechanical shock and higher visibility in low and moderate ambient lighting. Since the LED linear array is a light emitting device, it is more effective at getting the viewer's attention than a panel meter. Red, yellow and green LEDs can be used to quickly identify the proper limits of instrument operation. Distinct switching thresholds can be selected to allow the linear array to

simplify the machine operator's decision. Finally, in consumer oriented equipment, the LED linear array provides a new and distinctive selling feature for the product. Depending on the number of LEDs in the array, if the input signal is in analog form, the linear array decoder can be one or more operational amplifiers, a monolithic analog decoder, or a low cost analog to digital converter followed by a simple digital decoder. If binary or BCD information is available, such as from an analog to digital converter or a microprocessor, then only a simple digital decoder is required. Examples of several of these circuits will be shown.

When analog information is available, operational amplifiers or voltage comparators can be used to interface to the LED linear array. Since the LED lamps operate at low currents and low voltages, most operational amplifiers can drive LEDs without output buffering. Figure 2.4.4.4-2 shows examples of a bar graph display and a position indicator display. In both circuits, a five resistor voltage divider determines the switching thresholds of the LED lamps. The resistor values are not critical but the resistance ratio

2.30

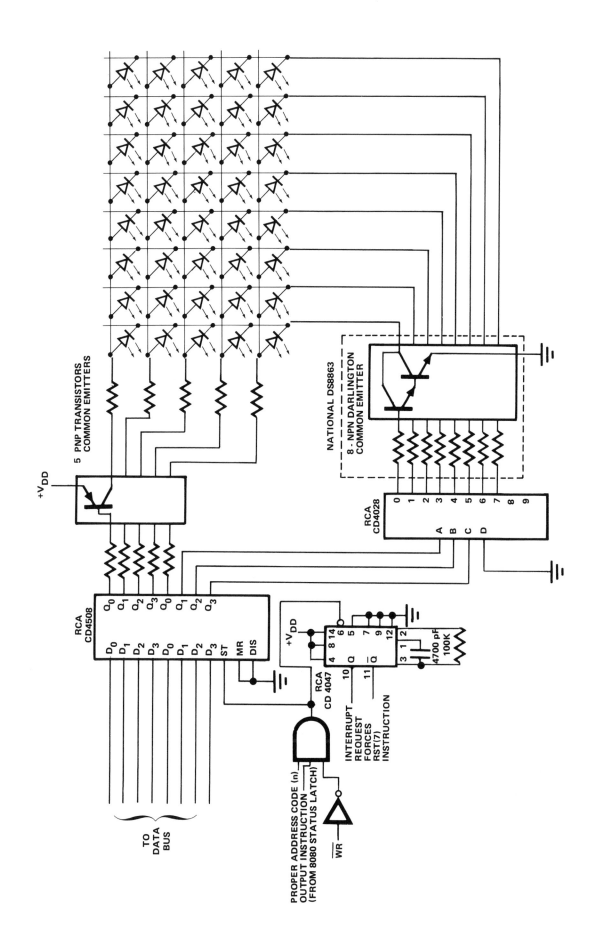

Figure 2.4.4.3-1 A Microprocessor Controlled X-Y Addressable Array.

ADDRESS	OP CODE	CLOCK CYCLES	COMMENTS
$(0038)_{16}$	PUSH PSW	11	
	PUSH HL	11	
	LHLD	16	HL = POINTER
	A_L	----	
	A_H		
	MOV A, M	7	A = (HL)
	OUT	10	STORES NEW ROW AND COLUMN INFORMATION
	(n)	----	
	INX HL	5	HL = HL + 1
	MOV A, M	7	A = (HL)
	OUT	10	STORES SECOND BYTE OF INFORMATION
	(n+1)	----	
ADDED FOR N EIGHT BIT LATCHES	. . .		
	INX HL	5	
	MOV A, M	7	
	OUT	10	STORES Nth BYTE OF INFORMATION
	(n+N-1)	----	
	MOV A, L	5	A = L
	CPI	7	COMPARES A TO $(17)_{16}$
	$(17)_{16}$	----	$(0F+8N)_{16}$ FOR N EIGHT BIT LATCHES
	JNZ	10	A ≠ $(17)_{16}$ JUMP TO LOOP
	$LOOP_L$	----	
	$LOOP_H$		
	MVI A	7	A = $(10)_{16}$
	$(10)_{16}$	----	
	STA	13	$POINTER_L$ = $(10)_{16}$
	A_L	----	
	A_H		
	POP HL	10	
	POP PSW	10	
	EI	4	ENABLE INTERRUPT FLAG
	RET	10	RETURN
LOOP	INX HL	5	HL = HL + 1
	SHLD	16	POINTER = POINTER + 1
	A_L	----	
	A_H		
	POP HL	10	
	POP PSW	10	
	EI	4	ENABLE INTERRUPT FLAG
	RET	10	RETURN

IN GENERAL,

FOR C COLUMNS, N EIGHT BIT LATCHES, THE TOTAL NUMBER OF CLOCK CYCLES REQUIRED TO REFRESH THE ENTIRE DISPLAY ONCE EQUALS:

CLOCK CYCLES = C (121 + 22N) - 1

= 1143 CLOCK CYCLES TO REFRESH THE DISPLAY IN FIGURE 2.4.4.3-1

MEMORY ADDRESS	CONTENTS (D_7-D_0)	COMMENTS
$A_H A_L$ $A_H A_L + 1$	$POINTER_L$ $POINTER_H$	POINTER = NEXT COLUMN TO BE DISPLAYED
$(X X 1 0)_{16}$	0 0 0 $Y_5 Y_4 Y_3 Y_2 Y_1$	COLUMN 1
$(X X 1 1)_{16}$	0 0 1	COLUMN 2
$(X X 1 2)_{16}$	0 1 0 5 BIT	COLUMN 3
$(X X 1 3)_{16}$	0 1 1 ROW	COLUMN 4
$(X X 1 4)_{16}$	1 0 0 INFOR-	COLUMN 5
$(X X 1 5)_{16}$	1 0 1 MATION	COLUMN 6
$(X X 1 6)_{16}$	1 1 0	COLUMN 7
$(X X 1 7)_{16}$	1 1 1	COLUMN 8

POINTER →

X = DON'T CARE

Figure 2.4.4.3-2 Intel 8080A Microprocessor Program used to Strobe the X-Y Addressable Array Shown in Figure 2.4.4.3-1.

between resistors will specify the threshold, V_1 to V_4, as shown in Figure 2.4.4.4-2. As shown, the linear displays can detect an analog signal greater than zero. In circuit 2.4.4.4-2a (bar graph), if $V_{IN} < V_1$, where V_1 is the switching threshold of the first lamp in the array, then all

A. NULL METER THAT COMPARES INPUT VARIABLE TO REFERENCE

B. INDOOR THERMOMETER

C. DIGITAL METER THAT OFFERS 5% RESOLUTION

D. LIQUID LEVEL INDICATOR

Figure 2.4.4.4-1 Typical Applications for Position Indicator and Bar Graph Arrays.

lamps in the array will be off. As V_{IN} increases above each switching threshold, V_1 to V_4, the corresponding output of each operational amplifier conducts current to ground which turns the LED on. R is selected to limit the current through the LED to the desired value. Since each LED lamp has its own current limiting resistor, standard red, high efficiency red, yellow, and green lamps can be mixed in the same array. Voltage comparators with open collector outputs can also be used in the circuit. In circuit 2.4.4.4-2b (position indicator) only one LED lamp in the array will be on. If $V_{IN} < V_1$, the outputs of all operational amplifiers will be pulled up to V_{CC}. U_1 will source current to LED_1, turning it on but LED_2 through LED_5 will be turned off with approximately zero bias. As V_{IN} increases above V_1, the output of U_1 will conduct current to ground, turning LED_1 off but turning LED_2 on. Each LED lamp will turn on sequentially as V_{IN} increases above the specified thresholds. Either circuit can be expanded in size to accomodate any desired bar graph display. The circuits shown in Figure 2.4.4.4-2 can be operated from a single polarity power supply providing that only positive signals are to be detected. Both positive and negative signals can be detected if a dual polarity power supply is used. A dual operational amplifier with three LEDs can be connected as shown in circuit 2.4.4.4-2b to make a solid state null detector circuit, that compares V_{IN} to a specified V_{REF} determined by the resistor divider network. The LEDs would indicate $V_{IN} < (V_{REF} - \Delta)$, $(V_{REF} - \Delta) < V_{IN} < (V_{REF} + \Delta)$, and $V_{IN} > (V_{REF} + \Delta)$, or negative, zero or positive if V_{REF} is equal to zero and Δ is small. Such a device could replace a conventional analog null meter.

For longer LED linear arrays, IC decoders are available that decode the analog input signal into a bar graph or a position indicator display. Figure 2.4.4.4-3 shows an example of an LED bar graph display that uses the Siemens UAA-180 decoder and a position indicator display that uses the Siemens UAA-170 decoder. For both devices, a voltage divider network consisting of R_3, R_4, and R_5 determine

$$V_1 = \frac{R_1}{R_1+R_2+R_3+R_4+R_5} \, V_{CC}$$

$$V_2 = \frac{R_1+R_2}{R_1+R_2+R_3+R_4+R_5} \, V_{CC}$$

$$V_3 = \frac{R_1+R_2+R_3}{R_1+R_2+R_3+R_4+R_5} \, V_{CC}$$

$$V_4 = \frac{R_1+R_2+R_3+R_4}{R_1+R_2+R_3+R_4+R_5} \, V_{CC}$$

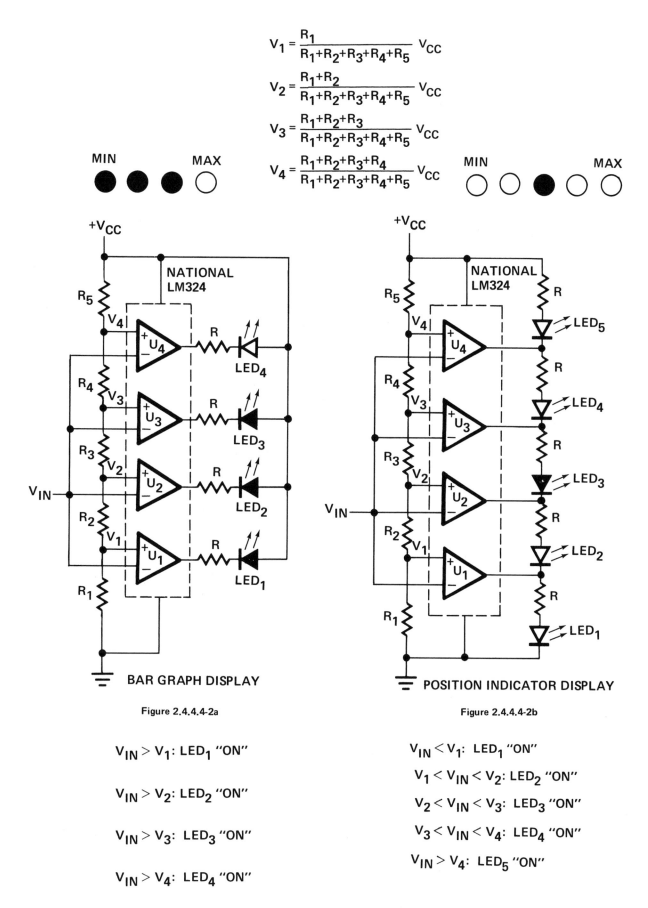

BAR GRAPH DISPLAY

Figure 2.4.4.4-2a

POSITION INDICATOR DISPLAY

Figure 2.4.4.4-2b

$V_{IN} > V_1$: LED$_1$ "ON"

$V_{IN} > V_2$: LED$_2$ "ON"

$V_{IN} > V_3$: LED$_3$ "ON"

$V_{IN} > V_4$: LED$_4$ "ON"

$V_{IN} < V_1$: LED$_1$ "ON"

$V_1 < V_{IN} < V_2$: LED$_2$ "ON"

$V_2 < V_{IN} < V_3$: LED$_3$ "ON"

$V_3 < V_{IN} < V_4$: LED$_4$ "ON"

$V_{IN} > V_4$: LED$_5$ "ON"

**Figure 2.4.4.4-2 Operational Amplifiers or Voltage Comparators
Used to Decode an Analog Signal into a Bar Graph
or Position Indicator Display.**

the switching range of the linear array. Using the UAA-180 decoder, if V_{IN} is less than V_{MIN}, all LED lamps will be off. If V_{IN} is greater than V_{MIN}, the input voltage will be decoded as a bar graph array with all LEDs on when V_{IN} is greater than V_{MAX}. For the UAA-170 decoder when V_{IN} is less than V_{MIN}, then LED_1 will be on. As V_{IN} increases above V_{MIN} then LED_2 through LED_{16} will turn on sequentially. When V_{IN} is greater than V_{MAX}, LED_{16} will be on. Since the UAA-170 and UAA-180 have constant current drivers for the LEDs, no current limiting resistors are required. However, the constant current drivers are programmed externally by R_1 and R_2. If longer linear arrays are desired, additional UAA-170s or UAA-180s can be cascaded in series.

Digital information can also be decoded as an LED bar graph or a position indicator display. If an analog signal is available, it may be desirable and cost effective to use a low cost analog to digital converter and then decode the digital outputs as a linear array. The position indicator display can be decoded with a one of n decoder as shown in Figure 2.4.4.4-4. One of n decoders can be cascaded to form any size of LED position indicator display desired. Since only one LED in the position indicator display will be on at a time, an x-y addressable array can be used to simplify decoding. Figure 2.4.4.4-5 shows how two one of eight decoders and 17 external components can be used to address 64 LED lamps in a position indicator display. Using commercially available 1 of 4, 1 of 8, 1 of 10, or 1 of 16

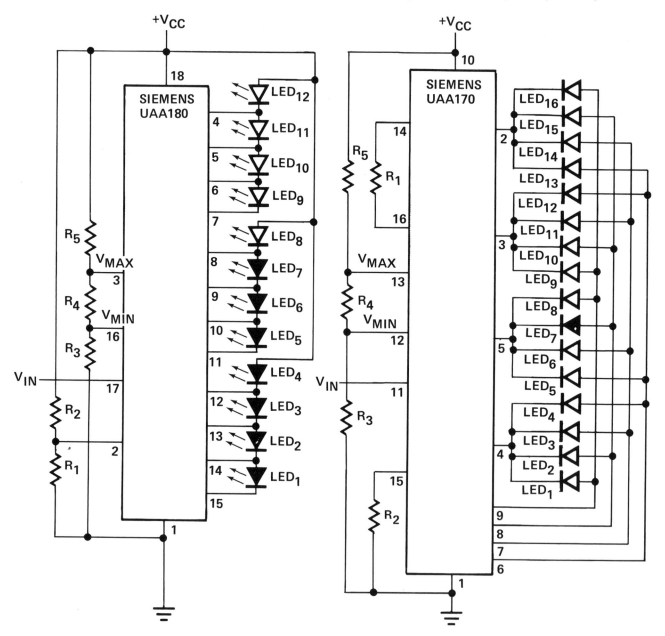

Figure 2.4.4.4-3 Use of Siemens UAA 170 and UAA 180 Position Indicator and Bar Graph Decoders.

2.34

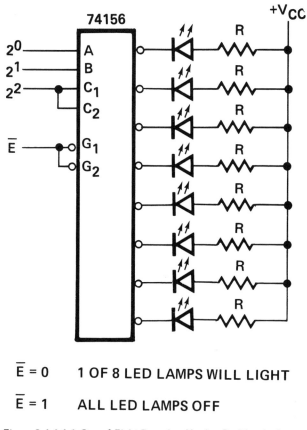

$\overline{E} = 0$ 1 OF 8 LED LAMPS WILL LIGHT

$\overline{E} = 1$ ALL LED LAMPS OFF

Figure 2.4.4.4-4 One of Eight Decoder Used as Position Indicator Decoder/Driver.

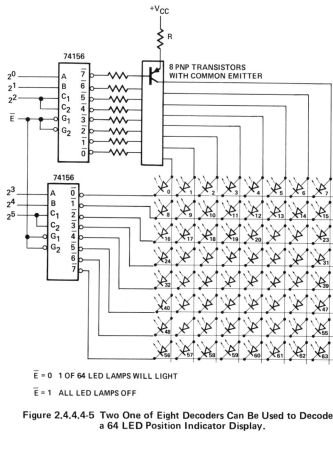

$\overline{E} = 0$ 1 OF 64 LED LAMPS WILL LIGHT

$\overline{E} = 1$ ALL LED LAMPS OFF

Figure 2.4.4.4-5 Two One of Eight Decoders Can Be Used to Decode a 64 LED Position Indicator Display.

decoders, almost any size LED position indicator array can be decoded. Using the technique shown in Figure 2.4.4.4-4, four 1 of 16 decoders, four inverters (4/6 package) and 64 external components would be used to address the same 64 LED lamps.

Digital logic can also be used to decode an LED bar graph display. Figure 2.4.4.4-6 shows how a one of eight decoder and an And gate network can be used to decode an eight LED bar graph display. The display can be expanded to any length with the proper decoder and an extended And gate network. For large displays, a strobed x-y addressable approach is simpler. Figure 2.4.4.4-7 shows an example of such an approach. The lower three bits of the input are connected to an eight output bar graph decoder. The upper three bits of the input are continuously compared to the outputs of a divide by eight counter. The output of the counter is decoded by a one of eight decoder which row strobes the 64 LED matrix on a 1/8 duty factor. If the value of the counter is less than the upper three bits of the input, then all LEDs in that row are turned on. When the value of the counter is equal to the upper three bits of the input, the lower three bits of the input are decoded by the bar graph decoder. For values of the counter that are greater than the input, all LED lamps for that row are turned off. This technique can also be used to decode BCD inputs by substituting a decade counter, one of ten

decoder, and ten line bar graph decoder for the devices specified. The maximum size of such an array is limited only by the requirement of a certain minimum duty factor to obtain the desired time averaged luminous intensity.

Microprocessors can be interfaced directly to any of the digital decoder circuits previously shown by adding a latch to hold the input information. The microprocessor would need to update the latch only when the information is to be changed. The techniques described in section 2.4.4.3 can also be used to implement a linear array. Unless the application is microprocessor time limited, these techniques are more cost effective than the bar graph decoder circuit described in Figure 2.4.4.4-7. The microprocessor would be required to decode the input information in software and continuously refresh the linear array. The circuit described in Figure 2.4.4.3-1 can be used to implement either a bar graph or a position indicator display.

2.4.5 Backlighting

Information can be more readily assimilated from a panel by an observer if only those symbols or characters relevant to a particular condition are visible. Ideally, the panel has a "dead front" appearance for all symbols except those required to be seen, and those are made visible with light

2.35

projected through them from a source in back of the panel; hence the term "backlighting", illustrated in Figure 2.4.5-1.

Red LEDs of earlier technology did not produce enough light to make backlighting practical. More recent technology not only makes red backlighting practical, but yellow and green as well.

Front lighting differs from backlighting in that it is used mainly as a substitute for or a supplement to ambient illumination. Because of scattered and stray light, it is difficult to selectively display symbols by a front lighting technique. LEDs can also be used in front lighting since the supplementary light is usually needed only when ambient light is subdued. Front lighting is, therefore, a simple design and does not require as much design attention as backlighting.

2.4.5.1 Fundamental Backlighting Requirements

There are only four parameters to consider. In the order of usual importance, they are:

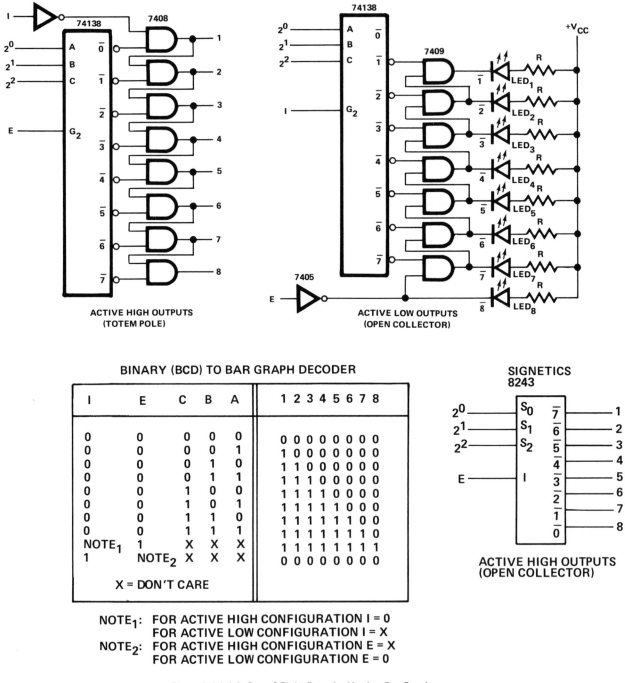

BINARY (BCD) TO BAR GRAPH DECODER

I	E	C	B	A	1 2 3 4 5 6 7 8
0	0	0	0	0	0 0 0 0 0 0 0 0
0	0	0	0	1	1 0 0 0 0 0 0 0
0	0	0	1	0	1 1 0 0 0 0 0 0
0	0	0	1	1	1 1 1 0 0 0 0 0
0	0	1	0	0	1 1 1 1 0 0 0 0
0	0	1	0	1	1 1 1 1 1 0 0 0
0	0	1	1	0	1 1 1 1 1 1 0 0
0	0	1	1	1	1 1 1 1 1 1 1 0
NOTE$_1$	1	X	X	X	1 1 1 1 1 1 1 1
1	NOTE$_2$	X	X	X	0 0 0 0 0 0 0 0

X = DON'T CARE

NOTE$_1$: FOR ACTIVE HIGH CONFIGURATION I = 0
FOR ACTIVE LOW CONFIGURATION I = X
NOTE$_2$: FOR ACTIVE HIGH CONFIGURATION E = X
FOR ACTIVE LOW CONFIGURATION E = 0

Figure 2.4.4.4-6 One of Eight Decoder Used as Bar Graph Decoder/Driver.

2.36

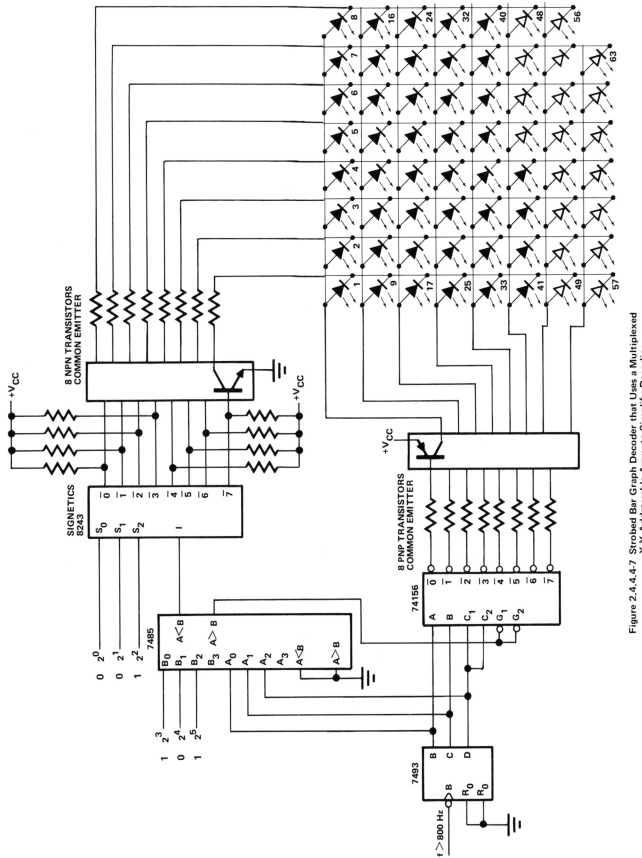

Figure 2.4.4.4-7 Strobed Bar Graph Decoder that Uses a Multiplexed X-Y Addressable Array to Simplify Decoding.

2.37

1. Contrast
2. Uniformity
3. ON Sterance*
4. Radiation Pattern

*Because this section deals only in photometric (no radiometric) quantities, the prefix luminous and subscript v may be dropped from the terms and symbols:

1. Luminous flux, ϕ_v
2. Luminous incidance, E_v
3. Luminous exitance, M_v
4. Luminous intensity, I_v
5. Luminous sterance, I_v

The radiation pattern of the backlighted symbol is ideally lambertian; but if the symbol complexity makes it unrecognizable anyway at large off-axis angles, the angular variation of intensity (or sterance) is of little importance.

Figure 2.4.5-1   Backlighting Used to Selectively Cause Special Symbols to Appear.

ON sterance describes the intensity per unit area of the symbol when backlighting is applied. For a particular background and symbol reflectance, the ON sterance determines the ambient light level where the backlighted symbol can be adequately distinguished. "Adequate" distinction of the symbol is not precisely definable because it bears a subjective relationship to the detail which must be resolved in the symbol.

Uniformity of sterance over the illuminated area is also somewhat subjective. In some applications, it has cosmetic value only, but in others, non-uniformity can cause loss of detail which may cause misrecognition of the symbol. Uniformity does, however, have an objective criteria for evaluation. Because eye response is logarithmic, a 3dB

sterance variation (2:1) is usually acceptable, and this is the basis of the developments described later.

Contrast, the most important consideration, is evaluated in terms of three contrast ratios:

a) ON/OFF CR
b) ON/BACKGROUND CR
c) OFF/BACKGROUND CR

ON/OFF CR is the ratio of ON sterance to OFF sterance of the symbol. The OFF sterance is the product of surface reflectance and ambient light, so ON/OFF CR varies inversely as the ambient light level. Similarly, ON/BACKGROUND CR is the ratio of the ON sterance of the symbol to the BACKGROUND sterance, which also is the product of surface reflectance and ambient light. Neither ON/OFF CR nor ON/BACKGROUND CR should always be "as high as possible". If ON/BACKGROUND CR is too high, eye fatigue may result. This is because the ON sterance results from radiated, not reflected, flux. Both ON/OFF CR and ON/BACKGROUND CR should, of course, be adequate and ideally they are equal.

If ON/OFF CR = ON/BACKGROUND CR, it follows that OFF/BACKGROUND CR is 1.00. When tradeoffs are being made, because of high ambient conditions and marginal flux availability, the usual optimization favors ON/BACKGROUND CR, but OFF/BACKGROUND CR must not be ignored.

Adequate ON/BACKGROUND CR requirements are subjective and BACKGROUND reflectances vary widely, but an approximation to the ON sterance requirement can be drawn from Figure 2.4.5-2. The curve is drawn from:

1. Assume normal ambient, E_A = 500 lx

2. Assume BACKGROUND surface reflectance, R_{BG}:

   $M_{BG}/E_A = R_{BG} = 2\%$

3. Assume lambertian radiation of flux reflected, so $M_{BG} = \pi L_{BG}$

4. Calculate BACKGROUND sterance, L_{BG}:

$$L_{BG} \left(\frac{cd}{m^2}\right) = \left[\frac{1}{\pi} \left(\frac{cd}{lm}\right)\right] [E_A \, (lx)] \left[\frac{R_{GB}(\%)}{100\%}\right]$$

$$= \left[\frac{1}{\pi} \left(\frac{cd}{lm}\right)\right] [500 \, lx] [0.02]$$

$$= 3.18 \frac{cd}{m^2}$$

(2.4.5-1)

5. Require ON/BACKGROUND CR, (L_{ON}/L_{BG}) to be within the limits $2.5 < (L_{ON}/L_{BG}) < 10$; choose $(L_{ON}/L_{BG}) = 5$ and calculate:

$$L_{ON} = L_{BG} \ (L_{ON}/L_{BG}) \qquad (2.4.5\text{-}2)$$

$$= 3.18 \ \frac{cd}{m^2} \ (5) = 15.92 \ cd/m^2$$

For the tolerable limits required in step 5, the curve in Figure 2.4.5.2 shows that an ambient ranging from 250 lx to 1000 lx can be accomodated. Less narrow requirements would, of course, broaden the range.

2.4.5.2 ON Sterance Design Considerations

There are two requirements for the ON sterance. It must be adequately high and uniform. To obtain uniform sterance requires a diffuser, as shown in Figure 2.4.5-3. The diffuser/LED combination, properly designed, gives a background of uniform sterance over which the "legend plate", bearing the symbol opening, is placed. Without proper diffusion, sterance would not be uniform over all portions of the symbol. A viewer would see the LED through only those portions of the symbol opening that lie between the viewer's eye and the LED.

A diffuser functions as diagrammed in Figure 2.4.5-3. Normal (perpendicular) incidence, E_v, on the back surface causes flux to be emitted at the front side, being scattered by the diffusant such that each increment of area on the front radiates in all directions. In an ideal (but not realizable) diffuser, each lumen impinging at the back would emerge from the front; that is, nothing would be reflected and nothing lost in the diffuser. Upon emerging, this lumen would be scattered; ideally, the scattering would be lambertian, so the ratio of total flux emitted to normal (perpendicular) intensity is π. Thus, in an ideal diffuser, an incidence of one lumen per square meter would cause emission of one lumen per square meter and normal intensity would be $1/\pi$ candelas per square meter. A "Diffusance Quotient" (similar to Intelligence Quotient) can be defined as the ratio:

$$D.Q. = \pi \ L_{VOUT}/E_{vIN} \qquad (2.4.5\text{-}3)$$

having a normal value of 1.00. D.Q. can be lowered by absorption and reflection loss, and can be raised by anisotropic (e.g., light pipe) effects in the diffuser.

For example, one type of diffusing film is described as having a transmission of 55% and a "gain" of 400%. For this film, the D.Q. is 0.55 x 4.00 = 2.2, and from equation (2.4.5-3) we can calculate:

$$\frac{L_{v \ OUT} \ (cd/m^2)}{E_{v \ IN} \ (lx)} = \frac{D.Q.}{\pi} = 0.70 \ (\frac{cd/m^2}{lx}) \qquad (2.4.5\text{-}4)$$

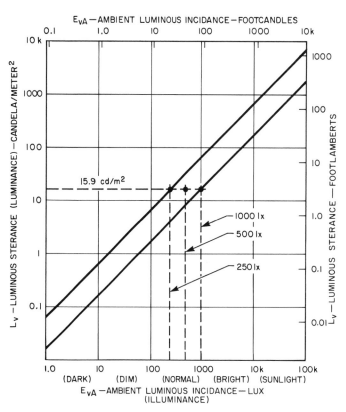

Figure 2.4.5-2   Approximate "ON" Luminous Sterance Required Under Given Ambient Light.

LAMBERTIAN LED: $\dfrac{E_v(x)}{E_v(0)} = \cos^4(\theta) = \left(\dfrac{1}{1+\left(\frac{x}{d}\right)^2}\right)^2$

NON-LAMBERTIAN: $\dfrac{E_v(x)}{E_v(0)} = \left(\dfrac{I_v(\theta)}{I_v(0)}\right) \cos^3\theta$

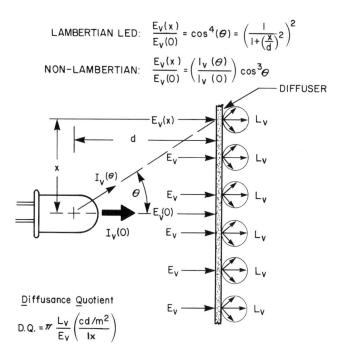

Diffusance Quotient

$$D.Q. = \pi \ \frac{L_v}{E_v} \left(\frac{cd/m^2}{lx}\right)$$

Figure 2.4.5-3   Diffuser Effect on Flux Direction as Related to Radiation Pattern.

2.39

Industry standards for characterization of diffusers have not yet appeared, although diffusers are available such as "Light Diffusing Film" from 3M Visual Products Division, and "Chromafuse" from Panelgraphic. Tradeoffs in diffuser selection relate to D.Q. -- a high D.Q. (>1) means the radiation pattern is narrower than lambertian because transmission loss cannot be zero. It is unlikely that the radiation pattern would be broader than lambertian, so a low D.Q. (<1) means a lossy diffuser, but a narrow radiation pattern can compensate. D.Q. is therefore the one single number that comes closest to relating diffuser specifications to backlighting performance.

Adequately high ON sterance may require a tradeoff with uniformity or the addition of extra LEDs. This can be seen by analyzing, in Figure 2.4.5-3, the incidence on the diffuser at points displaced a distance, x, from the axis. The analysis uses the inverse square law*:

$$E_v = I_v/d^2 \qquad (2.4.5-5)$$

　*accuracy of the inverse square law is within 5% for distance, d, as short as two LED diameters.

At $x = 0$, $\theta = 0$, so the incidence on the diffuser is:

$$E_v(0) = I_v(0)/d^2 \qquad (2.4.5-6)$$

At any other point, x, the incidence VECTOR is:

$$|E_v| \angle \theta = I_v(\theta)/(x^2 + d^2) \qquad (2.4.5-7)$$

but the NORMAL incidence component of the incidence VECTOR is the vector magnitude multiplied by $\cos \theta$, so:

$$E_v(x) = [I_v(\theta)/(x^2 + d^2)]\ \cos \theta \qquad (2.4.5-8)$$

The non-uniformity of incidence can be seen as the ratio $E_v(x)/E_v(0)$, found by taking the ratio of equation (2.4.5-8) to (2.4.5-6):

$$\frac{E_v(x)}{E_v(0)} = \frac{[I_v(\theta)/x^2 + d^2)]\ \cos \theta}{I_v(0)/d^2} \qquad (2.4.5-9)$$

$$= \frac{I_v(\theta)}{I_v(0)} = (\frac{d^2}{x^2 + d^2})\ \cos \theta$$

$$= \frac{I_v(\theta)}{I_v(0)}\ \cos^3 \theta$$

In equation (2.4.5-9), the ratio $[I_v(\theta)/I_v(0)]$ is recognized as the relative intensity vs. θ, which is the radiation pattern as given in most LED data sheets. Thus, if the LED radiation pattern is lambertian, $\cos \theta$ can be substituted for $[I_v(\theta)/I_v(0)]$ to give:

$$(2.5.4-10)$$

$$\frac{E_v(x)}{E_v(0)} = \cos^4 \theta \quad \text{FOR LAMBERTIAN RADIATION PATTERN}$$

A curve of equation (2.4.5-10) is given in Figure 2.4.5-4 along with curves for several non-lambertian LEDs. The curves for the non-lambertian LEDs were obtained by applying equation (2.4.5-9) to the radiation patterns in their data sheets.

Here is how to use these curves and equations for a single LED backlighting a legend, assuming no reflectors:

EXAMPLE:

Legend area: 10mm x 20mm

ON Sterance, L_v = 20 cd/m^2

Uniformity: corner $L_v \geqslant 0.5$ x center L_v

Assume diffuser mentioned earlier: D.Q. = 2.2

SOLUTION:

1.　Refer to Figure 2.4.5-3 and determine $E_v(0)$ from equation (2.4.5-3):

$$E_v(0) = \frac{\pi \times 20\ \text{cd/m}^2}{2.2} = 28.56\ \text{lx}$$

2.　Find the distance from the center to the corner:

$$x = \tfrac{1}{2} \sqrt{(10\text{mm})^2 + (20\text{mm})^2} = 11.18\text{mm}$$

3.　Choose an LED and either construct a curve according to equation (2.4.5-9) or use a curve from Figure 2.4.5-4. Assume now a type 5082-4650/55. Find θ at which $E_v/E_v(0) = 0.5$:

　Figure 2.4.5-4 gives $\theta = 25°$

4.　Compute LED-to-diffuser distance:

　$d = x/\tan \theta = 11.18\text{mm}/\tan 25° = 23.98\text{mm}$

5.　Compute LED intensity required, using equation (2.4.5-6) and results of step 1, step 4:

　$I_v(0) = 28.56\ \text{lx} \cdot (23.98\text{mm})^2 = 16.42\ \text{mcd}$

This is about five times the 3.0 mcd minimum specified for the 5082-4655, so a different LED is tried. Applying steps 3, 4, 5 to a 5082-4657/58:

2.40

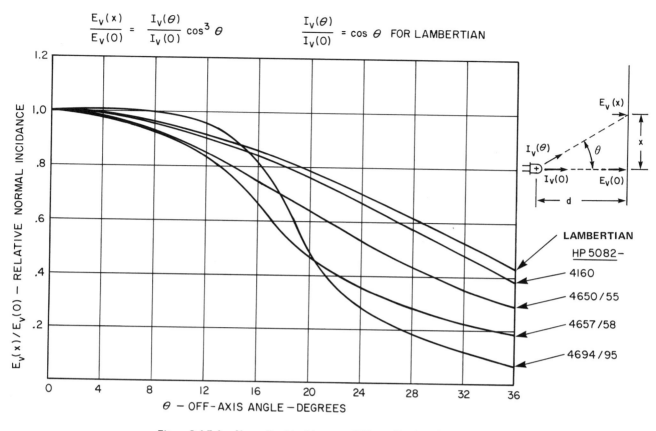

$$\frac{E_V(x)}{E_V(0)} = \frac{I_V(\theta)}{I_V(0)} \cos^3 \theta \qquad \frac{I_V(\theta)}{I_V(0)} = \cos \theta \quad \text{FOR LAMBERTIAN}$$

LAMBERTIAN
HP 5082–
—4160
—4650 / 55
—4657 / 58
—4694 / 95

θ — OFF-AXIS ANGLE — DEGREES

$E_V(x)/E_V(0)$ — RELATIVE NORMAL INCIDENCE

Figure 2.4.5-4   Normalized Incidence on Diffuser Obtained from Radiation Patterns.

3. From Figure 2.4.5-4, $\theta = 19°$

4. d = 11.18mm/tan 19θ = 32.47mm

5. $I_V(0) = 28.56$ lx x $(32.47\text{mm})^2 = 30.11$ mcd

This is marginal for the 5082-4657, but the 5082-4658 will do, with $I_F \cong 16$ mA.

Note that the exercise in trying the 5082-4350/55 could have been avoided by first computing the intensity of the required legend area:

$$I_{V\ OUT} = L_{V\ OUT} \times \text{LEGEND AREA} \qquad (2.4.5\text{-}11)$$

$$= 20 \text{ cd/m}^2 (10 \text{ mm x } 20 \text{ mm})$$

$$= 4 \text{ mcd}$$

A single LED would hardly be likely to product a legend area intensity greater than the basic LED intensity.

Note also that the solution with the 5082-4658 requires a distance behind the diffuser of 32.47 mm, which may be inconveniently large. By using more than one LED, the ON sterance and uniformity requirements over the entire legend area can be met with much less distance required behind the panel. The design procedure is a bit more complicated because it is necessary to take account of the overlapping incidence from adjacent LED(s). The 10mm x 20mm box should not be evenly subdivided (e.g. 2 LEDs 10mm apart) because the corners do not receive the overlapping

incidance. A little more "cut-and-try" on paper is much less costly, however, than "scrap-and-try-again" on the workbench.

Another POSSIBLE tradeoff would be to allow a slightly greater non-uniformity, i.e., $E_V(x)/E_V(0)<0.5$. This would permit a larger θ, and hence a smaller d and $I_V(0)$ would be possible. Remember that $I_V(0)$ requirement varies jointly as the SQUARE of d.

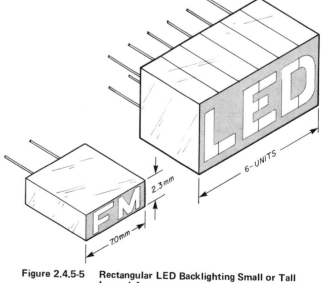

Figure 2.4.5-5   Rectangular LED Backlighting Small or Tall Legend Areas.

2.41

2.4.5.3 Backlighting Construction for Small-to-Medium Legend Areas

For illuminating small symbols, the simplest approach is the use of a rectangular LED, designed for the purpose and having built-in diffusant that meets most uniformity requirements. As shown in Figure 2.4.5-5, it can also be clustered to illuminate fairly large areas. At their specified input current levels, their ON sterance easily exceeds 40 cd/m^2. All that is needed with these is the symbol-bearing legend plate and perhaps a contrast enhancement filter. If it is desirable to have separately-illuminated areas in the cluster, thin black tape separators are adequate for crosstalk control. The bodies of these rectangular LEDs are translucent, so for good OFF/BACKGROUND CR they should be shielded from stray light behind the panel -- more black tape. Painting is not recommended.

Another simple approach uses LEDs made with undiffused encapsulant as in Figure 2.4.5-6. This "direct-illumination" method uses the basic design consideration of Section 2.4.5.2 to select the LED and spacing, d, for proper ON sterance and uniformity. This approach is typically used for fairly small legend areas (<10mm diameter); the large (10mm x 20mm) area was prescribed in the earlier examples for purposes of illustrating design tradeoffs.

The shortest distance behind the panel is achieved by using an "egg crate" reflector in which each LED is near the apex of a conically- or pyramidally-shaped reflector cavity. As seen in Figure 2.4.5-7, the reflector can be made to either combine more than one LED to backlight a large legend, or provide separation with a "web" to isolate separately energized areas. Egg-crate reflectors offer the additional benefit of re-directing side-emitted flux from the LED so that efficiency is raised and uniformity is improved. Because much of the flux reaching the diffuser comes *indirectly* from the reflector rather than directly from LED, this arrangement is described as "indirect" illumination.

Figure 2.4.5-7 also shows the recommended placement of the diffuser, legend plate, and (if used) contrast enhancement filter. Placing the legend plate behind the diffuser would "break" the symbol edges and lower the OFF/BACKGROUND CR; so unless fuzzy symbol edges are desired, the legend plate should be located as shown. Contrast enhancement filtering is always out in front.

Another method of indirect illumination uses a plastic "light pipe" as in Figure 2.4.5-8. Except for the hole to accept the LED, the plastic is solid and relies on greater-than-critical-angle reflections for its efficiency. Scratches and contact with materials other than air cause light losses at the surfaces, so these surfaces should be protected. Over a length of more than four diameters, multiple reflections cause enough diffusion to permit elimination of the separate diffuser; the front end of the "pipe" may be roughened if the length is marginal.

2.4.5.4 Backlighting Construction for Very Large Legend Areas

For very large symbols to be viewed from large distances (>10m) the uniformity requirement is sharply reduced. This permits the LEDs to be viewed directly through the symbol with NO DIFFUSER. To obtain highest possible efficiency (or ON sterance), each LED in the arrays of Figure 2.4.5-9 should be seated in a cell of an "egg crate"-type reflector. The average ON sterance is then found by dividing the single LED intensity by the area of a single cell. In Figure 3.4.5-9, notice that both arrays have the same number of LEDs, but different aspect ratios. Except for this, the "honeycomb" advantage over the "square" is rather small (a factor of only $2/\sqrt{3}$). Overlapping incidence is likely to leave smaller "cold" spots in the "honeycomb" than in the "square".

2.5 Communications and Signalling Applications

The terms "communications and signalling" are intended to categorize all those applications in which the primary function of the LED is to provide flux for detection by means other than human vision (but may include some of these as well). Such applications include:

- Card/tape reader (esp. low hole/no-hole ratio)
- Tape loop stabilizer (max and min loop sensors)
- End-of-tape sensor (reflective or transmissive)
- Optical tachometer (motor speed control)
- Assembly line monitor (parts counting/orientation)
- Bar code scanner (POS machine UPC)
- Opto-mechanical synchronizer (ignition timing)
- Safety interlock (with phase-lock loop in high ambient)
- Carriage travel sensor (beam break or reflect)
- Shaft position encoder (using arrayed devices)
- High voltage isolator (air gap or fiber optic)
- Smoke detector (both scattering- and obscuration-type)
- Densitometer (chemical analysis)
- Liquid level monitor (clear as well as opaque)

Their shorter wavelengths and higher modulation bandwidths give LEDs performance superior to IREDs in many of these applications, despite the higher quantum efficiencies of IREDs. Now new devices are available with wavelengths short enough to benefit from spectral considerations, but with quantum efficiencies so high (up to 1.5%) that they rival that of amphoteric IREDs. These new devices, emitting at 670 nm and 700 nm, are sufficiently visible that optical alignment can be done visually without the spectral viewing equipment required with IREDs.

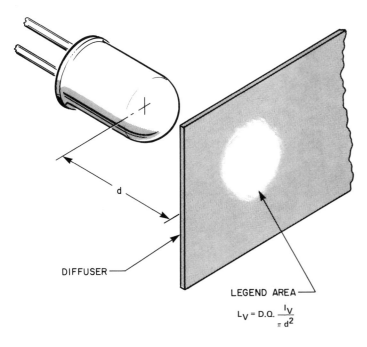

DIFFUSER

LEGEND AREA

$$L_V = D.Q. \frac{I_V}{\pi d^2}$$

Figure 2.4.5-6   Direct Illumination; Undiffused LED Makes Sharp Edges when "d" is Small.

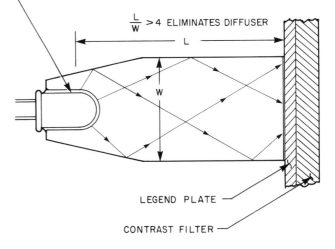

LED CEMENTED IN PLACE WITH UNDIFFUSED EPOXY

$\frac{L}{W} > 4$ ELIMINATES DIFFUSER

L

W

LEGEND PLATE

CONTRAST FILTER

Figure 2.4.5-8   Plastic Light Pipe; Reflects Side Emission and Diffuses Forward.

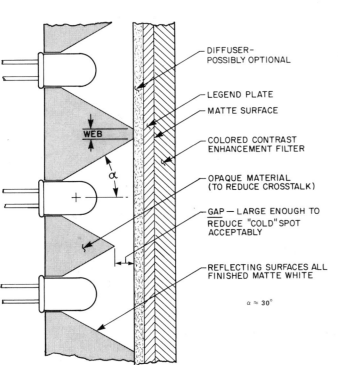

DIFFUSER—POSSIBLY OPTIONAL

LEGEND PLATE

MATTE SURFACE

COLORED CONTRAST ENHANCEMENT FILTER

OPAQUE MATERIAL (TO REDUCE CROSSTALK)

GAP — LARGE ENOUGH TO REDUCE "COLD" SPOT ACCEPTABLY

REFLECTING SURFACES ALL FINISHED MATTE WHITE

WEB

α

$\alpha \approx 30°$

Figure 2.4.5-7   Indirect Illumination; Egg-Crate Reflector Improves Uniformity.

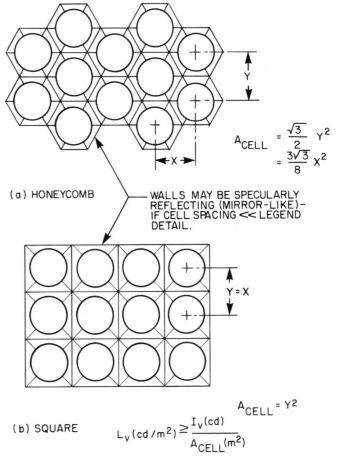

Y

X

(a) HONEYCOMB

$$A_{CELL} = \frac{\sqrt{3}}{2} Y^2 = \frac{3\sqrt{3}}{8} X^2$$

WALLS MAY BE SPECULARLY REFLECTING (MIRROR-LIKE)— IF CELL SPACING ≪ LEGEND DETAIL.

Y = X

(b) SQUARE

$$A_{CELL} = Y^2$$

$$L_V(cd/m^2) \geq \frac{I_V(cd)}{A_{CELL}(m^2)}$$

Figure 2.4.5-9   LED Clusters — Honeycomb or Square Patterns for Very Large Legend Area.

2.43

2.5.1 Device Characterization for Communications and Signalling

Specifications of luminous intensity or luminous sterance are not directly applicable in most signalling applications. However, they can be converted to radiometric quantities if the luminous efficacy, η_v, is known. If η_v is not given in the data sheet it can be found by the method described in Section 7.1. Equation 2.5.1-1 gives the relationship to be used in making the conversion:

$$\eta_v\left(\frac{lm}{W}\right) = \frac{\phi_v(lm)}{\phi_e(W)} = \frac{I_v(cd)}{I_e(W/sr)} = \frac{L_v(cd/m^2)}{L_e\left(\frac{W/sr}{m^2}\right)} \qquad \text{(2.5.1-1)}$$

The most common conversion requirement is for intensity. For example, if an LED has a luminous intensity $I_v = 3.5$ mcd and a luminous efficacy $\eta_v = 147$ lm/W, then, since 1 cd = 1 lm/sr:

$$I_e = \frac{3.5 \times 10^{-3}\ cd}{147\ lm/W} = 23.8\ \mu W/sr \qquad \text{(2.5.1-2)}$$

Conversely, if a device used in signalling is used also for visual effects, if η_v is known, the photometric properties can be obtained from given radiometric quantities.

Another important parameter in signalling is the radiation pattern. This is usually given in the form of $I_r(\theta)$, the relative intensity normalized at $\theta = 0$ so $I_r(0) = 1.0$. The intensity at any angle is then the product of the axial intensity multiplied by $I_r(\theta)$.

To obtain the amount of flux that is radiated into a cone of half-angle θ, or included-angle 2θ, it is necessary to evaluate the integral:

$$\phi(\theta) = \int_0^\theta I_e I_r(\theta)\,[2\pi\sin\theta]\,d\theta \qquad \text{(2.5.1-3)}$$

Here I_e is the axial intensity. The integral can be done mathematically if $I_r(\theta)$ is a reasonably describable function. If it is not, a summation is done by the method of Section 7.3.4.

When used with an optical system for which the **numerical aperture** (N.A.) is given, the amount of flux the LED radiates into the system is found from equation 2.5.1-3, since $\theta = \sin^{-1}(N.A.)$. To aid designers in quickly evaluating this flux, the data sheets for the new 670 nm and 700 nm devices give the results of equation 2.5.1-3 in normalized form:

$$\frac{\phi(\theta)}{I_e} = \int_0^\theta I_r(\theta)\,[2\pi\sin\theta]\,d\theta \qquad \text{(2.5.1-4)}$$

Thus the flux into a particular N.A. is the product of the specified axial intensity, multiplied by the normalized integral:

$$\phi(N.A.) = I_e \times \left.\left(\frac{\phi(\theta)}{I_e}\right)\right|_{\theta\,=\,\sin^{-1}(N.A.)} \qquad \text{(2.5.1-5)}$$

2.5.2 Flux Properties in Signalling

Wavelength compatibility can be of critical importance in dealing with mediums having sharply varying spectral transmittance or reflectance, or with detectors such as photoconductors, since they have relatively narrow spectral response. The effective flux coupling is found by spectral integration of the product of all spectrums involved in the system, as shown in Figure 2.5.2-1. In addition to those shown in the figure, there may also be other spectral effects, and these should be included in the product to be integrated.

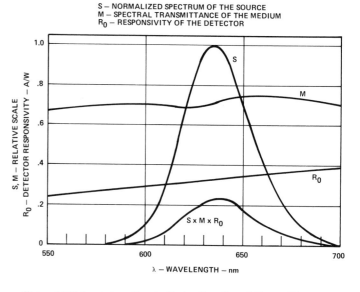

Figure 2.5.2-1   Integration to Derive Coupling of Spectral Source, Medium, and Detector.

Stability of the LED flux is sometimes also a matter of concern. It is unwise to depend on any randomly chosen LED or IRED to be a source of stable flux, even if the forward current is precisely regulated. LEDs used as optical standards would be no exception except that they are elaborately heat-sunk and are operated intermittently at a forward current that is far below rated maximum; even with these precautions, the ambient temperature must be noted and appropriate correction applied.

For applications requiring stable flux from potentially unstable LEDs (e.g. for photometer or radiometer transfer standard), the best technique is the use of a beam splitter

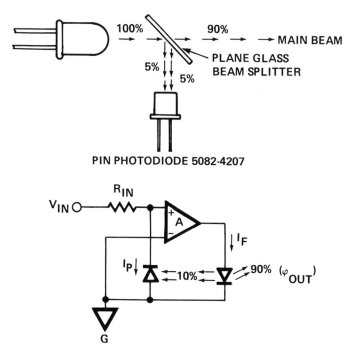

Figure 2.5.2-2 Servo Stabilization; Main Beam Flux Linearly Related to Input Voltage.

and a PIN photodiode in a servo system as shown in Figure 2.5.2-2. This arrangement is especially good at wavelengths below 800 nm where PIN photodiodes have a temperature coefficient of zero. If the LED used in this arrangement has a very narrow beam angle, the relative position of the photodiode should be adjusted to give the same numerical aperture (N.A.) of the optical system into which the main beam is directed. Obviously, mechanical movement of the components is not tolerable.

2.5.3 Lens System with LEDs

A fundamental principle of optics is that sterance is a constant. That is, as the image passes from one image position to another through successive lenses, the intensity per unit area of each image is reduced only by reflection loss (Fresnel loss) at the lens surface. The same is true for the image-to-object sterance ratio if the object is a lambertian emitter. It is also very nearly true when the object is an LED, as long as the cone angle for flux acceptance of the optical system is not so large that there is great deviation from a lambertian pattern. This deviation can be derived from the relative intensity, $I_r(\theta)$ of the radiation pattern, and expressed as:

$$\Delta_L = (1 - I_r(\theta)/\cos\theta) \qquad (2.5.3-1)$$

Regardless of how non-lambertian the radiation pattern is, the quantity of flux coupled into an optical system of given numerical aperture (N.A.) can be calculated from the LED radiation pattern. "Numerical aperture" is defined in Figure

2.5.3-1 and the appropriate flux integration method can be selected from Section 7.3.4.

Although N.A. is usually used in characterizing a finitely-focussed optical system, a handy relationship using the f/no relates the incidance at a target, E_T, to the sterance of a source:

$$\frac{E_T}{L_S} = \frac{\pi\tau}{4(f/no)^2(1 + \frac{d_T}{d_S})^2} \qquad (2.5.3-2)$$

In equation 2.5.3-2 the transmittance, τ, depends only on reflection losses and is usually greater than 0.7. The source-to-lens distance, d_S, and lens-to-target distance, d_T, are, respectively, the object distance, d_O and image distance, d_i relating to focal length, f, and magnification, m.

$$\frac{1}{d_O} + \frac{1}{d_i} = \frac{1}{f} \qquad (2.5.3-3)$$

$$m = \frac{d_i}{d_O} \qquad (2.5.3-4)$$

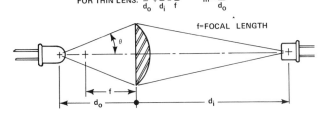

FOR THIN LENS: $\frac{1}{d_O} + \frac{1}{d_i} = \frac{1}{f}$ $m = \frac{d_i}{d_O}$

f=FOCAL LENGTH

(a) BASIC DEFINITION OF NUMERICAL APERTURE, N.A. = sinθ

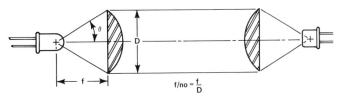

(b) FOR INFINITY-FOCUSED LENS, N.A. = $\frac{1}{2(f/no)}$

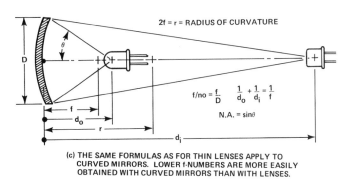

2f = r = RADIUS OF CURVATURE

f/no = $\frac{f}{D}$ $\frac{1}{d_O} + \frac{1}{d_i} = \frac{1}{f}$

N.A. = sinθ

(c) THE SAME FORMULAS AS FOR THIN LENSES APPLY TO CURVED MIRRORS. LOWER f-NUMBERS ARE MORE EASILY OBTAINED WITH CURVED MIRRORS THAN WITH LENSES.

Figure 2.5.3-1 Simple Thin Lens Formulas and Definition of "Numerical Aperture".

The ratio of focal length to lens diameter is the f/no.

The relationships of equations 2.5.3-2, -3, -4 are not adequate for designing precision optics but are good enough for most signalling applications of LEDs.

One application where a lens (or lenses) enhance LED performance is in aperture or edge sensing as in Figure 2.5.3-2. The edge moves down to cut the beam between source and sensor (target). With no lens, the beam diameter is as large as the LED and the "sharpness" of the cutting depends on how small the LED or the sensor can be. Moreover, the source-to-sensor coupling is at the mercy of the inverse square law.

The ideal situation for edge or hole sensing is to have an extremely tiny beam diameter at the beam cutting edge and a large flux coupled through to the detector when the edge is not breaking the beam. Thus two lenses are ideal. The "source lens" images the LED at the plane of the beam-breaking edge, then the "target lens" images the LED image on the sensor.

Confocal lens adjustment is obtained whenever two optical systems having the same N.A. are focused on the same point from opposite directions. All the flux entering the confocal point from one lens passes through it and enters the other lens. In Figure 2.5.3-2b, if the "target" lens had a shorter focal length, it could still be confocal, but d_T would be shorter.

Applying equation 2.5.3-2 to the arrangement of Figure 2.5.3-2b, with $d_S = d_T$ yields the incidance to sterance ratio (E_T/L_S) shown in the figure. By comparing this result to that of the NO-LENS arrangement in Figure 2.5.3-2a, the coupling improvement factor of 5.26 is found. Not only is the coupling improved, but the beam diameter at the beam-cutting edge is reduced by the ratio $d_S/d_{SL} \approx 3$.

In reflective pattern sensing, it is possible to operate without lenses, but, as in edge sensing, the signal and resolution are improved by the use of lenses, as in Figure 2.5.3-3. The LEDs used to irradiate the bar code should not be imaged at the plane of the code; although this would increase the flux coupled to the photodiode, it might cause interference patterns from interaction between the code and details of the LED image. That is the reason a cluster should be used for LED 1; a single LED 1 could cause an undesirable interference pattern. That is also the reason that LED 2 (if it is used instead of LED 1) should be imaged by LENS 2 at the plane of LENS 1, or slightly to the left of it. With the image of LED2 focused at the plane of LENS1, it will be defocussed at the bar code plane and the possibility of pattern interference is reduced.

The line-resolution of the code is the diameter of the photodiode multiplied by d_{oD}/d_{iD}. In accordance with equation 2.5.3-2 the d_{iS}/d_{oS} ratio should be made as small as possible, consistent with the imaging requirements.

2.5.4 No-Lens Signalling

In some applications adequate performance is obtained without lenses to modify the size and radiation pattern of the LED optical port. Two examples are discussed here: smoke detection and tachometry.

Smoke detection by the scattering or reflective method requires only proper baffling and placement of the LED and photodetector, as shown in Figure 2.5.4-1. In both the coaxial and radial arrangement, the photodetector and baffle are so positioned that in the absence of smoke, no flux from the LED reaches the photodetector. Not shown in Figure 2.5.4-1 are the supporting walls of the smoke-sensing chamber, but these must also be designed with regard to the possibility of stray flux -- either from the LED or outside. If the stray flux reaching the photodetector is steady, sensitivity is reduced only by the square root of the stray flux amplitude. However, if the stray flux is variable, sensitivity is linearly reduced.

For room-style smoke detectors either the coaxial or the radial arrangement can be used, but for smoke detection in a flue, smoke-stack, or air-duct, the radial arrangement is preferred because there is no obstruction by the baffle to passage of air or smoke. In very large diameter ducts or stacks, projection lenses for both the LED and the photodiode should be used. The optical arrangement should maximize the volume defined by the intersection of the beam projected by the LED with the reception beam of the photodetector. This intersection volume must, of course, exclude the walls of the duct or flue.

In tachometry, a beam-breaking edge can be used, but without lensing the optical ports of the LED and photodetector might be too large to give adequate resolution. For example, if a 25mm diameter disc is to resolve a shaft rotation in one-degree increments, the line spacing at the perimeter would be:

$$x = \frac{\Delta\theta \cdot \pi \cdot D}{360^\circ} = 0.22 \text{ mm} \qquad (2.5.4\text{-}1)$$

By placing next to the rotating disc a stationary "shutter" disc having the same line spacing, when the rotating disc moves there are alternate periods of flux transmittance and no transmittance at intervals as small as whatever line spacing can be achieved. This allows the optical ports of the LED and the photodetector (on opposite sides of the disc pair) to have arbitrarily-sized optical ports. If each disc has

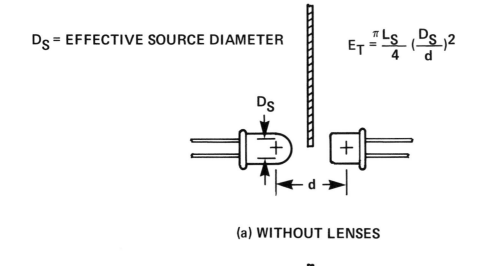

D_S = EFFECTIVE SOURCE DIAMETER

$$E_T = \frac{\pi L_S}{4}\left(\frac{D_S}{d}\right)^2$$

D_S

d

(a) WITHOUT LENSES

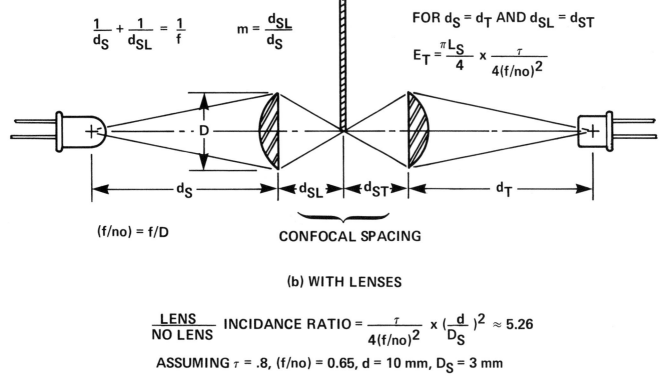

$$\frac{1}{d_S} + \frac{1}{d_{SL}} = \frac{1}{f} \qquad m = \frac{d_{SL}}{d_S} \qquad \text{FOR } d_S = d_T \text{ AND } d_{SL} = d_{ST}$$

$$E_T = \frac{\pi L_S}{4} \times \frac{\tau}{4(f/no)^2}$$

D

d_S d_{SL} d_{ST} d_T

$(f/no) = f/D$

CONFOCAL SPACING

(b) WITH LENSES

$$\frac{\text{LENS}}{\text{NO LENS}} \quad \text{INCIDANCE RATIO} = \frac{\tau}{4(f/no)^2} \times \left(\frac{d}{D_S}\right)^2 \approx 5.26$$

ASSUMING τ = .8, (f/no) = 0.65, d = 10 mm, D_S = 3 mm

Figure 2.5.3-2   Edge Sensing; Lenses Increase Coupling and Sharpen Precision.

n lines per revolution, the photodetector output will rise and fall n times per revolution. This gives shaft speed information, but no directional information.

If directional information is desired, a stationary disc with $(n+1)$ lines next to a moving disc with n lines per revolution will produce a Moire pattern rotating n times per shaft revolution in a direction opposite that of the shaft. With equal alternating light/dark lines, the Moire pattern will be dark on one side of the shaft and have 50% transmittance diametrically opposite. If the stationary disc has $(n+2)$

lines, the Moire pattern has two cycles per revolution and rotates $n/2$ times per shaft revolution in opposite direction. With $(n+3)$ lines, the Moire has three cycles per revolution and rotates $n/3$ times per revolution, again in opposite direction. To obtain Moire rotation in the same direction as the shaft, the stationary disc line spacing should be $(n-1)$, $(n-2)$, etc. Obviously, the stationary disc pattern need not go all the way around the shaft; it needs to cover only 180° of the Moire pattern cycle to allow LED/photodetector pairs to be spaced at 90° of the Moire cycle.

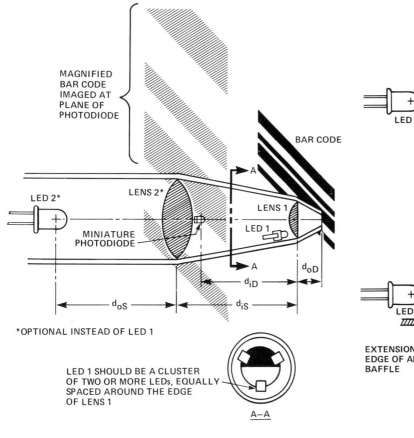

MAGNIFIED BAR CODE IMAGED AT PLANE OF PHOTODIODE

BAR CODE

LED 2*

LENS 2*

MINIATURE PHOTODIODE

LENS 1

LED 1

d_{oD}

d_{iD}

d_{oS}

d_{iS}

*OPTIONAL INSTEAD OF LED 1

LED 1 SHOULD BE A CLUSTER OF TWO OR MORE LEDs, EQUALLY SPACED AROUND THE EDGE OF LENS 1

A–A

Figure 2.5.3-3   Reflective Pattern Sensing with Coaxial Lens Arrangement.

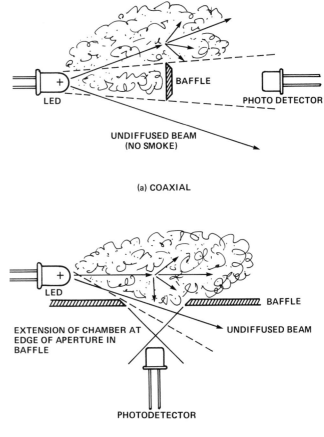

SMOKE-DIFFUSED BEAM

BAFFLE

LED

PHOTO DETECTOR

UNDIFFUSED BEAM (NO SMOKE)

(a) COAXIAL

LED

BAFFLE

EXTENSION OF CHAMBER AT EDGE OF APERTURE IN BAFFLE

UNDIFFUSED BEAM

PHOTODETECTOR

(b) RADIAL

Figure 2.5.4-1   Reflective-Type Smoke Detector Optical Arrangements.

The electrical circuit to be used in detecting the direction and speed of the shaft rotation is very simple, as seen in Figure 2.5.4-2. The dual J-K flip-flop in Figure 2.5.4-2(a) is the basic direction detecting element giving one pulse per Moire cycle at Q_1 for clockwise (CW) rotation and one pulse per Moire cycle at Q_2 for anticlockwise (ACW) rotation. While Q_1 pulsates, Q_2 is steady low, and vice versa, so direction and velocity information is available. If Q_1 and Q_2 are connected to the UP and DOWN inputs, respectively, of an up/down counter, the count will indicate shaft position providing there are no extraneous pulses. That is the reason for the $\overline{Q}$-to-J feedback. After a CW clocking of Q_1 (CK_1), Q_1 cannot return to zero until the Moire pattern has moved $90°$, either forward (NORMAL CR_1) or backward (BACKWARD CR_1). It is therefore essential that BACKWARD CR_1 be accompanied by clock (CK_2) of Q_2. This will allow a DOWN count to cancel an UP count if the shaft is vibrating through more than $90°$ of the Moire cycle.

To make sure there is no race at the CK and CR inputs, hysteresis should be used in the photodetector amplifier/comparator, as in Figure 2.5.4-2(b). The feedback diodes, D_3 and D_4 are mainly for high speed operation to restrict the input and output voltage excursions. Hysteresis

can (and should) be used if they are omitted, but the values will be different. $I_{P,LH}$, the photocurrent needed to cause an L-to-H transition at the output will be the same:

$$I_{P,LH} = \frac{V_{CC}}{R_3} \qquad (2.5.4\text{-}2)$$

but the photocurrent at which an H-to-L transition occurs will be:

$$I_{P,HL} = \frac{V_{CC}}{R_3} \left(\frac{R_2}{R_1 + R_2}\right) \qquad (2.5.4\text{-}3)$$

Also, V_{OH} will rise to within a few millivolts of V_{CC}, but $V_{OL} \approx 0$.

The line patterns forming the Moire patterns need not be integrally related. Although the Moire pattern will always shift through $360°$ for a one-cycle shift of the moving pattern, the physical spread of the Moire pattern can be made whatever is necessary using the relationships in Figure 2.5.4-3(a) for radially aligned bars. Linear motion, of course, can be interpreted by differentially spaced parallel bars, using the same relationships.

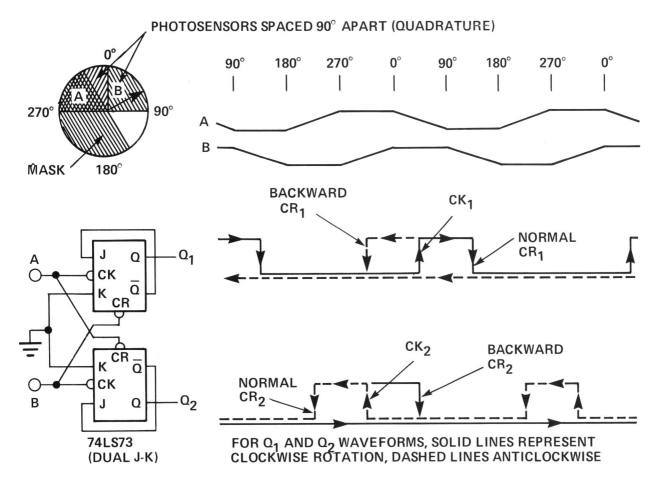

PHOTOSENSORS SPACED 90° APART (QUADRATURE)

**74LS73
(DUAL J-K)**

FOR Q_1 AND Q_2 WAVEFORMS, SOLID LINES REPRESENT
CLOCKWISE ROTATION, DASHED LINES ANTICLOCKWISE

(a) BASIC DIRECTIONAL SENSING. FOR CLOCKWISE ROTATION, Q_1 PULSATES;
FOR ANTICLOCKWISE ROTATION, Q_2 PULSATES.

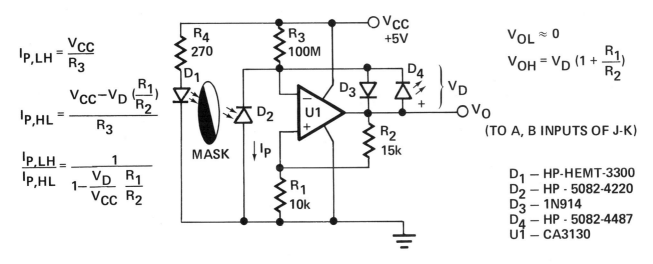

$$I_{P,LH} = \frac{V_{CC}}{R_3}$$

$$I_{P,HL} = \frac{V_{CC} - V_D \left(\frac{R_1}{R_2}\right)}{R_3}$$

$$\frac{I_{P,LH}}{I_{P,HL}} = \frac{1}{1 - \frac{V_D}{V_{CC}} \frac{R_1}{R_2}}$$

$$V_{OL} \approx 0$$

$$V_{OH} = V_D \left(1 + \frac{R_1}{R_2}\right)$$

(TO A, B INPUTS OF J-K)

D_1 — HP-HEMT-3300
D_2 — HP - 5082-4220
D_3 — 1N914
D_4 — HP - 5082-4487
U1 — CA3130

(b) COMPARATOR WITH HYSTERESIS TO PREVENT TROUBLE WITH RACES AT
J-K INPUTS WHEN BACKWARD CLEAR OCCURS.

Figure 2.5.4-2   Quadrature Phase Detector for Tachometry with
Direction Sensing.

2.49

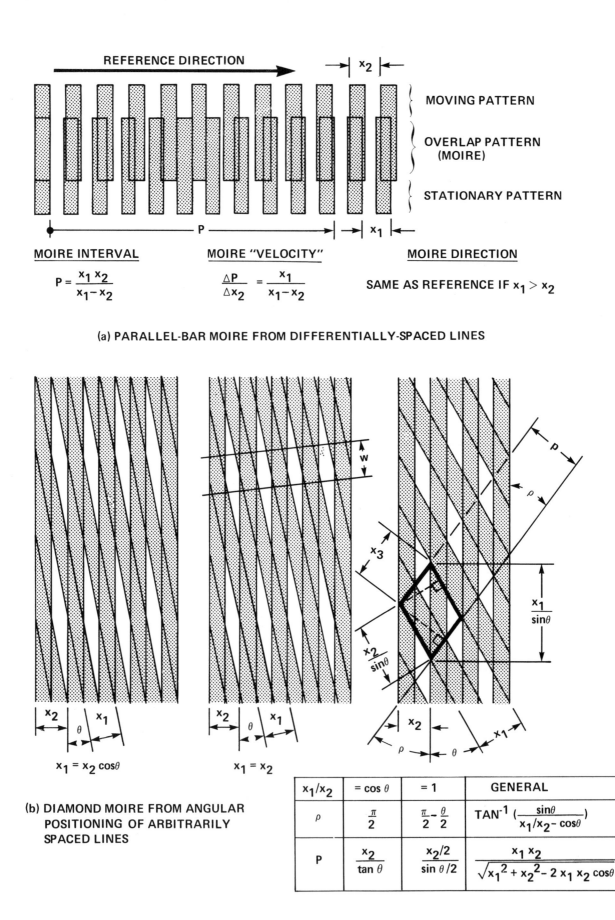

REFERENCE DIRECTION

MOVING PATTERN

OVERLAP PATTERN (MOIRE)

STATIONARY PATTERN

MOIRE INTERVAL

$$P = \frac{x_1 x_2}{x_1 - x_2}$$

MOIRE "VELOCITY"

$$\frac{\Delta P}{\Delta x_2} = \frac{x_1}{x_1 - x_2}$$

MOIRE DIRECTION

SAME AS REFERENCE IF $x_1 > x_2$

(a) PARALLEL-BAR MOIRE FROM DIFFERENTIALLY-SPACED LINES

$x_1 = x_2 \cos\theta$

$x_1 = x_2$

(b) DIAMOND MOIRE FROM ANGULAR POSITIONING OF ARBITRARILY SPACED LINES

x_1/x_2	$= \cos\theta$	$= 1$	GENERAL
ρ	$\dfrac{\pi}{2}$	$\dfrac{\pi}{2} - \dfrac{\theta}{2}$	$TAN^{-1}\left(\dfrac{\sin\theta}{x_1/x_2 - \cos\theta}\right)$
P	$\dfrac{x_2}{\tan\theta}$	$\dfrac{x_2/2}{\sin\theta/2}$	$\dfrac{x_1 x_2}{\sqrt{x_1^2 + x_2^2 - 2 x_1 x_2 \cos\theta}}$

Figure 2.5.4-3   Moire Patterns Used to Obtain Direction Information in Tachometry.

Moiré patterns for quadrature direction sensing can also be produced using an angularly positioned stationary pattern of arbitrary line spacing, as in Figure 2.5.4-3(b). The geometrical derivation of the Moire cycle interval, P, and angle, ρ, from the line spacing x_1 and x_2 at angle θ is seen from the parallelogram. The parallelogram has one side $(x_2/\sin\theta)$ and a diagonal $(x_1/\sin\theta)$ with included angle θ, so the other side (x_3) of the parallelogram is found from the Law of Cosines:

$$x_3 = \sqrt{(\frac{x_1}{\sin\theta})^2 + (\frac{x_2}{\sin\theta})^2 - 2(\frac{x_1 x_2}{\sin^2\theta})\cos\theta} \qquad (2.5.4\text{-}4)$$

The two right triangles in the parallelogram are similar so:

$$\frac{P}{x_2/\sin\theta} = \frac{x_1}{x_3} \qquad (2.5.4\text{-}5)$$

Solving equations 2.5.4-4,5 for P gives the general formula in Figure 2.5.4-3(b). Notice that if $\theta = 0$, this formula becomes the same as that for the differentially spaced bars in Figure 2.5.4-3(a). The pattern angle, ρ, is derived from the triangle formed in the right half of the parallelogram by applying the Law of Sines:

$$\frac{x_2 / \sin\theta}{\sin\rho} = \frac{x_1 / \sin\theta}{\sin(\pi-\theta-\rho)} \qquad (2.5.4\text{-}6)$$

which, solved for ρ, gives the general formula in Figure 2.5.4-3(b).

For either parallel bar or angular Moire pattern, the recommended detector "window" width W is less than 90° of the Moire cycle, that is P/4. Making it larger does not give appreciably higher photocurrent and requires overlap of 90° spaced photodetectors. Taking the photocurrent for W = P/4 as the normal:

W/P	RELATIVE "LIGHT" CURRENT	"LIGHT / DARK" RATIO
0.25	1.000	7.01
0.10	.434	19.00
0.05	.223	39.00

A straight-line, angularly positioned stationary pattern can also be used with a radial moving pattern, yielding curved Moire patterns. However, if the radius of curvature is large enough, relative to the line spacing, quadrature spaced sensors with simple geometry can still be used.

2.5.5 Signalling Over Long Distances

Straightforward signal transmission as well as reflective pattern sensing at large distances (>1m) can be done with LEDs, but auxiliary optics are required.

The relationship between the sterance of the LED source and the incidance at the target was given in equation 2.5.3-2 and is repeated in Figure 2.5.5-1(a). By applying equation 2.5.3-3 the (f/no) drops out and the incidance/sterance ratio is:

$$\frac{E_T}{L_S} = \tau (\frac{\pi}{4} D_S^2) \frac{1}{d_T^2} \qquad (2.5.5\text{-}1)$$

Notice that the term in parentheses is just the area of the lens. It is as if an enlarged source, having an area equal to the lens and having the sterance of the LED were radiating according to the inverse square law $1/(d_T)^2$ toward the receiving detector. If another lens of diameter D_R, and focal length f_R, is placed in front of the receiver and focused on the apparent source, then again applying equation (2.5.3-2) the net result gives the incidance E_R at the receiver:

$$\frac{E_R}{L_S} = \tau_S(\frac{\pi}{4}D_S^2) \frac{1}{d_T^2} \times \tau_R(\frac{\pi}{4} D_R^2) \frac{1}{f_R^2} \qquad (2.5.5\text{-}2)$$

With a source imaged on a diffusely reflecting target, the image becomes a source which can then be imaged on a receiver. The expression for the incidance to sterance ratio in such a situation is given in Figure 2.5.5-1(b). Except for the reduction (R_T/π) due to target reflectance, the expression is very nearly the same as equation 2.5.5-2; that is, the incidance/sterance ratio varies as the product of lens areas and inversely as the square of the product of focal length times separation distance. Applying this principle yields the result in Figure 2.5.5-1(c) for coaxial mirror optics. The flat secondary mirror is not the best arrangement, but serves to illustrate the principle. In practice, the secondary mirror (the back of the source mirror) would be slightly convex to increase the focal distance to the detector so it would not require as broad a reception angle.

Both the systems of Figure 2.5.5-1(b and c) can be used for such applications as reflective scanning, remote obstruction sensing, etc. Focussed on an empty space, with low background reflectance, they can be used for smoke detection.

For straightforward signal transmission, lenses are limited by the inverse square law applied to the separation distance. Also they require a rigid mechanical structure. For these reasons, fiber optics are much more convenient and, literally, more flexible.

The principle of light transmission by a fiber optic is shown in Figure 2.5.5-2. Snell's law states:

$$n_1 \sin \theta_1 = n_2 \sin \theta_2 \qquad (2.5.5\text{-}3)$$

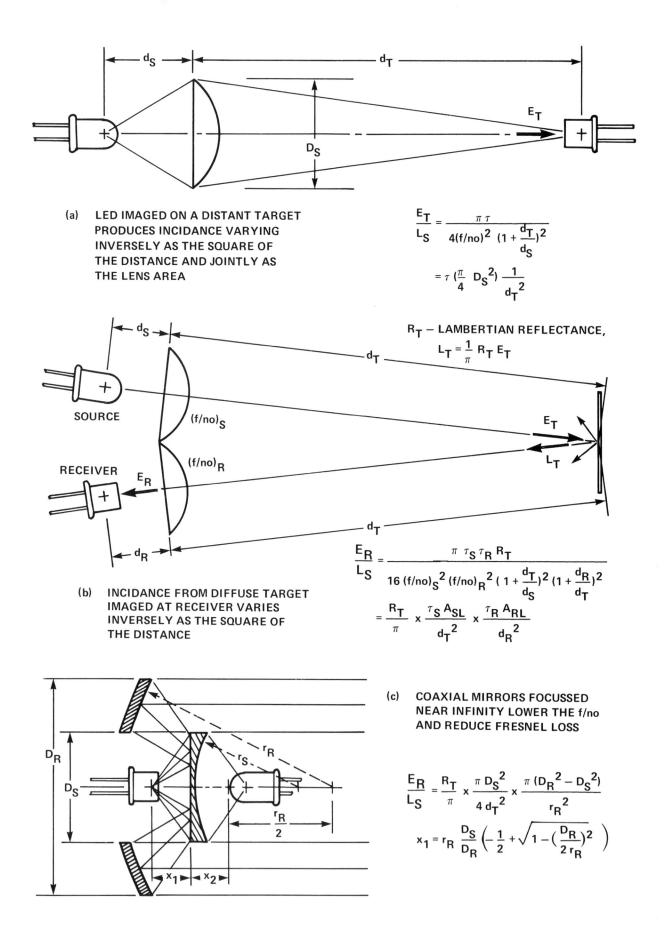

(a) LED IMAGED ON A DISTANT TARGET PRODUCES INCIDANCE VARYING INVERSELY AS THE SQUARE OF THE DISTANCE AND JOINTLY AS THE LENS AREA

$$\frac{E_T}{L_S} = \frac{\pi\, \tau}{4(f/no)^2\,(1 + \frac{d_T}{d_S})^2}$$

$$= \tau\,(\frac{\pi}{4}\,D_S^2)\,\frac{1}{d_T^2}$$

R_T – LAMBERTIAN REFLECTANCE,

$$L_T = \frac{1}{\pi}\,R_T\,E_T$$

(b) INCIDANCE FROM DIFFUSE TARGET IMAGED AT RECEIVER VARIES INVERSELY AS THE SQUARE OF THE DISTANCE

$$\frac{E_R}{L_S} = \frac{\pi\,\tau_S\,\tau_R\,R_T}{16\,(f/no)_S^2\,(f/no)_R^2\,(1 + \frac{d_T}{d_S})^2\,(1 + \frac{d_R}{d_T})^2}$$

$$= \frac{R_T}{\pi} \times \frac{\tau_S\,A_{SL}}{d_T^2} \times \frac{\tau_R\,A_{RL}}{d_R^2}$$

(c) COAXIAL MIRRORS FOCUSSED NEAR INFINITY LOWER THE f/no AND REDUCE FRESNEL LOSS

$$\frac{E_R}{L_S} = \frac{R_T}{\pi} \times \frac{\pi\,D_S^2}{4\,d_T^2} \times \frac{\pi\,(D_R^2 - D_S^2)}{r_R^2}$$

$$x_1 = r_R\,\frac{D_S}{D_R}\left(-\frac{1}{2} + \sqrt{1 - \left(\frac{D_R}{2\,r_R}\right)^2}\,\right)$$

Figure 2.5.5-1   Long-Range Direct and Reflective Signalling Using Lenses and Mirrors.

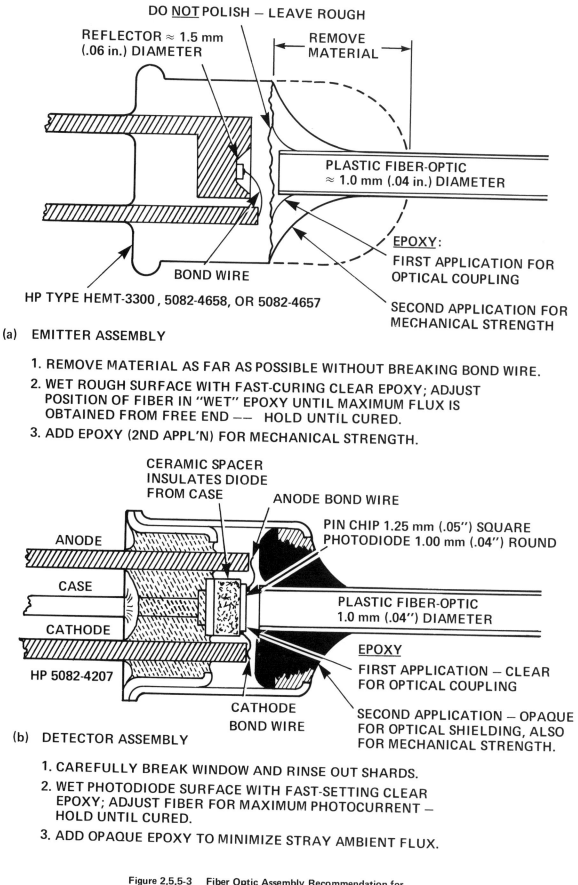

DO NOT POLISH — LEAVE ROUGH

REFLECTOR ≈ 1.5 mm (.06 in.) DIAMETER

REMOVE MATERIAL

PLASTIC FIBER-OPTIC ≈ 1.0 mm (.04 in.) DIAMETER

EPOXY:

FIRST APPLICATION FOR OPTICAL COUPLING

SECOND APPLICATION FOR MECHANICAL STRENGTH

BOND WIRE

HP TYPE HEMT-3300 , 5082-4658, OR 5082-4657

(a)   EMITTER ASSEMBLY

1. REMOVE MATERIAL AS FAR AS POSSIBLE WITHOUT BREAKING BOND WIRE.

2. WET ROUGH SURFACE WITH FAST-CURING CLEAR EPOXY; ADJUST POSITION OF FIBER IN "WET" EPOXY UNTIL MAXIMUM FLUX IS OBTAINED FROM FREE END —— HOLD UNTIL CURED.

3. ADD EPOXY (2ND APPL'N) FOR MECHANICAL STRENGTH.

CERAMIC SPACER INSULATES DIODE FROM CASE

ANODE BOND WIRE

PIN CHIP 1.25 mm (.05") SQUARE PHOTODIODE 1.00 mm (.04") ROUND

ANODE

CASE

PLASTIC FIBER-OPTIC 1.0 mm (.04") DIAMETER

CATHODE

EPOXY

FIRST APPLICATION — CLEAR FOR OPTICAL COUPLING

HP 5082-4207

SECOND APPLICATION — OPAQUE FOR OPTICAL SHIELDING, ALSO FOR MECHANICAL STRENGTH.

CATHODE BOND WIRE

(b)   DETECTOR ASSEMBLY

1. CAREFULLY BREAK WINDOW AND RINSE OUT SHARDS.

2. WET PHOTODIODE SURFACE WITH FAST-SETTING CLEAR EPOXY; ADJUST FIBER FOR MAXIMUM PHOTOCURRENT — HOLD UNTIL CURED.

3. ADD OPAQUE EPOXY TO MINIMIZE STRAY AMBIENT FLUX.

Figure 2.5.5-3   Fiber Optic Assembly Recommendation for Coupling LED to PIN Photodiode.

2.53

For a ray passing from a medium of refractive index n_1 into a medium n_2, and incident at an angle θ_1 with the surface vector of the medium boundary, θ_2 is the refraction angle. For $n_2 < n_1$ as θ_1 is increased, there is some angle, θ_c, at which $\theta_2 = 90°$, and if $\theta_1 > \theta_c$ the ray will be totally reflected. Rays entering at angles less than θ_o can propagate; those entering at larger angles are lost in the cladding of the fiber. There is, therefore, a numerical aperture for fiber optics and the relationship of N.A. to the indices of refraction of the core (n_1) and cladding (n_2) is derived in Figure 2.5.5-2.

In coupling LEDs to fiber optics, there are two principles to bear in mind:

1.   lenses are usually not much help

2.   make the source diameter less than the fiber diameter and the fiber diameter less than the detector diameter.

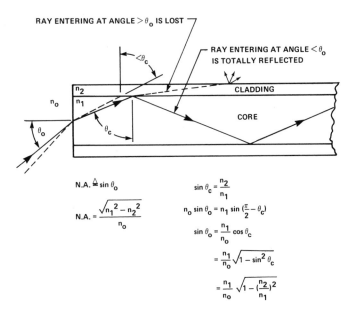

Figure 2.5.5-2   Fiber Optic Meridional Reflection Derivation of Numerical Aperture.

Lenses are not much help because, as seen in equation 2.5.3-2 the flux density cannot be made larger than it is at the surface of the source. The end of the fiber should, therefore, be as close as possible to the LED chip (die).

If the fiber diameter is less than the emitting area of the LED, flux will be lost. Even if the fiber diameter is larger than the LED emitting area, some flux is lost due to the N.A. limit. That is, fibers with large N.A. couple with lower insertion loss.

Again, at the receiving end, flux exiting from the fiber and missing the detector is lost, so a suitably large area detector should be used.

These principles are applied in the fiber optic assembly illustrated in Figure 2.5.5-3. The reason for selecting a plastic fiber optic is mostly cost, but also simplicity of assembly. Large N.A. plastic fibers of the kind shown in the figure are available from DuPont and from International Fiber Optics.

Although the plastic fiber has a large N.A., (low insertion loss) transmission loss (in dB/km) tends to be rather high. As seen in Figure 2.5.5-4, the fibers with lower transmission loss have higher insertion loss; furthermore, such fibers are usually more costly. Also shown in Figure 2.5.5-4 is the inverse-square-law coupling loss of a lens system.

When joining LEDs and photodiodes to lower-loss fiber bundles with lower transmission loss, it is usually more effective to first attach a stub ($\approx$100mm) to the device, then use a connector, such as those developed by AMP Incorporated to join the stub to the bundle. To attach a bundle directly to the LED requires first potting the end of the bundle in a binder to keep the fibers from spreading during attachment to the LED. Such spreading would raise insertion loss even further.

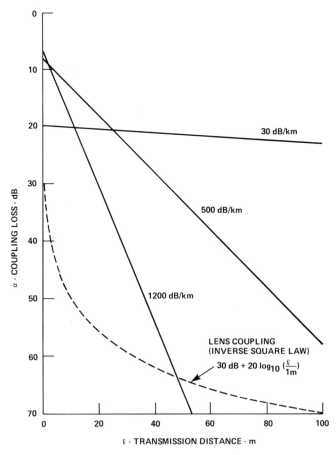

Figure 2.5.5-4   Insertion and Transmission Loss of Fiber Optics Compared with Lenses.

2.5.6 Film Exposure

In exposing photographic film with radiation from LEDs, it may or may not be necessary to take careful account of the spectral response of the emulsion. If there is doubt, the film data should be checked. The use of colored filters may be regarded as modifying the spectral response of the film.

Emulsion having a flat response over the spectrum of the LED requires only proper use of the typical exposure equation:

$$H = E t \qquad (2.5.6\text{-}1)$$

The units of E and t depend on how the film exposure requirement is given. If H is in "meter candle seconds" (mcs), then E is in lux (lx) and t is in seconds. If H is in "ergs per square centimer" (erg/cm^2) a conversion can be made to relate this to radiant incidance, E_e in $\mu w/cm^2$, and exposure time:

$$
\begin{aligned}
1\ erg/cm^2 &= 10^{-7}\ joule/cm^2 \\
&= 10^{-7}\ w \cdot sec/cm^2 \\
&= 10^{-1}\ (\mu w/cm^2) \cdot sec
\end{aligned}
\qquad (2.5.6\text{-}2)
$$

Inserting appropriate units in equation 2.5.6-1 gives the exposure relationships:

$$H\ (mcs) = E_v\ (lx) \cdot t(sec) \qquad (2.5.6\text{-}3)$$

$$H\ (erg/cm^2) = \frac{10 \cdot E_e\ (\mu w/cm^2) \cdot t\ (sec)}{10} \qquad (2.5.6\text{-}4)$$

The value of either E_v or E_e depend on how the LED is coupled to the film. With no lens, and the LED a distance, d, from the film:

$$E_v\ (lx) = 1000\ \frac{I_v\ (mcd)}{d^2\ (mm^2)} \qquad (2.5.6\text{-}5)$$

$$E_e\ (lx) = 100\ \frac{I_e\ (\mu w/sr)}{d^2\ (mm^2)} \qquad (2.5.6\text{-}6)$$

If a lens is used, i.e., a camera, the relationship in Figure 2.5.5-1(a) applies, with d_T/d_S being negligibly small, so that:

$$E_v(lx) = \frac{\pi \tau}{4\ (f/no)^2} \cdot L_v\ (cd/m^2) \qquad (2.5.6\text{-}7)$$

$$E_e\ (\mu w/cm^2) = \frac{\pi \tau}{4(f/no)^2} \cdot L_e\ (\frac{\mu w/sr}{cm^2}) \qquad (2.5.6\text{-}8)$$

Where spectral response can be ignored, the ASA film speed is related to minimum exposure requirement, H_m by:

$$H_m = 0.8/S_{ASA}* \qquad (2.5.6\text{-}9)$$

*ref: ANSI PH2.5–1972

In equation 2.5.6-9, H_m is in lux seconds (or meter candle seconds) and can be used directly in equation 2.5.6-3 because spectral response is not considered. Combining equation 2.5.6-3, -7, -9 and assuming $\tau = 0.8$ yields the result:

$$\frac{(f/no)^2}{t(sec)} = \pi\ L_v\ (cd/m^2) \cdot S_{ASA} \qquad (2.5.6\text{-}10)$$

for exposure to a density 0.1 above base-plus-fog density. In most applications, however, this is a marginal exposure. For a satisfactory film record of an LED or LED display, the exposure should be 10 to 100 times the value obtained from equation 2.5.6-10, either by lowering the f/no, or increasing the time, t, the sterance L_v, or the film speed used in recording. The exposure requirement in erg/cm^2 for use in equation 2.5.6-4 is obtained from the film characteristic curve (D-H curve) for wavelength ranges in which the film response is nearly constant over the LED spectrum.

To account for spectral effects, the incidance and exposure time requirements must be found by integrating:

$$\frac{E_e\ (\mu w/cm^2) \cdot t\ (sec)}{10} = \frac{\int \varphi_r\ (\lambda)\ d\lambda}{\int S(\lambda) \cdot \varphi_r(\lambda)d\lambda}\ (erg/cm^2) \qquad (2.5.6\text{-}11)$$

The spectral sensitivity, $S(\lambda)$ is the spectral inverse of the exposure requirement for a particular density, as given by the film D-H curves and spectral response curves. $S(\lambda)$ has the units $(erg/cm^2)^{-1}$. $\varphi_r(\lambda)$ is the relative spectral output of the LED and has no units. If a spectral filter is used, its spectral transmittance should be included as a factor in the integrand of the integral in the denominator of equation 2.5.6-11.

NOTES

Section 3

Optoisolators

3.0   OPTO-ISOLATORS

3.1   Optoisolator Theory

An opto isolator consists of a photon emitting device whose flux is coupled through optically transparent insulation to some sort of photodetector. The photon emitting device may be an incandescent or neon lamp, or an LED. The transparent insulation may be air, glass, plastic, or fiber-optic. The photodetector may be a photoconductor, photodiode, phototransistor, photoFET, or an integrated combination photodiode/amplifier. Various combinations of these elements result in a wide variety of input characteristics, output characteristics, and coupled characteristics. This discussion will be limited to the sort of optoisolator having an LED input, with a thin layer of transparent insulation separating it from a solid-state photodetector. The construction of such an optoisolator is shown in Figure 3.3.1-1, and the schematic representations are shown in Figure 3.3.1-2.

3.1.1   Photo Emitter

In the design of a photoemitter for an optoisolator, the main concern is optimization of the coupling to the photodetector. The parameters to be optimized are gain, bandwidth, optical port, and electrical characteristics.

A low series resistance is desirable, so a GaAs-based photoemitter is the best choice. A low forward voltage is also desirable, but not as important as optimization of the gain and bandwidth.

For GaAs-based photoemitters, with $GaAs_{1-x}P_x$ epitaxy, adjustment of x affects the wavelength, efficiency (gain), and speed of response (bandwidth). Gain, G, and bandwidth, B, normalized relative to their values at x = 0, are shown in Figure 3.1.1-1, along with a curve of the gain-bandwidth product, GB. It is clear that the gain-bandwidth product is optimized for x ≈ 30%. For this reason, all HP optoisolators use photoemitters in which the $GaAs_{1-x}P_x$ epitaxy is produced with x ≈ 30%, giving a wavelength of λ ≈ 700 nm.

With x ≈ 30%, the forward voltage is very nearly that of a "standard" red LED, which is a convenience in some digital applications (Section 3.6.1).

Optical port considerations for the photoemitter of an optoisolator differ considerably from those of an LED. LEDs are made with an annular emitting region around a centered bonding pad to give a large ratio of apparent-to-actual emitting area. For an optoisolator the emitting area is as small as possible, consistent with current density consideration, and the bonding pad is offset. The offset bonding pad allows minimal obscuration (shadowing)

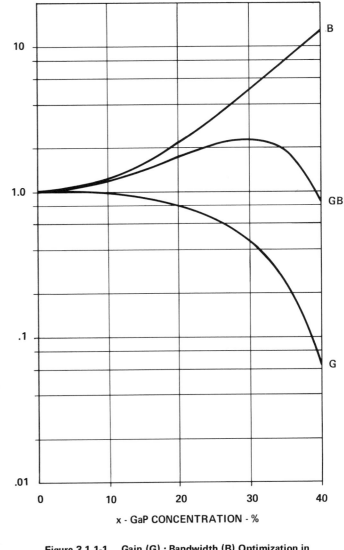

Figure 3.1.1-1   Gain (G) · Bandwidth (B) Optimization in Direct Bandgap Photoemission.

of the emitting area and allows the detector to be closely coupled. The small size reduces the flux loss at the edges and minimizes the variability of coupling due to variability in the spacing and alignment of the photodector's sensitive region.

3.1.2   Optical Medium

In selection of the optical medium, insulation is a vital consideration if the emitter-to-detector distance is very small (e.g., "sandwich" construction). If the distance is very large, such as through fiber-optics, lenses, or other medium (reflective or transmissive), the insulation is less important but spectral transmittance becomes significant, especially where plastics are involved. The effect of spectral transmittance as related to spectral properties of the emitter and the detector is discussed in Section 2.5.2.

3.1

Most optoisolators use a transparent junction coating to reduce Fresnel loss at the surfaces of the emitter and the detector. Since junction coating compounds are non-conductors, this provides insulation as well. In addition, HP optoisolators have a layer of FEP film (transparent Teflon) betweeen emitter and detector to insure good insulation (see Section 3.3).

Fresnel losses result from reflection when flux passes from one material to another having a different index of refraction. The fraction reflected is given by:

$$R = \left(\frac{n_2 - n_1}{n_2 + n_1}\right)^2 \qquad (3.1.2\text{-}1)$$

The fraction transmitted is then derived as:

$$\tau = \frac{4}{2 + \dfrac{n_2}{n_1} + \dfrac{n_1}{n_2}} \qquad (3.1.2\text{-}2)$$

For an optoisolator having a GaAsP emitter (n = 3.6) and a silicon detector (n = 3.5), the importance of having a coupling medium other than air (n = 1) is shown in Figure 3.1.2-1, applying equation 3.1.2-2 to both the changes of refractive index.

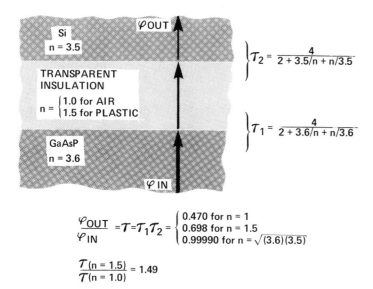

$$\frac{\varphi_{OUT}}{\varphi_{IN}} = \tau = \tau_1 \tau_2 = \begin{cases} 0.470 \text{ for } n = 1 \\ 0.698 \text{ for } n = 1.5 \\ 0.99990 \text{ for } n = \sqrt{(3.6)(3.5)} \end{cases}$$

$$\frac{\tau(n = 1.5)}{\tau(n = 1.0)} = 1.49$$

Figure 3.1.2-1   Refractive Index Effects in Optical Coupling Medium of Optoisolators.

3.1.3   Photodetector

The basics of photodetection by photodiodes are discussed in Section 4.1. Silicon photodiodes are very good detectors but usually require ancillary gain elements to produce adequate signal levels. To add gain with elements external to the optoisolator package is an inconvenience, raises the package count in the system, and degrades performance (CMR, speed). For these reasons, it is desirable to include

the gain element within the optoisolator package. This can be done in either of two ways:

(a)   Hybrid - costly, but permits separate optimization of the photodiode and its amplifier;

(b)   Integrated - reduces assembly cost but compromises performance.

There are two ways to integrate the photodetector and amplifier. One of these is the use of a phototransistor, in which the photodetecting region is the collector-base junction. The other is the use of a photodiode whose photocurrent is amplified by a transistor separately integrated on the same chip. The phototransistor is less complicated to produce but has some inherent performance disadvantages relative to the photodiode/transistor: poor linearity and low speed of response.

A linearity comparison is shown in Figure 3.1.3-1. The phototransistor's non-linearity results from the flow of collector current in the collector-base junction, thus causing reduction in the collector-base depletion region which in turn reduces its responsivity. Notice that raising the collector voltage restores the depletion region and reduces the non-linearity. In the photodiode/transistor, the collector current does not flow in the photodiode -- even if the cathode of the photodiode is connected to the collector in what is called the "phototransistor connection" (because it can be operated as a two-terminal device). Best linearity, however, is obtained by keeping fixed the voltage across the photodiode.

The speed of response of a phototransistor is inherently slow because the collector-base junction must be large in order to capture photons, and therefore has a large ($\approx$20 pF) junction capacitance amplified by Miller effect (see Figure 3.1.3-2). In the photodiode/transistor, the photodiode capacitance can be made lower ($\approx$10 pF) because the junction thickness does not affect the gain of the separately integrated transistor. More importantly, however, the collector-base capacitance to which Miller effect applies is extremely small ($\approx$0.5 pF). Furthermore, as seen in Figure 3.1.3-2, much of this is contributed by capacitance between external pins and pin connections. For this reason, if no base connection is required, the rise/fall time of HP 5082-4350/51 optoisolators is improved 30% to 40% by removal of the base pin. If this is impractical, or if a base connection is required, the board layout should minimize capacitance between the base and collector leads.

Because the HP photodiode/transistor isolator is an integrated (not hybrid) device, the voltage at the emitter of the output transistor must always be at or below the voltage at any other point in the isolator IC. For example, if the emitter is driving some above-ground point, such as

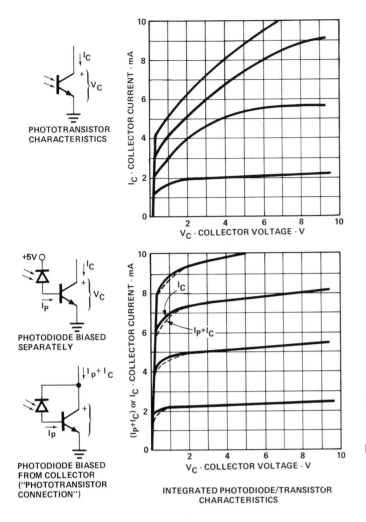

PHOTOTRANSISTOR
CHARACTERISTICS

PHOTODIODE BIASED
SEPARATELY

PHOTODIODE BIASED
FROM COLLECTOR
("PHOTOTRANSISTOR
CONNECTION")

INTEGRATED PHOTODIODE/TRANSISTOR
CHARACTERISTICS

Figure 3.1.3-1   Linearity of Photodiode/Transistor Compared
to that of Phototransistor.

another transistor, and if base bypass resistors are used to enhance the speed of response, the resistor to the isolator transistor base cannot be grounded. Correct connection is shown in Figure 3.1.3-3.

Nor can the emitter be permitted to float. For example, if the photodiode alone is to be employed, the emitter should be connected to the base. Failure to do so yields the results shown in Figure 3.1.3-4. The collector may float but must not be permitted to become negative with respect to the emitter, so it also should be connected to the emitter.

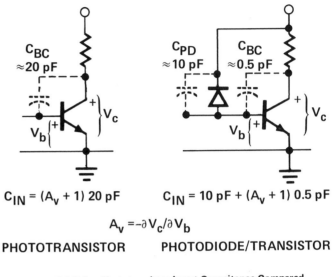

$$C_{IN} = (A_v + 1)\ 20\ pF$$

$$C_{IN} = 10\ pF + (A_v + 1)\ 0.5\ pF$$

$$A_v = -\partial V_c / \partial V_b$$

PHOTOTRANSISTOR   PHOTODIODE/TRANSISTOR

Figure 3.1.3-2   Phototransistor Input Capacitance Compared
to Photodiode/Transistor.

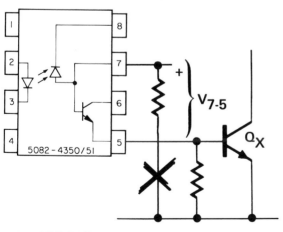

INCORRECT

RESIDUAL CHARGE ON BASE OF
Q_X MAY MAKE $V_{7-5} < 0$ AND
CAUSE EXTREMELY SLOW
OPERATION

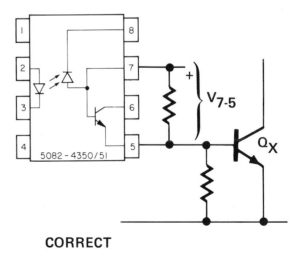

CORRECT

V_{7-5} CANNOT BECOME NEGATIVE

Figure 3.1.3-3   Correct Connection of Base Bypass Resistor
with Emitter Above Ground.

NOTE CHANGE OF SCALE FOR $V_{CE} > 0$

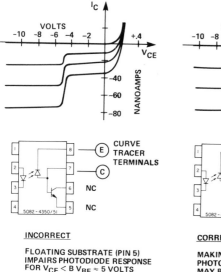

INCORRECT

FLOATING SUBSTRATE (PIN 5) IMPAIRS PHOTODIODE RESPONSE FOR $V_{CE} < 8 V_{BE} \approx 5$ VOLTS

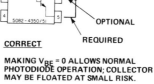

CORRECT

MAKING $V_{BE} = 0$ ALLOWS NORMAL PHOTODIODE OPERATION; COLLECTOR MAY BE FLOATED AT SMALL RISK.

Figure 3.1.3-4   Floating Substrate Impairment of Integrated Photodiode Performance.

3.1.4   Amplifier Options

Integration of a photodiode on the same chip with amplifying transistor(s) entails some tradeoffs. The silicon resistivity cannot be as high as that for an optimal PIN photodiode because it would then be impossible to integrate reasonable gain/speed transistors. On the other hand, silicon resistivity low enough to make optimal transistors would not permit a reasonably good photodiode. Within the constraints of this tradeoff, ($C/A < 90$ pF/mm^2 for the photodiode and $GB > 500$ MHz for the transistor), a variety of detector-amplifier configurations can be executed.

Analog amplifiers in HP optoisolators are of two types, shown in Figure 3.1.4-1. One is the single transistor with the photodiode anode connected to the base and the cathode separated for reverse bias connection (e.g. V_{CC}). The other type is called a split-darlington; the emitter of the first transistor is connected to the base of the second transistor in the usual darlington manner, but the collector of the second is separated (split) from the collector of the first transistor to allow the output collector to drop to a lower V_{CE} in saturation. In the usual darlington circuit, the collectors are common, and the lowest possible output V_{CE} is the sum of V_{BE} of the second transistor and $V_{CE(SAT)}$ of the first. This results in a minimum output V_{CE} of ≈ 800 mV. With the *split*-darlington configuration, the output V_{CE} is not held up by the V_{BE} of the output transistor and may drop to less than 100 mV. This is especially important in digital applications requiring a low output V_{CE} for good noise immunity. The base of the second transistor is available for strobing and for speed enhancement with resistive bypassing -- handy features in

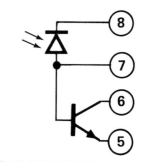

SINGLE TRANSISTOR AMPLIFIER

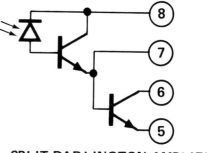

SPLIT-DARLINGTON AMPLIFIER

$$V_{CE2(SAT)} = V_{BE2} + V_{CE1(SAT)}$$

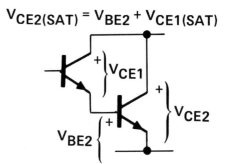

ORDINARY DARLINGTON AMPLIFIER

Figure 3.1.4-1   Analog-Type Photocurrent Amplifiers.

digital applications. The high gain makes the split-darlington amplifier useful in many low power applications and as a gain element in a closed loop where the input to the main amplifier must be isolated from the reference comparator (e.g. analog power supply regulation). However, because the base of the *first* transistor is not available for feedback connection, the split-darlington amplifier is not optimal for linear applications.

For digital applications at moderate data rates, the analog types can be used if their gain/output-current capabilities are adequate. For high data rates, the transistors must be operated at lower closed-loop gain in order to achieve the required bandwidth. HP detector/amplifiers for digital applications have a high speed linear amplifier driving a Schottky-clamped output transistor, as shown in Figure 3.1.4-2. Bias for the photodiode is decoupled from V_{CC} to reduce the possibility of "chatter" (oscillatory transition

3.4

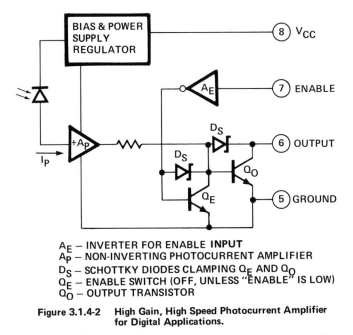

A_E — INVERTER FOR ENABLE **INPUT**
A_P — NON-INVERTING PHOTOCURRENT AMPLIFIER
D_S — SCHOTTKY DIODES CLAMPING Q_E AND Q_O
Q_E — ENABLE SWITCH (OFF, UNLESS "ENABLE" IS LOW)
Q_O — OUTPUT TRANSISTOR

**Figure 3.1.4-2   High Gain, High Speed Photocurrent Amplifier
for Digital Applications.**

from one logic state to another due to regenerative coupling via the power supply line). The linear amplifier has a tendency to be unstable if the high-frequency impedance of the power supply is not low enough. For this reason, a low-inductance bypass capacitor (0.01 μF ceramic disc) should be installed adjacent to each isolator of this type. This, and other chatter-suppression techniques are discussed in Section 3.3.

The Schottky clamp is a metal-silicon diode in parallel with the base-collector junction of the output transistor. A metal-silicon (Schottky) diode has a lower turn-on voltage than a P-N junction, so when the transistor is driven into saturation, the Schottky diode bypasses the current which would otherwise enter the base-collector junction. With reduced current entering the base-collector junction, there is a proportionate reduction in the charge to be removed when the transistor is to be turned off, and thus the Schottky clamp reduces the turn-off delay. The attendant drawback is the 400 mV higher $V_{CE\,(SAT)}$ for a Schottky clamped transistor.

The "enable" input has threshold voltage and input current levels resembling a TTL input. However, it is not necessary to apply a pullup resistor to insure its remaining high. Unless the enable is connected to a strobe, it may simply be left open. The strobe applied to the enable input may be either open-collector or active-pullup.

With the enable high, analog operation is possible because there is no hysteresis. However, the dynamic range is limited. The lower limit is the threshold input current for operation in the active region -- this threshold may be as high as 4 mA. The upper limit is the maximum dissipation rating on the output. Because of a touchy bias situation, analog separation is not recommended for designs to be mass produced.

The reason for omitting hysteresis was not to permit analog operation, but rather to permit maximum data rate. With hysteresis, there would be a higher immunity to both differential- and common-mode noise but the shifting threshold would reduce the data rate capability. The effect of hysteresis on data rate is the opposite of peaking.

3.2   Parameter Characterization

As seen in Section 3.1, a variety of input, output, and coupled characteristics are possible, depending on the choices of input and output devices and the manner in which they are optically coupled. Figure 3.2-1 illustrates a type of optoisolator with respect to which all the important parameters, analog as well as digital, can be visualized and described. Listed in order of their importance:

1.   Isolation (Common Mode Rejection, CMR)

2.   Insulation (maximum $V_{I\text{-}O}$)

3.   Speed (modulation bandwidth, propagation delay)

4.   Reverse coupling (ground looping)

5.   Forward coupling (Current Transfer Ratio, CTR; fan-out)

3.2.1   Isolation

The fundamental purpose of an isolator, whether optically, electrically, or magnetically coupled, is to enhance, in the output, the ratio of differential-mode to common-mode signals. Optical coupling is superior to electric or magnetic coupling because the photons that carry the differential mode signal do not carry any charge or require a magnetic flux to support their movement. Thus, the only means by which the common-mode signal can appear in the output are by:

(a)   modulating the input current and

(b)   stray capacitive coupling.

Means (a) is not really a property of the optoisolator and can be eliminated by impedance balancing, as seen in Figure 3.2-1, where R_{P1} and R_{P2} can be selected or adjusted to make $\partial I_C/\partial e_{CM} = 0$. Isolation characterization is therefore examined with respect to (b) stray capacitive coupling.

Analog isolation is seen simply as the ratio of the relative effects of differential-mode voltage, e_{DM}, and common-mode voltage, e_{CM}, on output current, I_C, expressed as the Common-Mode Rejection Ratio, CMRR:

$$CMRR = \left(\frac{\partial I_C/\partial e_{DM}}{\partial I_C/\partial e_{DM}}\right) \left(\frac{e_{DM}}{e_{CM}}\right) \qquad (3.2.1\text{-}1)$$

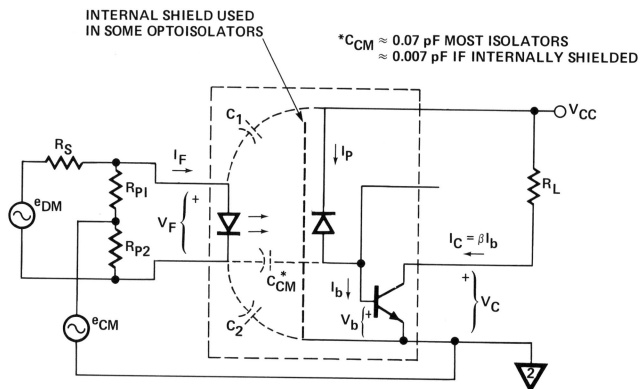

INTERNAL SHIELD USED
IN SOME OPTOISOLATORS

*$C_{CM} \approx 0.07$ pF MOST ISOLATORS
≈ 0.007 pF IF INTERNALLY SHIELDED

- ISOLATION:

  ANALOG: $CMRR \triangleq \dfrac{\partial I_C / \partial e_{DM}}{\partial I_C / \partial e_{CM}}$ $CMR = 20 \log_{10} (CMRR)$

  DIGITAL: MAX. TOLERABLE COM. MODE FOR PROPER DIGITAL OUTPUT

  $CMRV \triangleq$ MAX. TOLERABLE e_{CM} $CMTR \triangleq$ MAX. TOLERABLE de_{cm}/dt

  CM_H: CMRV or CMTR for $V_C \geqslant V_{OH}$ (INPUT DIODE OFF)

  CM_L: CMRV or CMTR for $V_C \leqslant V_{OL}$ (INPUT DIODE ON)

- INSULATION: $V_{I-O} \triangleq e_{CM}$ ABOVE WHICH DAMAGE MAY OCCUR.

- SPEED:

  ANALOG: 3 dB BANDWIDTH FOR dV_C/de_{CM}

  DIGITAL: DELAY IN PROPAGATION OF CHANGE IN LOGIC STATE, $e_{DM} \rightarrow V_C$

- REVERSE COUPLING: $C_{I-O} = C_1 + C_2 + C_{CM}$

- FORWARD COUPLING: CURRENT TRANSFER RATIO, $CTR \triangleq \dfrac{I_C}{I_F} \times 100\%$

Figure 3.2-1 Important Optoisolator Parameters; Illustration and
Brief Description

CMRR is often expressed in dB, in which case it is called the Common Mode Rejection, CMR:

$$\text{CMR} = 20 \log_{10} (\text{CMRR}) \qquad (3.2.1-2)$$

CMR cannot be specified without reference to the input circuit design. This is seen by analyzing equation 3.2.1-1:

$$\frac{\partial I_C}{\partial e_{DM}} = \beta \frac{\partial I_b}{\partial e_{DM}} = \beta \frac{\partial I_p}{\partial e_{DM}} = \beta \frac{\partial I_p}{\partial I_F} \frac{\partial I_F}{\partial e_{DM}} \qquad (3.2.1-3)$$

$$= \beta \frac{\partial I_p}{\partial I_F} \left(R_S + \frac{dV_F}{dI_F} \right)^{-1}$$

$$\frac{\partial I_C}{\partial e_{CM}} = \beta \frac{I_b}{\partial e_{CM}} = \beta \frac{I}{\partial e_{CM}} \partial \left(C_{CM} \frac{de_{CM}}{dt} \right) \qquad (3.2.1-4)$$

$$= \beta \, C_{CM} \, (2 \pi f_{CM})$$

Taking the ratio of equation 3.2.1-3 to equation 3.2.1-4 gives

$$\text{CMRR} = \frac{\beta (I_p/I_F)}{\beta \, (2\pi f_{CM} \, C_{CM}) \, \left(R_S + \dfrac{dV_F}{dI_F} \right)} \qquad (3.2.1-5)$$

The photocurrent-to-input current ratio (I_p/I_F) and common-mode coupling capacitance, C_{CM}, are properties of the optoisolator, however, CMRR can be made arbitrarily high by making R_S and dV_F/dI_F small. Note also that CMRR decreases as the frequency, f_{CM}, of the common mode signal rises. There is a limit to this effect. The cutoff frequency of the amplifier limits response to both e_{DM} and e_{CM} making the ratio unimportant. Because of the non-isolator variables affecting CMR, any meaningful description of this parameter must give conditions for I_F, R_S and f_{CM}. Notice that CMRR is independent of the amplifier gain, but does depend on the diode-to-diode current transfer ratio, I_p/I_F. This ratio is fairly constant: $I_p/I_F \approx 0.0015$ for all HP optoisolators and is very nearly the same also for non-HP types in which the "photodiode" is the base-to-collector junction of a phototransistor. Notice also that C_{CM} is only a small part of the total input-to-output capacitance, which is given in data sheets as $C_{I\text{-}O}$. C_{CM} also is fairly constant; $C_{CM} \approx 0.07$ pF in most HP isolators, but in the internally shielded types $C_{CM} \approx 0.007$ pF.

At present, the internal shield is offered in only the very high speed optoisolator, 5082-4361, in which the broader bandwidth would, without the shield, allow CMRV to dip lower as in Figure 3.2.1-1, or shift downward and leftward the curve in Figure 3.2.1-2.

For the unshielded single-transistor analog types, CMR can be improved by adding a neutralizing capacitor between the collector and either of the input pins. The value of this capacitor should be $\beta \times C_{CM} \approx 7$ pF. Neutralization can also be used with dual isolators of the single-transistor type, providing the neutralizing capacitor couples each collector to its corresponding input diode. Obviously, the neutralizing capacitor must have a voltage rating compatible with the application.

Digital operation requires that the output remain in proper logic state despite interference from e_{CM}. The extent to which such interference is tolerable can be described in either of two ways. For sinusoidal e_{CM}, there is a maximum tolerable amplitude, called the Common Mode Rejection Voltage, CMRV, which, if exceeded, will cause the output to change. However, a single value of CMRV does not describe e_{CM} tolerance. Since e_{CM} is capacitively coupled, CMRV varies inversely with frequency, as in Figure 3.2.1-1, up to the amplifier cutoff frequency, beyond which the slope is positive. For non-sinusoidal e_{CM}, i.e. transients, it is more convenient to describe the maximum tolerable rate of change, $\partial e_{CM}/\partial t$, in terms of Common Mode Transient Rejection, CMTR. Just as for CMRV, a single value of CMTR does not describe $\partial e_{CM}/\partial t$ tolerance. The abscissa over which CMTR is examined may be either the amplitude, e_{CM}, or the duration, t_{TR}, of the transient, as in Figure 3.2.1-2 because:

$$e_{CM} = e'_{CM} \times t_{TR} \qquad (3.2.1-6)$$

Observations have shown that e'_{CM} is very nearly a hyperbolic function of e_{CM} according to:

$$(e_{CM} - e_{CMO})(e'_{CM} - e'_{CMO}) \approx \text{CONSTANT} \qquad (3.2.1-7)$$

The "CONSTANT" is a function of the gain and speed of the amplifier in the isolator output. The asymptotes, e_{CMO} and e'_{CMO}, describe what is intuitively clear. That is, there is some transient excursion amplitude, e_{CMO} which is small enough that no matter how high the rate of rise may be, the isolator output will remain in its proper logic state. Similarly, for a sufficiently small rate of rise e'_{CMO}, the excursion amplitude e_{CM} is not limited (except by the insulation, Section 3.2.2). These asymptotes can be described in terms of current and voltage increments Δi_b and Δv_b at the amplifier input. To some extent, therefore, there are some circuit design choices that affect the isolator's defense against common mode transients.

Notice in Figures 3.2.1-1 and 3.2.1-2 that different curves are shown, CM_H and CM_L, representing common mode voltage tolerance with the output in logic HIGH state and LOW state, respectively. In general, reducing the value of R_L (see Figure 3.2-1) raises CM_H and lowers CM_L. Raising the level of input current, I_F, raises CM_L. For sinusoidal common mode voltage, polarity is of no concern because the rate of rise equals the rate of fall. For non-sinusoidal transients, it is worth noting that if the output is in a logic low state, a positively sloped transient has no effect. Thus in Figure 3.2.1-2, the CM_H curve applies only to positive

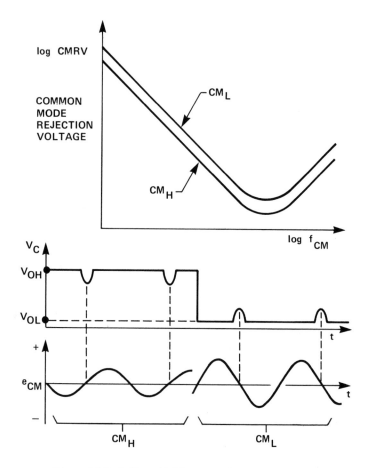

Figure 3.1.1—1 Sinusoidal Common Mode Voltage Rejection Property, CMRV.

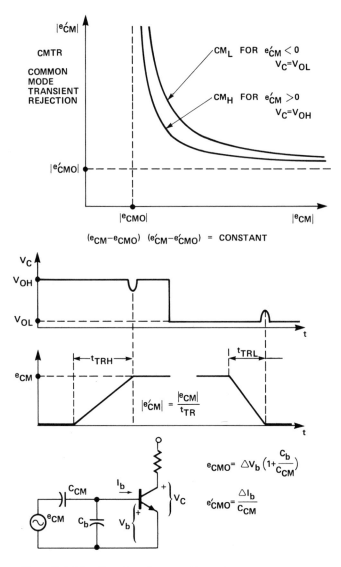

Figure 3.2.1-2 Transient Common Mode Voltage Rejection Property, CMTR.

transients. Conversely, in similar fashion, the CM_L curve applies only to negative transients. Therefore, if observation or other evidence suggests that transients predominantly have a higher rate of change in one direction than in the other (e.g. a sawtooth), R_L and I_F can be adjusted, as above, to selectively improve CM_H and CM_L. Such unequal rates of change are common in circuits having diodes or other nonlinear elements.

The curves in Figures 3.2.1-1 and 3.2.1-2 characterize only the static immunity to common mode interference. Common mode sinusoids and transients can also affect the dynamic performance adversely. For example, a transient occuring during a differential-mode logic transition can affect the rate of rise or fall in the output circuit, thus preventing a short bit from causing a proper logic transition at the output. Also, high frequency common mode interference can cause more than one output transition to occur during a single transition of the differential-mode signal. (See Section 3.6.2 and 3.6.6 for defensive techniques.)

Characterization of dynamic immunity to any common-mode interference can be done quite simply with an up/down decade counter as in Figure 3.2.1-3. With a reference data stream at the "up" input and the data

transmitted through the isolator at the "down" input, the counter should only toggle between two adjacent states -- either 4, 5, 4, 5 or 5, 6, 5, 6. Preset at 5 re-sets both R-S latches, causing LEDs 2 and 4 to glow. Should the isolator fail occasionally to make an output transition, the count will advance until a "carry" output sets the "carry" latch. Should extraneous transitions occur, the counter will decrement until a "borrow" output sets the "borrow" latch. Perfect operation is indicated as long as LEDs 1 and 3 remain off in the presence of e_{CM}.

CAUTION: getting a clean stream of reference data to the "up" input in the presence of e_{CM} *may pose enough difficulty to require extreme patience and heroic measures, especially if very high data rates are involved.*

The scheme shown in Figure 3.2.1-4 has the best chance of yielding valid results. Since the differential count is all that matters, the use of two ÷ x counters allows the reference

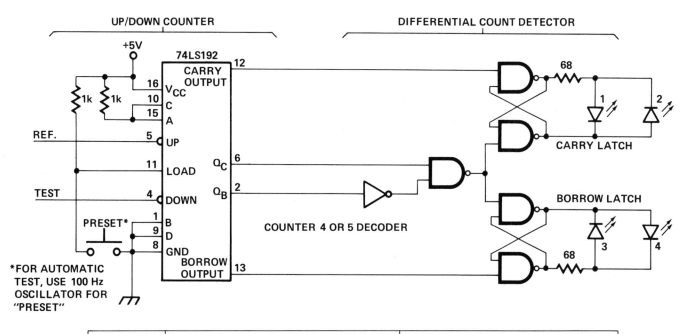

CONDITION		INDICATOR NUMBER (LED)				DESCRIPTION OF CONDITION
		1	2	3	4	
PASS	A	OFF	ON	OFF	ON	PERFECT OPERATION
FAIL	B	ON	ON	OFF	ON	TEST INPUT MISSING SOME COUNTS
	C	OFF	ON	ON	ON	EXTRANEOUS COUNTS AT TEST INPUT
	D	ON	ON	ON	ON	HIGH-RATE MIXTURE, CONDITIONS B&C
	E	ON	OFF	ON	OFF	PROBABLY OK — CHECK PRESET
	F	FLICKER	FLICKER	OFF	ON	OCCASIONAL CONDITION B
	G	OFF	ON	FLICKER	FLICKER	OCCASIONAL CONDITION C
	H	FLICKER	FLICKER	FLICKER	FLICKER	LOW-RATE MIXTURE, CONDITIONS B&C

Figure 3.2.1-3   Up/Down Counter, Differential Count Detector Sense Transmission Error.

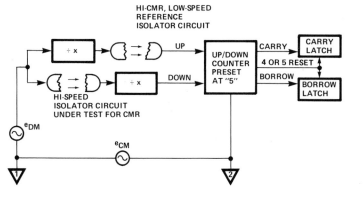

Figure 3.2.1-4   "Heroic Measures" to Assure Clean Reference Count in Dynamic CMR Test.

isolator to operate at a lower data rate so speed can be traded for better CMR. The isolator circuit under test must operate at the full data rate in the presence of the same interference. If the test is to be done with e_{DM} being a randomly coded data stream, the two $\div x$ counters should be pre-set at the same count (e.g. 0000) with e_{DM} at a fixed logic state. For most consistent results, inverters should be used, as needed, so that at "preset" all counter inputs are "zero" if they are negative-edge triggered or "one" if they are positive-edge triggered.

3.2.2   Insulation

A common, but usually erroneous, assumption is that insulation can be operated at any voltage up to that at which it breaks down. Many designers are unaware of corona and its effect on insulation. Of those who are aware of corona, some believe it occurs only at exposed terminals.

Corona (also known as "partial discharge") can occur within insulation materials, particularly those having an abundance of "microvoids". Due to inhomogenous electric fields within the material, the local field across a "microvoid" can rise to a level at which there is a local breakdown resulting in a partial discharge. This partial discharge at one microvoid shifts the field so it builds up across some other microvoid. Each partial discharge causes ions that locally degrade the insulation. The cumulative effect over long periods of time (weeks or months) is to lower the terminal-to-terminal breakdown voltage.

There is a voltage, called Corona Inception Voltage, CIV, above which corona can be observed using specialized equipment capable of responding to each partial discharge. As the test voltage is raised above CIV, there is an increase in the rate of occurrence of these partial discharges. Correspondingly, there is an increase in the rate at which the insulation is degraded.

To maintain a high quality of insulation, HP optoisolators are given a special treatment called "backfilling". After molding, they are placed in a silicone oil bath in a chamber which is evacuated until the gases from the microvoids have escaped. Then the chamber is pressurized, forcing silicone oil to fill the microvoids. With this treatment, CIV is believed to be well above the rated V_{I-O}, but 100% production testing for corona is not practical; with the methods and equipment now available, it is too time consuming.

However, to insure rejection of devices with initially defective insulation, there is 100% production testing with 3000 Vdc applied for 5 seconds at 45% relative humidity, rejecting for leakage greater than 1.0 μA.

The production test assures that each part meets the Underwriters Laboratories requirements for 220 Vac, 50/60 Hz operation. The UL formula for prescribing the test voltage:

line voltage = 220 Vrms
times 2 = 440 Vrms
plus 1000 Vrms = 1440 Vrms 50/60 Hz for 1 minute
or plus 20% = 1728 Vrms 50/60 Hz for l second
or times $\sqrt{2}$ = 2444 Vdc for l second

Except for those in hermetic packaging, all HP optoisolators easily meet UL requirements for status as a "Recognized Component" under File No. E55361.

Another important property of insulation is its leakage resistance. In HP optoisolators R_{I-O} is typically 10^{12} ohms. This appears to be an unreasonable claim when compared with the 1.0 μA leakage allowed in the 3000 Vdc test. Here again, it is a matter of practicality. With 3000 Vdc applied to 10^{12} ohms, the resulting current is only 3 nA. Precise observation of so small a current in a 5 second test is impractical, so the $R_{I-O} = 10^{12}$ ohms in given as typical, with periodic verification done on randomly selected samples. Both kinds of testing (R_{I-O} and V_{I-O}) are performed with all input leads connected to one side of the test voltage and all output leads to the other side -- a two-terminal test.

3.2.3 Speed of Response

For both analog and digital operation, the speed of response is highly dependent on the circuit as well as on characteristics of the optoisolator. In both analog and digital modes, the speed of response can be enhanced by use of feedback and peaking. These techniques are discussed in Section 3.5.6 for analog operation, and in Section 3.6.3 for digital operation. In this section, discussion centers on isolator properties affecting speed of response, and how the speed is characterized.

Analog operation requires the isolator to operate in its "active region", i.e., with the output collector neither cut off nor saturated over the required excursion range. In the circuit of Figure 3.2-1, e_{DM}, R_S, R_{PI}, and R_2 should be selected for $I_F = (I_{Fdc} \pm \Delta I_F)$ such that $0 \ll V_C \ll V_{CC}$ between the excursion limits of I_F. Speed of response is then characterized either in terms of 10%-90% rise time if ΔI_F is a step function, or in terms of 3 dB bandwidth if ΔI_F is a sinusoid.

In the isolator itself, the principal bandwidth-limiting elements are the photoemitter, the amplifying transistor, and the capacitance of the photodiode. In the photodiode, the photoelectrons are created within a few picoseconds after photons enter. The resulting photocurrent flows to the base with a rise time constant as seen at the base. The base time constant depends not only on the transistor in the isolator but also on the circuit used with the transistor. If the transistor is operated common-emitter as in Figure 3.2-1, then the base time constant:

$$\tau_B = R_B \, (C_{PD} + C_{BC}) + \beta \, R_L \, C_{BC} \qquad (3.2.3\text{-}1)$$

where:

$C_{PD} \approx 10 \text{ pF}$ is the photodiode capacitance

C_{BC} = Base-to-collector capacitance = 0.5 pF plus stray external capacitance between collector and base connections

R_L = Load resistance

R_B = Dynamic resistance to ground at the base

If no external resistance is added to bypass the base, then R_B is just the dynamic resistance at the base:

$$(3.2.3\text{-}2)$$

$$R_B = \frac{\partial V_B}{\partial I_b} = \frac{25 \text{ mV}}{I_b} = \beta \left(\frac{25 \text{ mV}}{I_C} \right) = \beta \left(\frac{25 \text{ mV}}{\frac{CTR}{100\%} \times I_{Fdc}} \right)$$

where CTR = Current Transfer Ratio (see Section 3.2.5). Then in equation 3.2.3-1

$$(3.2.3\text{-}3)$$

$$\tau_B = \beta \left(\frac{25 \text{ mV}}{\frac{CTR}{100\%} \times I_{Fdc}} \right) (C_{PD} + C_{BC}) + R_L C_{BC}$$

Substituting typical values, CTR = 20% and $I_{Fdc} = 15 \text{ mA}$ yields

$$(3.2.3\text{-}4)$$

$$\tau_B = \beta C_{BC} [R_L + 8.33\Omega \, (1 + \frac{C_{PD}}{C_{BC}})] = \beta C_{BC} [R_L + 175\Omega \,]$$

3.10

Equation 3.2.3-4 shows the importance of a low value of C_{BC}. It also shows that the base time constant is limited by R_L only if $R_L > 175\Omega$. Compare this with a phototransistor for which C_{BC} (≈ 20 pF) is larger by 40 times; also, $C_{PD} = 0$ and the base time constant is limited by $R_L > 8.33\Omega$.

Although this is not recommended for high speed operation with HP optoisolators, the transistor can be operated "common collector", as in Figure 3.2.3-1. This, incidentally, IS the recommended circuit for optimizing the speed of a phototransistor. Here the base-to-collector capacitance is not Miller-effect multiplied so the base time constant is:

$$\tau_B = (C_{PD} + C_{BC})\,[R_B + (\beta + 1)\,R_L] \qquad (3.2.3\text{-}5)$$

Assuming the same values for CTR and I_F in the expression of equation 3.2.3-2 for R_B, equation 3.2.3-5 becomes:

$$\tau_B = (\beta + 1)\,(C_{PD} + C_{BC})\,[R_L + \frac{R_B}{\beta + 1}] \qquad (3.2.3\text{-}6)$$

$$= (\beta + 1)\,(C_{PD} + C_{BC})\,[R_L + \frac{8.33\Omega}{1 + 1/\beta}]$$

$$= (\beta + 1)\,(C_{PD} + C_{BC})\,[R_L + 8.25\Omega]$$

Notice now that the base time constant is limited by $R_L > 8.25\Omega$ regardless of the relative values of C_{PD} and C_{BC}. Comparing equation 3.2.3-4 and equation 3.2.3-6, it is clear that if $C_{BC} \ll C_{PD}$ as in HP isolators, the common-emitter circuit is preferred for superior bandwidth, whereas for phototransistor types the common-collector circuit is better, particularly if the gain requirement makes $R_L \gg 8.25\Omega$.

Although in HP optoisolators the photoemitter is not usually the limiting element in speed of response, it should be mentioned. When it is current-source driven, the step response (10%-90% rise/fall time) is approximately 20 ns. This corresponds to a 3 dB bandwidth greater than 15 MHz.

For a given junction area and optical port, the remaining chip design parameters that raise the efficiency also lower the speed. This is because the photons are produced by electron-hole recombination. The photon emission rate is, therefore, proportional to the recombination rate, which is proportional to minority carrier density, which in turn is proportional to junction charge, Q_j. Thus:

$$\phi_e = \text{radiant flux} \propto Q_j \qquad (3.2.3\text{-}7)$$

Since Q_j is the product of forward current I_F, and minority carrier lifetime, τ_j:

$$\phi_e \propto Q_j = I_F\,\tau_j \qquad (3.2.3\text{-}8)$$

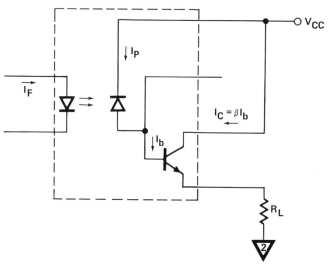

Figure 3.2.3-1 Common-Collector Operation of Single-Transistor Optoisolator.

It is clear that, for a fixed forward current, the speed (inverse of lifetime) can be increased only at the expense of efficiency.

If the photoemitter is the speed limitation in a particular circuit, some improvement can be obtained by peaking. In analog operation, peaking the photoemitter drive can also compensate for speed deficiency in the output circuit.

Digital operation requires the isolator to switch from one logic state to another. For the low state I_F (Figure 3.2-1) must be large enough to bring V_C well below some defined threshold, and for the high state I_F must be low enough to allow V_C to rise well above that threshold. Speed of response is therefore defined with respect to the time required for V_C to reach the threshold in response to a change of logic state at the input, i.e., switching I_F from one state to the other. Characterization of speed is either in terms of propagation delay or data rate.

Propagation delay is the time required for a change of logic state to propagate through the isolator and cause a change of logic state in its load. t_{PHL} is the propagation delay in causing the isolator output to drop from the high state to a specified threshold. t_{PLH} is the propagation delay in causing the output to rise from the low state to the threshold. t_{PHL} and t_{PLH} are shown in Figure 3.2.3-2.

While isolator characteristics (CTR, β, C_{PD}, C_{BC}, etc.) influence t_{PHL} and t_{PLH}, propagation delay is also influenced by the circuit. Raising I_F reduces t_{PHL} by causing a large collector current, but raises t_{PLH} by causing the output to be more deeply saturated and thus increasing the "storage" time. Raising the value of R_L also reduces t_{PHL} by reducing the current it sources to the V_C node but unfortunately this also raises t_{PLH} by reducing the current available to pull V_C up again when the input logic state is

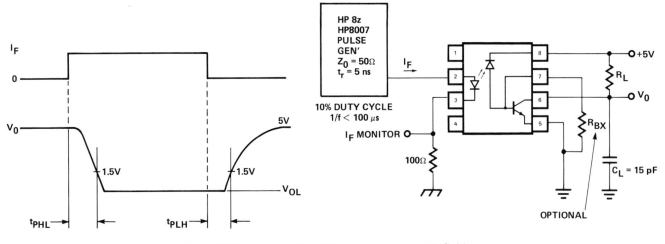

Figure 3.2.3-2 Propagation Delays, t_{PHL} and t_{PLH}; Definition and Measurement.

changed. A resistor, R_{BX} (Figure 3.2.3-2) in parallel with the base-to-emitter junction aids in turning off the transistor to reduce t_{PLH}; adding R_{BX} will require additional input current to keep the output current as high -- and t_{PHL} as low -- as it was before R_{BX} was added. The input current increment, I_{FX}, is required to make up for the current bypassed by R_{BX}, so its value depends on what type of isolator is used. As a general rule:

for single-transistor types:

$$I_{FX} = \left(\frac{V_{be}}{R_{BX}}\right)\left(\frac{I_F}{I_P}\right); I_{FX} \text{ (mA)} \approx \frac{700(V)}{R_{BX} \text{ (k}\Omega)} \qquad (3.2.3-9)$$

for split-darlington types:

$$\qquad\qquad\qquad\qquad\qquad\qquad\qquad (3.2.3-10)$$

$$I_{FX} = \frac{V_{be}}{R_{BX}} \quad \frac{I_F}{I_P} \quad \frac{I}{\beta 2} \; ; \; I_{FX} \text{ (mA)} \approx \frac{7(V)}{R_{BX} \text{ (k}\Omega)}$$

Making I_F very large, then adding R_{BX} reduces both t_{PHL} and t_{PLH}.

Data rate characterization of speed can be directly related to t_{PHL} and t_{PLH}, as shown in Figure 3.2.3-3. Data rate is defined for NRZ (non return to zero) data as the maximum rate for error-free transmission of a random data pattern. Random data implies that following an arbitrarily long string of consecutive ones or zeroes, there may be a string of ones and zeroes alternating at the maximum rate. The minimum duration of the zero or one level is the reciprocal of the maximum data rate in bits per second (b/s) whereas the maximum duration is arbitrary.

The pattern in Figure 3.2.3-3a produces the response in Figure 3.2.3-3b when observed with a 'scope triggered from the slow clock (t_2). If the scope is triggered at every step of

the fast clock (t_1), the result is the classical "eye" picture in Figure 3.2.3-3c. Data are recoverable as long as the "eye" is "open" above and below the threshold for a period of time greater than the "set-up" time required before clocking. Good practice would be to design for an "open eye" duration about twice the "set-up" time. Applying this rule gives the maximum data rate:

$$f_{NRZ(MAX)} \leqslant \frac{1}{t_{SET\text{-}UP} + t_{P(MAX)}} \qquad (3.2.3-11)$$

where $t_{P(MAX)}$ is either t_{PHL} or t_{PLH} -- whichever is the greater. The technique applies also to system characterization; that is, for the entire system (drivers, lines, wiring, etc.) a $t_{PHL(SYS)}$ and $t_{PLH(SYS)}$ can be observed and applied in equation 3.2.3-11.

With RZ (return to zero) data, or other self-clocking patterns, the concept of random data does not apply. Each bit interval includes a period of time at the high state and a period of time at the low state, as in Figure 3.2.3-4. With such a pattern, the longest time spent in either logic state cannot exceed one bit interval and the shortest is never less than half a bit interval.

Characterization of the RZ data rate in terms of t_{PHL} or t_{PLH} is valid, indeed equivalent, for systems with linear transient response and balanced threshold. However, optoisolators are usually non-linear. Furthermore, a transmission line long enough to affect the rise time will exhibit non-linear response. The characterization of RZ data rate should therefore be done with RZ data. A safe rule in judging the data rate that can be expected is:

$$f_{RZ(MAX)} < \frac{1}{2 \, t_{P(MAX)}} \qquad (3.2.3-12)$$

where $t_{P(MAX)}$ is either t_{PLH} or t_{PHL}, whichever is the greater.

3.12

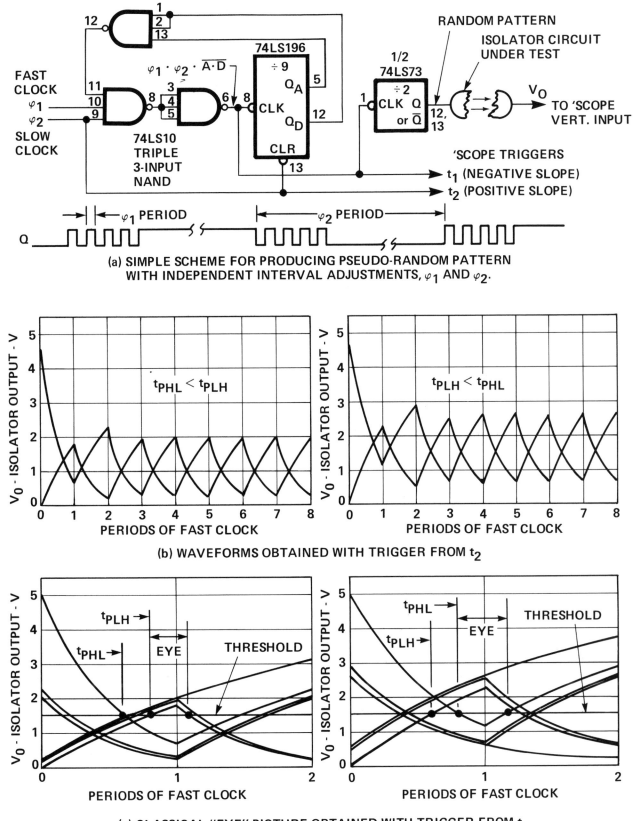

(a) SIMPLE SCHEME FOR PRODUCING PSEUDO-RANDOM PATTERN WITH INDEPENDENT INTERVAL ADJUSTMENTS, φ_1 AND φ_2.

(b) WAVEFORMS OBTAINED WITH TRIGGER FROM t_2

(c) CLASSICAL "EYE" PICTURE OBTAINED WITH TRIGGER FROM t_1

Figure 3.2.3-3   Pseudo-Random Code Generator and Observation of NRZ Data Rate.

3.13

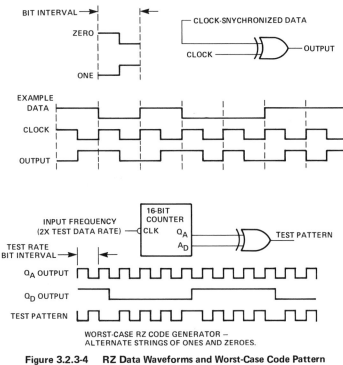

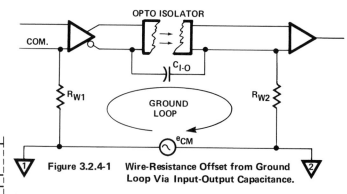

Figure 3.2.4-1 Wire-Resistance Offset from Ground Loop Via Input-Output Capacitance.

Figure 3.2.3-4 RZ Data Waveforms and Worst-Case Code Pattern Generator

3.2.4 Reverse Coupling

An important consideration in many applications of isolators is the degree to which they prevent the flow of ground loop current. Indeed, this is often the basis for use of *optically* coupled isolators in preference to other kinds. Ground loop current can be troublesome not only in the module generating the signal to be transmitted, but also in the module receiving the signal. As seen in Figure 3.2-10, voltage drops in connecting wires can cause annoying offsets.

Optoisolators virtually eliminate dc ground loops ($R_{I-O} \approx 10^{12}$ ohms), but they do allow some ac ground loop current. Internal capacitance between the input and output circuits permit ac ground loop current according to

$$i_{GL} = C_{I-O} \frac{de_{CM}}{dt} \qquad (3.2.4\text{-}1)$$

Note that the capacitance here is not merely the 0.07 pF of C_{CM} in Figure 3.2-1. The internally shielded optoisolators have a lower C_{CM}, but C_{I-O} is the same in both kinds. There are no circuit tricks that can be done to reduce C_{I-O} or its effects.

In applications where the $C_{I-O} \approx 1$ pF of most optoisolators is intolerably high, the only recourse is the selection of a type that has a larger physical separation between the input and output circuits. Such types are usually more costly because they require lenses or fiber optics to obtain adequate optical coupling between widely separated circuits.

With $C_{I-O} = 1$ pF, the ground loop current that flows from a 60 Hz e_{CM} is only 533 pA per volt rms. Even with 220 V rms, the ground loop current is less than 120 nA. If several isolators are used, however, or if e_{CM} has a high frequency or voltage, the resulting ground loop current should be considered.

3.2.5 CTR (or Gain)

With respect to analog types of optoisolators (Figure 3.2-1), the gain is simply defined as the ratio of the output current to the input current. This property is called the Current Transfer Ratio (CTR) and is usually given in percent. If an analog optoisolator is to be used in a digital application, it is worth noting that the CTR is given for a very low collector voltage. This assures a designer that for the specified level of input current, the collector voltage will be adequately low if the current available from the load (I_C in Figure 3.2-1) does not exceed I_F x CTR ÷ 100%.

For optoisolators intended mainly for digital applications, the CTR may be given differently. Rather than being described as ratio of output/input current, it may be described as the "fan out" capability relative to some particular logic family. For example, specification of a "fan out" of 8 TTL gates means that the output of the optoisolator can "sink" a current of 8 x 1.6 mA while maintaining an adequately low output voltage. Where an external pullup resistor is used, the current available from the pullup resistor must be considered in the design.

A more common way of specifying CTR for a digital application is to give the maximum value of the output voltage for given levels of input and output currents. This voltage, V_{OL}, is the output voltage at logic low. For example, if V_{OL} = 0.4V max at I_F = 1.6 mA and I_O = 4.8 mA, this means that with V_O = 0.4V, the minimum CTR is 300% and as long as the available load current does not exceed 4.8 mA, the output voltage will be less than 0.4V with an input current of 1.6 mA or more.

CTR is not constant for all levels of input current. This is due partly to the supralinearity of the photoemitter, but a larger portion of this variation is due to the change in the gain, h_{fe}, of the output amplifier, especially if it has more than one transistor. This variation is described by giving either the output current or the CTR as a function of input current.

3.3   Isolator Packaging

3.3.1   Packaging of Plastic DIP Isolators

Figure 3.3.1-1 shows the mechanical construction of Hewlett-Packard's 8 pin optically coupled isolators. The GaAsP emitter and silicon detector are die attached and wire bonded to separate four-pin lead frames. They are then covered with an inert silicone junction coating. An insulating film is sandwiched between the emitter and detector lead frames. Teflon FEP film is used as the insulating medium between the lead frames. This insulating film assures the excellent low leakage, high voltage insulation between input and output. Finally, the entire assembly is encapsulated in epoxy to insure package integrity.

The same 8 pin package can also be used for dual channel isolators. Mechanical construction of a dual is similar to the construction of a single channel isolator. Due to pin limitations on the dual, V_{CC} and GND of each detector are normally connected together and brought out to two of the four output pins. The remaining two output pins are used for the output of each detector. This configuration retains the high speed performance obtained by biasing the photodiode at a constant reverse voltage. Since the base of the output transistor is unconnected, the base to collector capacitance is lowered and the dual isolator has a somewhat shorter propagation delay than the corresponding single channel isolator. However, circuits that require feedback applied to the base or an external resistor from base to emitter to improve speed, cannot use an 8 pin dual channel isolator. The pinouts of Hewlett-Packard's 8 pin single and dual channel isolators are shown in Figure 3.3.1-2.

3.3.2   Packaging of High Reliability Isolators

Some applications require higher reliability than can be obtained with a plastic encapsulated opto isolator. In general, if an application requires integrated circuits in ceramic packages and metal can transistors for higher reliability in severe environments, then a hermetically packaged isolator should be used. Presently, hermetic isolators are either packaged in a metal can or in a 16-pin ceramic package. Figure 3.3.2-1 shows the mechanical construction of Hewlett-Packard's 16-pin hermetic optically coupled isolator. Silicon detectors are die attached and wire bonded inside a 16 pin ceramic package. GaAsP emitters are die attached and wire bonded to a separate ceramic insert. Solder preforms are applied between the ceramic package and the insert is then soldered in place. Next, the assembly is potted with transparent insulating material to improve optical coupling and electrical insulation between the emitter and detector. Finally, a metal lid is attached to the package to insure a hermetic seal. The finished optically coupled isolator can withstand storage

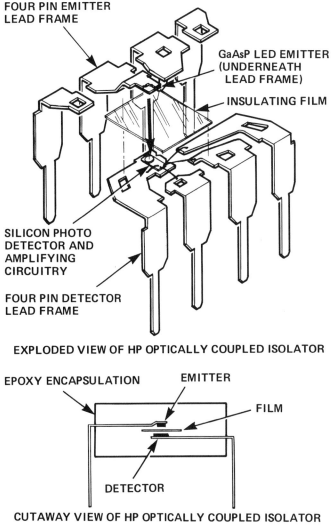

EXPLODED VIEW OF HP OPTICALLY COUPLED ISOLATOR

CUTAWAY VIEW OF HP OPTICALLY COUPLED ISOLATOR

Figure 3.3.1-1   Mechanical Construction of Hewlett Packard's Plastic Encapsulated Opto Isolators.

temperatures from -65°C to +150°C, operating temperature extremes from -55°C to +125°C and 98% relative humidity at 65°C without failure.

These high reliability isolators are also available with additional high reliablity screening for military applications. Hewlett Packard has established two standard high reliability test programs, patterned after MIL-M-38510, Class B. These programs are known as the TX and TXB programs.

3.3.3   Compatibility of Six and Eight Pin Isolators

In many respects, six and eight pin optically coupled isolators have direct pin compatibility. Figure 3.3.3-1 illustrates this compatibility for six pin phototransistor or photodarlington isolators and for eight pin photodiode-transistor or photodiode-split darlington isolators. The eight pin package allows a separate V_{CC} bias supply that provides high speed operation and further circuit versatility while still maintaining direct mechanical compatibility.

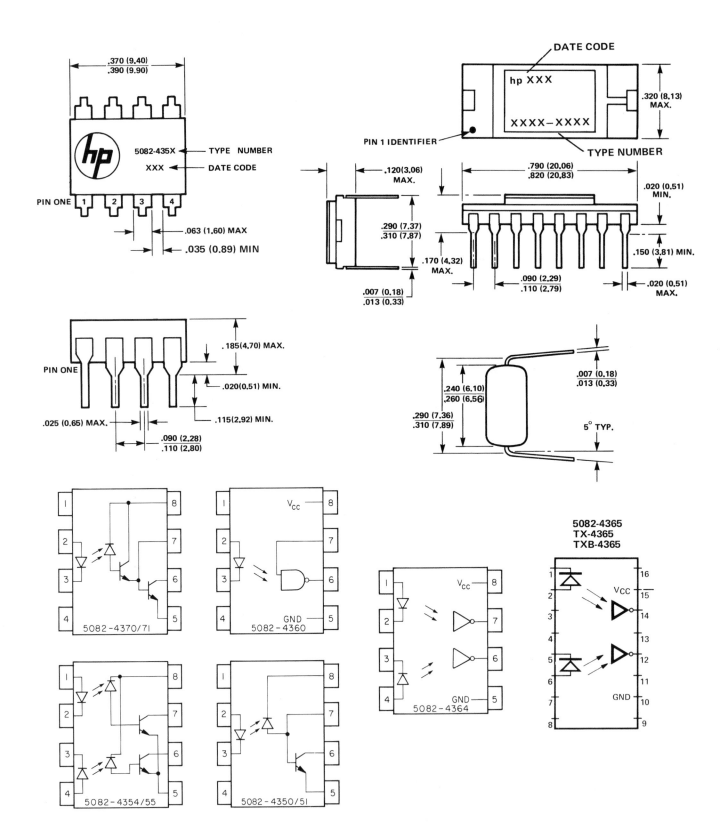

Figure 3.3.1-2   Outline Drawings and Pinouts of Hewlett Packard's
                  Opto Isolators.

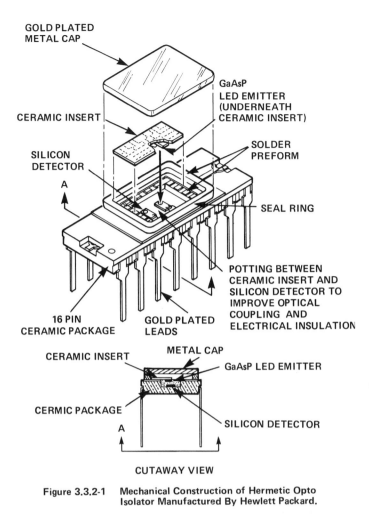

GOLD PLATED METAL CAP

GaAsP LED EMITTER (UNDERNEATH CERAMIC INSERT)

CERAMIC INSERT

SOLDER PREFORM

SILICON DETECTOR

A

SEAL RING

POTTING BETWEEN CERAMIC INSERT AND SILICON DETECTOR TO IMPROVE OPTICAL COUPLING AND ELECTRICAL INSULATION

16 PIN CERAMIC PACKAGE

GOLD PLATED LEADS

CERAMIC INSERT

METAL CAP

GaAsP LED EMITTER

CERMIC PACKAGE

SILICON DETECTOR

A

CUTAWAY VIEW

Figure 3.3.2-1   Mechanical Construction of Hermetic Opto Isolator Manufactured By Hewlett Packard.

The only extra cost of this interchangeability is an allowance for two additional holes on the printed circuit. These holes are used for pins 1 and 8 of the eight pin isolator and left open for a six pin isolator. To provide proper bias for the photodiode, pin 8 should be connected to V_{CC}. Since six pin isolators have an overall length of 8.89 mm (.350 inches) and eight pin isolators have an overall length of 9.91 mm (.390 inches), both isolators require 10.16 mm (.400 inch) spacing on standard 2.54 mm. (.100 inch) printed circuit pad spacing. Using dual channel eight pin isolators, four isolator channels can be realized in the same space as one 16 pin integrated circuit.

This direct pin compatibility holds for all of Hewlett-Packard's eight pin single channel optically coupled isolators. The 5082-4350 family and the 5082-4370 family offer switching speed advantages over phototransistor and photodarlington isolators. The split darlington configuration of the 5082-4370 family allows the output transistor to saturate as low as .1V for improved noise margin with TTL circuitry. For a V_{CC} of 5 volts, the 5082-4360 family is also directly pin compatible. The 5082-4360 family has extremely fast switching speeds, low input current requirements and a TTL compatible enable input.

3.3.4 Layout Considerations for Optically Coupled Isolators

Optically coupled isolators have excellent common mode rejection characteristics. However, for optimum common mode rejection, special considerations should be given to the circuit layout. Stray capacitance between the input circuit and the output circuit should be minimized. This can be done by physically separating all input and output circuitry on the circuitboard. For line receiver applications, the shielded cable should also be dressed properly to minimize stray capacitance. Since the base of the output transistor is especially susceptible to common mode transients, a special ground trace under the isolator can serve as a shield. To further minimize capacitive coupling into the base, pin 7 can be clipped off at the side of the package. A simpler approach in an application requiring more than one isolator is to use dual packages, because the base is unconnected.

Circuit layout can also effect propagation delays through the optically coupled isolators. Since most opto isolators have an open collector output, excessive shunt capacitance on pin 6 can limit rise and fall times of the isolator. For this reason, the logic gate being driven by the isolator should be as close to the isolator as possible and the pullup resistor mounted in close proximity to the isolator. An external resistor can be connected between pin 5 and pin 7 to reduce t_{PLH} of the 5082-4370 isolator. If this technique is used, capacitance between pin 5 and pin 6 should be minimized because of the "Miller" effect. Figure 3.3.4-1 illustrates some of these circuit layout considerations.

3.3.5   Bypass Capacitor Requirements

Some isolators such as the 5082-4360 series family require an external bypass capacitor to prevent internal oscillations. An isolator that is not functioning properly due to internal oscillations exhibits the following symptoms: the isolator has an extremely low current transfer ratio which requires an excessively high input LED current to cause the output collector to saturate. If noise is coupled into the isolator from its load, the isolator may have multiple transitions for a single input pulse. To prevent these internal oscillations, a low inductance .01 μF ceramic capacitor should be placed between pins 5 and 8 as close to the device as possible. For optimum results, the bypass capacitor should be connected in such a manner as to minimize the coupling of noise generated by the isolator load into the V_{CC} and ground connections of the isolator. This can be accomplished by using separate V_{CC} and ground lines for the isolator or isolators than those used by the digital logic. A second technique to minimize noise coupling is to lay out the printed circuit with a topology that connects the V_{CC} and ground for the digital logic between the power supply and the V_{CC} and ground of the opto isolator. These techniques

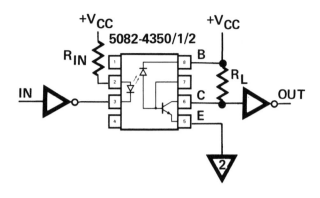

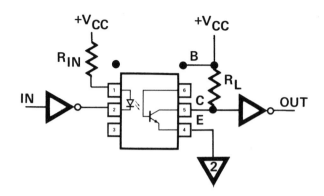

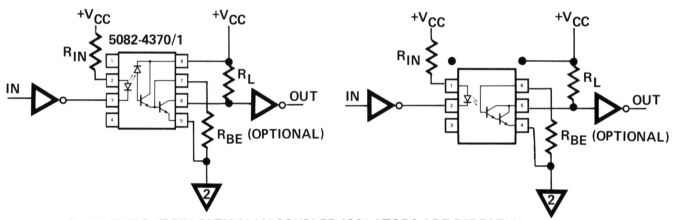

SIX AND EIGHT PIN OPTICALLY COUPLED ISOLATORS ARE DIRECTLY
PIN COMPATIBLE WHEN TWO EXTRA HOLES AND SHORT V$_{CC}$ TRACE
ARE ADDED TO PRINTED CIRCUIT BOARD.

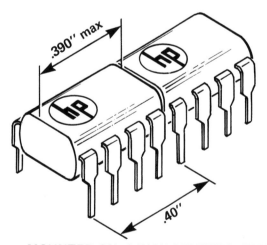

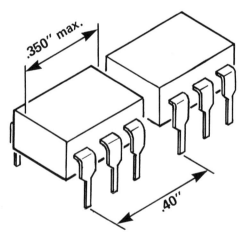

MOUNTED ON .1 INCH CENTERS, SIX PIN OPTO ISOLATORS REQUIRE
THE SAME SPACE AS EIGHT PIN OPTO ISOLATORS.

Figure 3.3.3-1   Pinout Compatibility and End Stackability of 6
and 8 Pin Isolators.

3.18

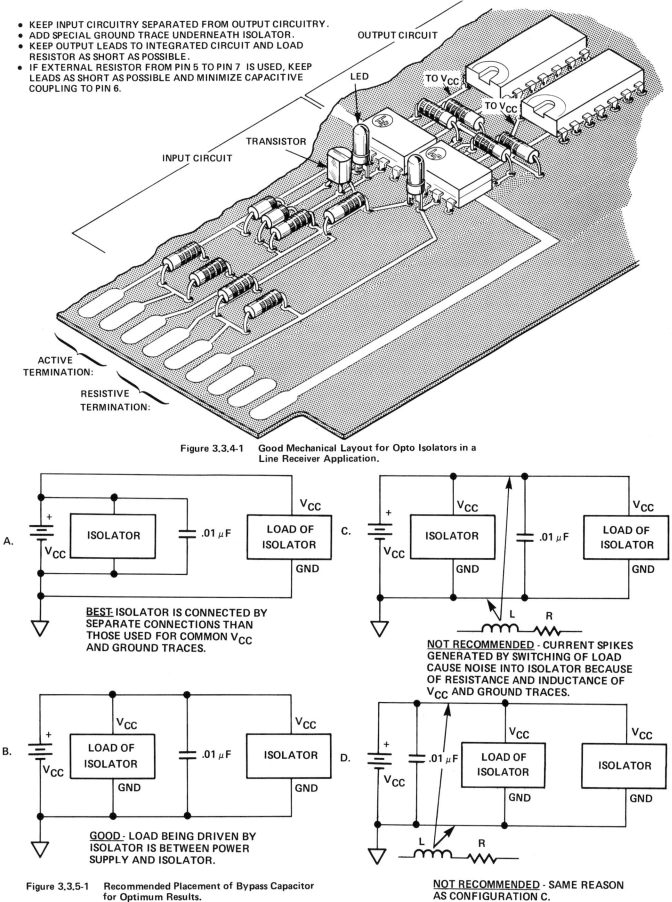

- KEEP INPUT CIRCUITRY SEPARATED FROM OUTPUT CIRCUITRY.
- ADD SPECIAL GROUND TRACE UNDERNEATH ISOLATOR.
- KEEP OUTPUT LEADS TO INTEGRATED CIRCUIT AND LOAD RESISTOR AS SHORT AS POSSIBLE.
- IF EXTERNAL RESISTOR FROM PIN 5 TO PIN 7 IS USED, KEEP LEADS AS SHORT AS POSSIBLE AND MINIMIZE CAPACITIVE COUPLING TO PIN 6.

Figure 3.3.4-1   Good Mechanical Layout for Opto Isolators in a Line Receiver Application.

A. **BEST**- ISOLATOR IS CONNECTED BY SEPARATE CONNECTIONS THAN THOSE USED FOR COMMON V_{CC} AND GROUND TRACES.

B. **GOOD**- LOAD BEING DRIVEN BY ISOLATOR IS BETWEEN POWER SUPPLY AND ISOLATOR.

C. **NOT RECOMMENDED** - CURRENT SPIKES GENERATED BY SWITCHING OF LOAD CAUSE NOISE INTO ISOLATOR BECAUSE OF RESISTANCE AND INDUCTANCE OF V_{CC} AND GROUND TRACES.

D. **NOT RECOMMENDED** - SAME REASON AS CONFIGURATION C.

Figure 3.3.5-1   Recommended Placement of Bypass Capacitor for Optimum Results.

3.19

are illustrated in Figure 3.3.5-1. When several isolators are used on a single printed circuit board, a separate bypass capacitor should be connected across pins 5 and 8 of each isolator. These same bypassing techniques also apply to the 5082-4365 isolator with the exception that V_{CC} and ground are on pins 15 and 10. When several isolators are end stacked, a very clean mechanical layout can be accomplished by connecting the pullup resistor to the V_{CC} of an adjacent isolator.

In noisy environments, bypassing may also be required for the 5082-4350 series and for the 5082-4370 series isolators to prevent V_{CC} transients from being coupled into the photodiode. The voltage transients can be large enough to cause multiple transitions of the output while the collector is switching from one state to another. When extremely high common mode transients are present, sufficient energy might be coupled into the photodiode and overstress or cause a catastrophic failure of the detector. A resistor connected between V_{CC} and pin 8 of the isolator will limit these current surges to safe levels but will not affect the normal operation of the opto isolator. This technique is illustrated in Figure 3.6.2-4.

3.4   CTR Degradation

3.4.1   Introduction

It is an established fact that the total photon flux emitted by an optoelectronic device diminishes slightly over the operating lifetime of the device. Barring catastrophic failures or overstressing of the optoelectronic device, this change of photon emission is almost imperceptible for many tens of thousands of hours in visual applications. However, this change in light output can be measured with a sensitive photodetector. Figure 3.4.1-1 shows the average change in light output vs. time for a 655 nm GaAsP lamp manufactured by Hewlett-Packard. These lamps were stressed at 50 mA DC for 40,000 hours. At lower stress currents, the change of light output vs. time is reduced. Since the total flux emitted by the LED normally diminishes over time, this change is often referred to as a degradation of light output, although in some instances, the light output of an LED has actually increased over time.

An optically coupled isolator is an optoelectronic emitter-detector pair. Any degradation of light output of the emitter will cause a change in the apparent gain of the entire device. Current gain of an isolator is normally specified as a ratio of output current to input current expressed as a percent for a specified input current and called the current transfer ratio (CTR). The change in gain of the isolator can be expressed as a change in current transfer ratio over time and is commonly called CTR degradation.

CTR degradation is important because an excessive amount of degradation or a marginally designed system can cause a reduction in performance and eventual system failure unless an allowance is made for it. Figure 3.4.1-2 shows a simple optically isolated logic interface that can be used to illustrate this problem. The available output current, I_O is equal to the input current, I_F, times the CTR of the isolator. Keeping in mind the I_{OL} capabilities of the logic family driving the isolator LED, I_F is picked so that the resulting I_O will be large enough to sink the I_{IL} of the logic gate being driven by the isolator and the current flowing through the pull-up resistor.

$$I_o \geq m \, |I_{IL}| + \frac{V_{CC2} - V_{OL}}{R_L} \qquad (3.4.1\text{-}1)$$

where: m is the number of logic inputs being driven by the isolator, I_{IL} is the maximum input current required by the logic gate, V_{CC2} is the power supply voltage, V_{OL} is the maximum output voltage for the logic gate being driven by the isolator, and R_L is the pull-up resistor.

For example, if $V_{CC2} = 5V$, $R_L = 5.6K$, $I_{IL} = -1.6 \, \text{mA}$, $V_{OL} = .4V$ and m = 1, then I_O should be greater than or equal to 2.4 mA. This I_O corresponds to an isolator driven at 16 mA I_F with a CTR $\geq 15\%$. A reduction of CTR over time that lowers the CTR to less than 15% will reduce I_O below 2.4 mA and the output voltage of the isolator will increase above .4V. The noise margin of the circuit will then degrade until the output voltage rises higher than the switching threshold of the gate (about 2 Vbe for TTL families) and the circuit ceases to function. Clearly, the designer must anticipate this CTR degradation and either start with a CTR higher than 15% or design the circuit such that a lower CTR will still allow the circuit to function.

Variations in V_{CC} and resistor tolerances can also contribute to system failures. A well designed circuit using optically coupled isolators should allow some margin for CTR degradation as well as consider the worst cased effects of temperature, component tolerances, and power supply variations. Also when using isolators in analog circuits, degradation may cause a change in the gain or offset of the overall system.

3.4.2   Cause of CTR Degradation

An optically coupled isolator can be modeled by the system block diagram shown in Figure 3.4.2-1. Potential causes of CTR degradation are a reduction in efficiency (η) of the emitter, a decrease in the transmission of the optical path (K), a reduction in responsivity (R) of the photodetector, or a change in gain (β) of the output amplifier. It is generally accepted that CTR degradation is caused primarily by the reduction in total efficiency (η) of the optoelectronic emitter. Also, since the gain of the output amplifier (β) is related to its input current, CTR degradation may be compounded by the change in β due to a decrease in photocurrent (I_p) caused by a reduction in η. The current transfer relationship for the isolator shown in Figure 3.4.2-1 is:

$$\frac{I_O}{I_F} = K\,R\,\eta\,(I_F,\,t)\,\beta\,(I_P) \qquad (3.4.2\text{-}1)$$

where: I_O is the collector current, I_F is the input current η is the quantum efficiency of the LED expressed as photons emitted per electron of input current, K is the total transmission of the optical path, R is the responsivity of the photodetector in terms of electrons of photocurrent per photon, I_P is the photocurrent, and β is the gain of the output amplifier.

If the gain of the output amplifier β is constant, than the CTR degradation of the isolator is equal to the degradation of light output of the optoelectronic emitter. In general, the gain of the output amplifier is a function of its input current so the rate of CTR degradation may be different than the rate of LED degradation. Assuming that K, R, and I_F remain constant, and that β varies as a function of photocurrent, then the change in CTR over time can be related to the change in emitter efficiency by equation 3.4.2-2.

$$\frac{\Delta CTR}{CTR} = \frac{\Delta I_o}{I_o} = \frac{\Delta\eta}{\eta}\left[1 + (1 + \frac{\eta}{\Delta\eta})\,\frac{\Delta\beta}{\beta}\right] \qquad (3.4.2\text{-}2)$$

$$I_P(\text{final}) = (1 + \frac{\Delta\eta}{\eta})\,I_P\,(\text{initial})$$

where: Δ is defined as the final value minus the initial value — i.e., $\Delta\beta = \beta(I_P(\text{final})) - \beta(I_P(\text{initial}))$; CTR, I_0, η, β are the initial values; and $\Delta\eta/\eta$ is the normalized change in light output for a specified operating time.

Typical values of I_P (I_F) and $\beta(I_P)$ for the 5082-4350 and 5082-4370 series of optically coupled isolators are shown in Figures 3.4.2-2 and 3.4.2-3. The amount of CTR degradation can then be calculated from Equation 3.4.2-2 by knowing the amount of LED degradation and the initial value of I_F. For example, suppose a 5082-4350 isolator is operated at 16 mA I_F and the LED degrades by 15% so that $\Delta\eta/\eta = -.15$. Referring to Figure 3.4.2-2, I_P is typically 24 μA so that β is typically about 145. Since $\Delta\eta/\eta = -.15$, then I_P (final) is 20 μA which corresponds to a β of 150. Thus $\Delta\beta/\beta = 5/145$ so that $\Delta CTR/CTR = -.12$. In general, $|\Delta CTR/CTR| < |\Delta\eta/\eta|$ if $\Delta\beta/\beta > 0$ and $|\Delta CTR/CTR| > |\Delta\eta/\eta|$ if $\Delta\beta/\beta < 0$. However, unless $\Delta\beta/\beta$ is large, the rate of CTR degradation will be approximately equal to the rate of emitter degradation. The effect of the extra stage of gain in the 5082-4370 series of isolators is to

increase the sensitivity of the detector to much smaller values of I_F. This allows the 5082-4370 type of isolator to be used at much lower input currents, thus reducing the rate of emitter degradation. In addition, $\Delta\beta/\beta > 0$ in this operating region which tends to further minimize the effects of emitter degradation.

Another type of isolator contains an output amplifier that switches rather abruptly at a certain threshold of I_P. An example of this type of isolator is the 5082-4360 series isolator. Typical I_P vs. I_O characteristics for the 5082-4360 series isolator are shown in Figure 3.4.2-4. Above this threshold, the output of the isolator has more than adequate gain to handle most types of digital circuits. The exact value of the switching threshold, I_{Pth} varies with $\eta(I_F)K\,R$ and somewhat on the processing of the photodetector IC. The effect of emitter degradation for this type of isolator is an increase in the amount of input current that corresponds to I_{Pth}. The design objective for this type of isolator is to insure that there is sufficient input current, I_F, to generate a value of I_P greater than I_{Pth} even with some emitter degradation.

Characterization of $\partial n/\partial t$ will then allow the effects of emitter degradation on overall CTR degradation to be determined for the various types of optically coupled isolators.

Hewlett Packard has tested a considerable number of 700 nm LED emitters packaged in glass topped TO-18 style hermetic packages for $d\eta/dt$ at 20 mA I_F. These emitters are used in all of the opto isolators manufactured by Hewlett Packard. Data was taken for both 25°C and 125°C so a direct comparison can be made. Figure 3.4.2-5 shows the mean, 10 and 90 percentiles of normalized efficiency vs. time for LED's from several different wafer and diffusion lots. These curves indicate that a decrease in intensity of 5% at 1000 hours and 8% in 5000 hours is to be expected when operating at 20 mA I_F and 25°C.

Figure 3.4.2-5 shows that operation of 125°C does not significantly alter the amount of emitter degradation over time. Temperature is, therefore, only a minor contributor to CTR degradation. Longer term degradation effects can be gleaned from the curve presented earlier for 40,000 hours of operation for the very similar 655nm GaAsP devices.

It is known that the rate of LED light output degradation is influenced by the materials and processing parameters used to manufacture the GaAsP LED, and the junction temperature of the LED in addition to the current density through the LED. The current flowing through the LED can be considered to be the sum of a "radiative" diffusion current and several components of "non radiative" currents. Light is generated only by the "radiative"

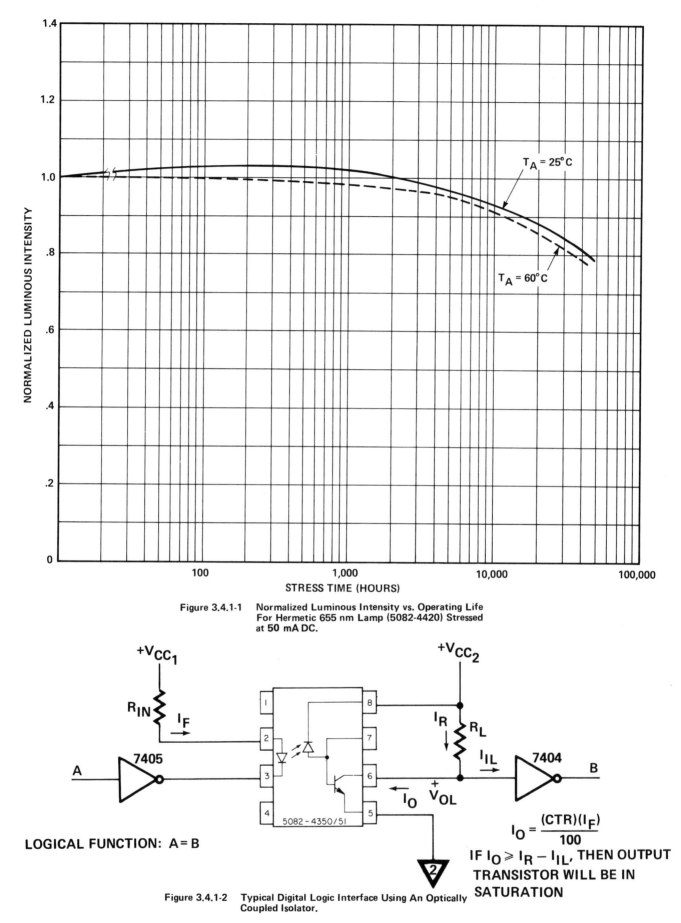

Figure 3.4.1-1   Normalized Luminous Intensity vs. Operating Life
For Hermetic 655 nm Lamp (5082-4420) Stressed
at 50 mA DC.

LOGICAL FUNCTION: A = B

$$I_O = \frac{(CTR)(I_F)}{100}$$

IF $I_O \geq I_R - I_{IL}$, THEN OUTPUT
TRANSISTOR WILL BE IN
SATURATION

Figure 3.4.1-2   Typical Digital Logic Interface Using An Optically
Coupled Isolator.

3.22

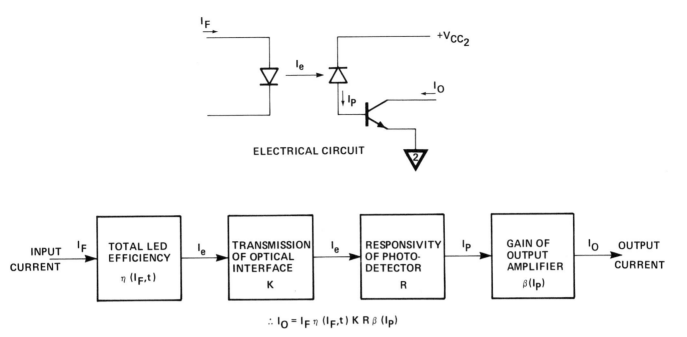

ELECTRICAL CIRCUIT

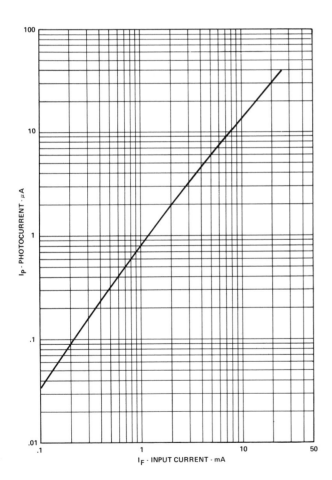

$$\therefore I_O = I_F \, \eta \, (I_F, t) \, K \, R \, \beta \, (I_P)$$

Figure 3.4.2-1   System "Model" for an Opto Isolator.

Figure 3.4.2-2   Photocurrent vs. Input Diode Forward Current for Hewlett Packard Plastic Encapsulated Opto Isolators.

diffusion current. If the "non radiative" current should increase, then less light will be emitted by the LED. The ratio of "radiative" to "non radiative" currents changes as a function of total current flowing through the LED. This accounts for the supralinearity of light emission vs. input current. Another important consideration is the magnitude of the LED "stress" current as compared to the current at which the CTR is measured. The proportion of "non-radiative" diode current to "radiative" diffusion current tends to increase at lower measurement currents. For example, given a stress current of 10 mA and measurements of CTR at 1 mA and 10 mA before and after stress, the percentage of CTR change will be greater for the 1 mA measurement than for the 10 mA measurement. For this reason, excessive peak transient LED currents should be avoided. Since the devices in Figure 3.4.2-5 were stressed at 20 mA DC and tested at 10 mA DC, the results shown reflect a somewhat worst case rate of degradation. At this time, the cause for this increase in "non radiative" current over time is not completely understood. Figure 3.4.2-6 shows the change in CTR vs. time for 5082-4370 type couplers. Note the significant difference between the 20 mA stress and 1.6 mA stess current curves -- both had CTR measurements made at I_F = 1.6 mA.

Typical results of degradation at 2K hours plus extrapolated 10K hour data is shown in Table 3.4.2-1. While no completely valid model exists for extrapolation, results to date suggest that a continuous exponential decay model is probably valid. The 10K extrapolated data is based on this assumption. Results of the 40,000 hour operating life test shown in Figure 3.4.1-1 on the very similar 655 nm GaAsP LEDs tend to substantiate this "exponential decay model".

3.23

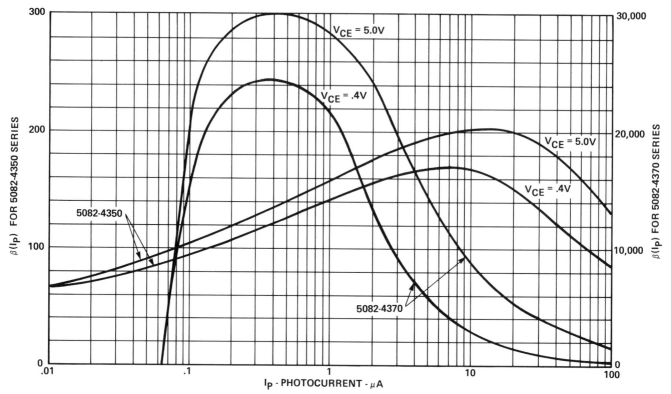

Figure 3.4.2-3   DC Current Gain for the Output Amplifier of the
5082-4350 Series and 5082-4370 Series OptoIsolators.
Isolators.

DEVICE TYPE	I_F STRESS T_A = MAX	2K HOURS	10K HOURS
5082-4350	5 mA	−7% $\triangle$CTR	−12% $\triangle$CTR
	20 mA	−10% $\triangle$CTR	−15% $\triangle$CTR
5082-4360	10 mA	−10% $\triangle I_F$	−15% $\triangle I_F$
5082-4365	20 mA	−10% $\triangle I_F$	−15% $\triangle I_F$
5082-4370	1.6 mA	−5% $\triangle$CTR	−10% $\triangle$CTR

TABLE 3.4.2-1 Typical CTR Degradation vs. Time for
Hewlett Packard Opto Isolators.

In a critical application, the user may want to multiply these "typical" figures by a factor of 2 or 3. The spread between mean, 10 and 90 percentile in Figure 3.4.2-5 shows this to be a reasonable assumption.

When using this CTR degradation information, three things should be kept in mind. This data is for continuous operation of the LED at maximum LED current. In most applications, the input LED is operated in a pulsed mode or in a mode in which it is on for only a fraction of the total system operating time. Since CTR degradation occurs only during the time that the LED is on, the required operating life of the isolator may only be a small fraction of the projected system operating lifetime. Thus the effects of CTR degradation can be minimized by phasing digital logic such that the isolator LED is normally off. Lower stress currents will also decrease the rate of degradation. Finally, this data only applies to the 700 nm emitters manufactured by Hewlett Packard. Isolator emitters manufactured at different wavelengths by different GaAsP or GaAs processes may have significantly different rates of degradation.

3.4.3 Worst Case Circuit Design to Allow for CTR Degration

Proper circuit design can alleviate most of the potential problems of CTR degradation. For the circuit of Figure 3.4.1-2, the minimum value of R_L was selected by equation 3.4.1-1. This equation can be modified to allow a guardband for CTR degradation and to worst case V_{CC1}, V_{CC2}, R_{IN}, and R_L variations. Every circuit design should account for these variations plus the effects of ambient temperature extremes on device parameters. The minimum value of R_L is determined by the minimum value of I_O resulting from a minimum CTR device at a given I_F. In this state, the isolator must guarantee a logical "0" to the input of the logic circuitry. The maximum value of R_L is determined by the maximum amount of leakage, I_{OH}, into the collector of the isolator at the maximum ambient temperature of the circuit. In this state, the isolator must guarantee a logical "1" to the input of the logic circuitry. R_L should be selected to satisfy these constraints:

$$R_L(MIN) \geq \frac{V_{CC2}(MAX) - V_{OL}}{\dfrac{I_F(MIN)\,CTR(MIN)\,(1 - x/100)}{100} - m\,|I_{IL}|} \quad (3.4.3\text{-}1)$$

3.24

$$R_L \text{ (MAX)} \leqslant \frac{V_{CC2}(MIN) - V_{OH}}{I_{OH}(MAX) + m\, I_{IH}} \qquad (3.4.3\text{-}2)$$

where: V_{OL}, V_{OH}, I_{IL}, I_{IH} are guaranteed minimums and maximums for the particular logic family, m is the total number of logic inputs connected to the isolator. CTR, I_{OH} are guaranteed isolator parameters. V_{CC2} and I_F are circuit parameters based on worst cased values for V_{CC1} and R_{IN}. x is the maximum % degradation allowed over the system operating life.

Within these limits for R_L, a tradeoff exists between R_L and propagation delay. t_{PLH} is determined by the saturation delay of the isolator plus an RC constant determined by R_L and the total shunt capacitance, C_L, at the collector. t_{PHL} is inversely proportional to I_F and generally independent of R_L. Figure 3.4.3-1a shows a worst case TTL design that uses equations 3.4.3-1 and 3.4.3-2 to allow for CTR degradation. In this example, several design choices are available to the designer and one or a combination of them an optimize this design for cost and effectiveness.

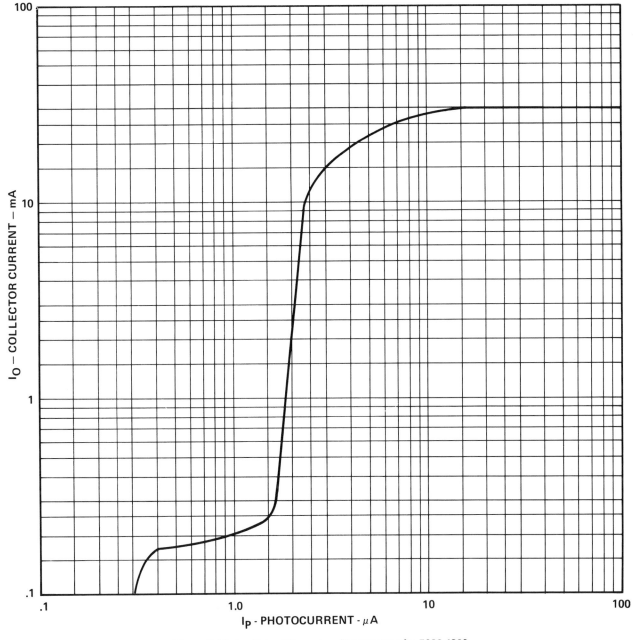

Figure 3.4.2-4   Output Current vs. Photocurrent for 5082-4360 Series Opto Isolators.

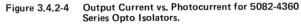

3.25

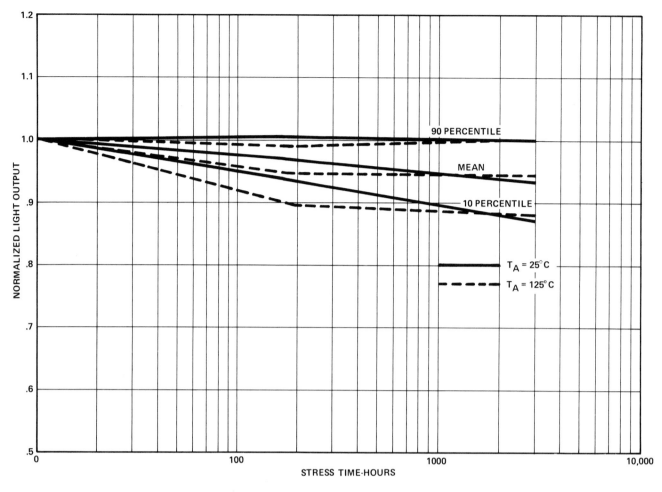

Figure 3.4.2-5 Normalized Luminous Intensity vs. Operational Life for 700 nm LED Emitters Used for Hewlett-Packard Optically Coupled Isolators.

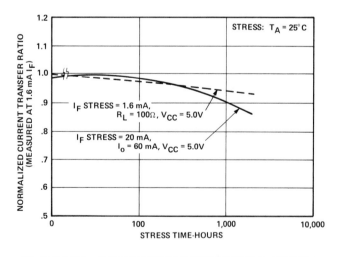

Figure 3.4.2-6 Normalized Current Transfer Ratio vs. Operating Life for 5082-4370 Series Opto Isolators.

1. V_{CC1}, V_{CC2} tolerances can be tightened (possible addition of zener diode).

2. R_{IN}, R_L tolerances can be tightened.

3. A TTL buffer gate can be substituted for the 7405 to increase I_F slightly.

4. A higher CTR isolator can be specified.

A simple and cost effective design technique that reduces the effects of CTR degradation and improves the overall performance of the system is to replace the gate being driven by the isolator with a gate having a lower I_{IL} requirement. For example, Figure 3.4.3-1b shows a TTL to LSTTL worst case design. This circuit is identical to Figure 3.4.3-1a with the exception that the 7404 gate is replaced with a 74LS04 gate. Since TTL and LSTTL have approximately the same propagation delay, the circuit of Figure 3.4.3-1b will have a significantly shorter t_{PLH} than the circuit of Figure 3.4.3-1a because R_L is smaller. Figure 3.4.3-1b can be optimized to allow more than one load, a larger amount of CTR degradation, or a lower CTR isolator (5082-4350) at lower cost.

3.26

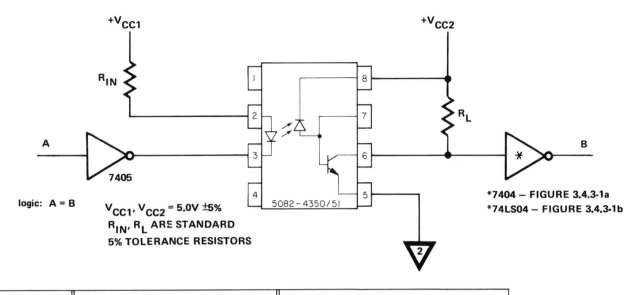

logic: A = B

V_{CC1}, V_{CC2} = 5.0V ±5%
R_{IN}, R_L ARE STANDARD
5% TOLERANCE RESISTORS

*7404 – FIGURE 3.4.3-1a
*74LS04 – FIGURE 3.4.3-1b

TTL FAMILY	I_{IL}	V_{IL}	I_{IH}	V_{IH}	I_{OL}	V_{OL}	I_{OH}	V_{OH}
74S	−2 mA	.8V	50 μA	2V	20 mA	.5V	−1000 μA	2.7V
74H	−2 mA	.8V	50 μA	2V	20 mA	.4V	− 500 μA	2.4V
74	−1.6 mA	.8V	40 μA	2V	16 mA	.4V	− 400 μA	2.4V
74LS	−.36 mA	.8V	20 μA	2V	8 mA	.5V	− 400 μA	2.7V
74L	−.18 mA	.7V	10 μA	2V	3.6 mA	.4V	− 200 μA	2.4V

FIGURE 3.4.3-1a (TTL TO TTL INTERFACE)

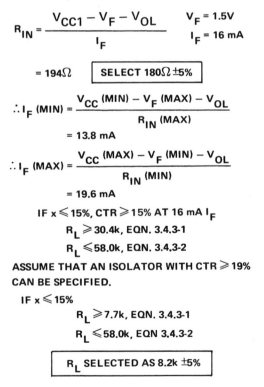

$$R_{IN} = \frac{V_{CC1} - V_F - V_{OL}}{I_F}$$

V_F = 1.5V
I_F = 16 mA

= 194Ω SELECT 180Ω ±5%

$$\therefore I_F (MIN) = \frac{V_{CC} (MIN) - V_F (MAX) - V_{OL}}{R_{IN} (MAX)}$$

= 13.8 mA

$$\therefore I_F (MAX) = \frac{V_{CC} (MAX) - V_F (MIN) - V_{OL}}{R_{IN} (MIN)}$$

= 19.6 mA

IF x ≤ 15%, CTR ≥ 15% AT 16 mA I_F

R_L ≥ 30.4k, EQN. 3.4.3-1
R_L ≤ 58.0k, EQN. 3.4.3-2

ASSUME THAT AN ISOLATOR WITH CTR ≥ 19%
CAN BE SPECIFIED.

IF x ≤ 15%

R_L ≥ 7.7k, EQN. 3.4.3-1
R_L ≤ 58.0k, EQN. 3.4.3-2

R_L SELECTED AS 8.2k ±5%

5082-4351

V_F = 1.5V (1.75V MAX) AT 20 mA I_F
CTR ≥ 15% AT 16 mA I_F, V_0 = .4V
I_{OH} ≤ .5 μA AT V_{CC} = 5.5V, 25°C

FIGURE 3.4.3-1b (TTL TO LS TTL INTERFACE)

R_{IN} SELECTED AS 180Ω ±5%

$\therefore I_F (MIN)$ = 13.8 mA
$\therefore I_F (MAX)$ = 19.6 mA

IF x ≤ 15%, CTR ≥ 15% AT 16 mA I_F

R_L ≥ 3.4k, EQN 3.4.3-1
R_L ≤ 100k, EQN 3.4.3-2

IF x ≤ 25%, CTR ≥ 15% AT 16 mA I_F

R_L ≥ 4.0k, EQN 3.4.3-1
R_L ≤ 100k, EQN 3.4.3-1

R_L SELECTED AS 4.3k ±5%

Figure 3.4.3-1   Worst Case Design of TTL to TTL/LSTTL
Interface Using 5082-4351 Isolator that Allows
for CTR Degradation.

Some optically coupled isolators such as the 5082-4360 family have a highly non linear current transfer function which is shown in Figure 3.4.3-2. This transfer function can be described more accurately by a minimum threshold of I_{Fth} that is required to turn the isolator on rather than a current transfer ratio. As the light output of the LED degrades, the net effect is a slight increase in the value of I_{Fth}. Allowance for a change in I_{Fth} due to LED degradation can best be accomplished by driving the isolator with a somewhat higher I_F than the minimum I_F required for a guaranteed V_{OL} (at a specified maximum I_{OL}). The calculation of the required I_F to allow LED degradation of X% is shown below:

$$I_P \propto (I_F)^n \qquad (3.4.3-3)$$

where $1.1 \leqslant n \leqslant 1.3$

$$\therefore \left[1 - \frac{X}{100}\right] = \left[\frac{I_{PH}}{I_P}\right] = \left[\frac{I_{FH}}{I_F}\right]^n \qquad (3.4.3-4)$$

$$\therefore I_F = \frac{I_{FH}}{\left[1 - \dfrac{X}{100}\right]^{1/n}} \qquad (3.4.3-5)$$

where: I_{FH} is the guaranteed minimum isolator input current for a specified I_{OL} and V_{OL}, n is the slope of I_P vs. I_F on logarithmic coordinates (Figure 3.4.2-2), and x is the maximum % degradation allowed over the system operating life.

Figure 3.4.3-3 shows a worst case LSTTL to TTL design using equation 3.4.3-5 to allow for CTR degradation. Since I_{OL} of this type of optically coupled isolator does not decrease with operating life, the equations for R_L can be simplified to:

$$R_L \geqslant \frac{V_{CC2}(MAX) - V_{OL}}{I_{OL}(MIN) - m|I_{IL}|} \qquad (3.4.3-6)$$

$$R_L \leqslant \frac{V_{CC2}(MIN) - V_{OH}}{I_{OH}(MAX) + mI_{IH}} \qquad (3.4.3-7)$$

where: V_{OL}, V_{OH}, V_{IL}, I_{IH} are guaranteed minimums and maximums for the particular logic family, m is the total number of logic inputs connected to the isolator. $I_{OL(MIN)}$ and $I_{OH(MAX)}$ are guaranteed isolator parameters. V_{CC2} is the power supply voltage.

Since it is very important to insure that a guardband exists for I_F and some applications, i.e., line receiver applications have a wide range of input currents, it may be necessary to "regulate" the isolator input current. One circuit for doing this is shown in Figure 3.6.1-6 and Figure 3.6.1-7.

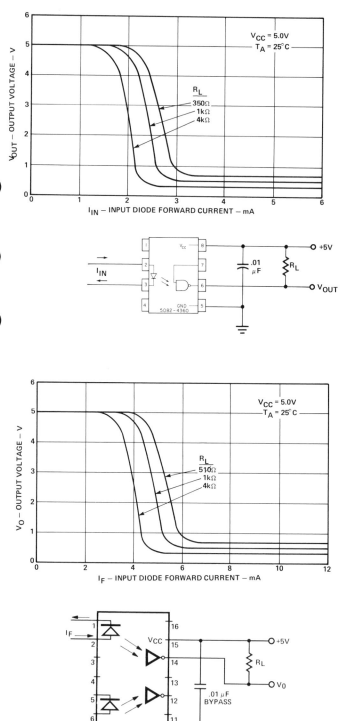

HP 5082-4365 INPUT-OUTPUT CHARACTERISTICS

Figure 3.4.3-2   Input vs. Output Characteristics for 5082-4360 Series Opto Isolators.

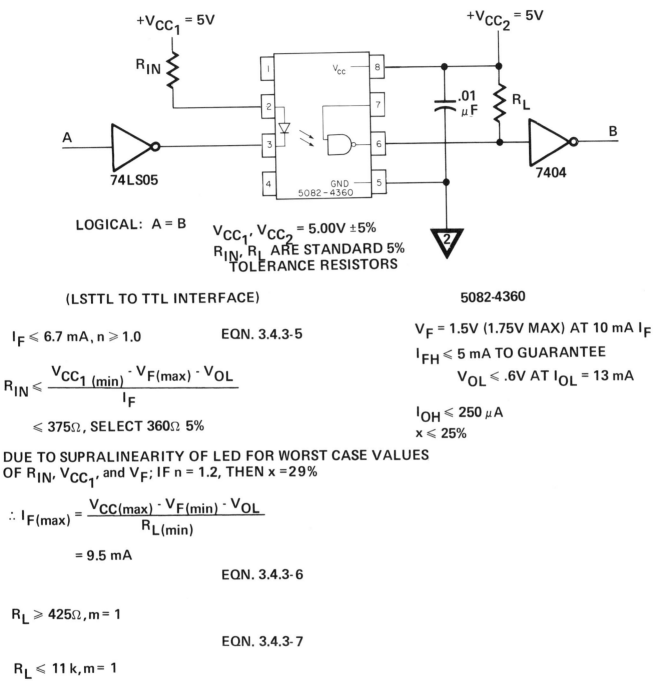

LOGICAL: A = B

V_{CC_1}, V_{CC_2} = 5.00V ±5%

R_{IN}, R_L ARE STANDARD 5% TOLERANCE RESISTORS

(LSTTL TO TTL INTERFACE)

$I_F \leqslant 6.7$ mA, n ⩾ 1.0   EQN. 3.4.3-5

$$R_{IN} \leqslant \frac{V_{CC_1\ (min)} - V_{F(max)} - V_{OL}}{I_F}$$

$\leqslant 375\Omega$, SELECT 360Ω 5%

DUE TO SUPRALINEARITY OF LED FOR WORST CASE VALUES OF R_{IN}, V_{CC_1}, and V_F; IF n = 1.2, THEN x =29%

$$\therefore I_{F(max)} = \frac{V_{CC(max)} - V_{F(min)} - V_{OL}}{R_{L(min)}}$$

= 9.5 mA

EQN. 3.4.3-6

$R_L \geqslant 425\Omega$, m = 1

EQN. 3.4.3-7

$R_L \leqslant 11$ k, m = 1

$\therefore R_L$ SELECTED AS 510Ω 5%, m = 1

5082-4360

V_F = 1.5V (1.75V MAX) AT 10 mA I_F

$I_{FH} \leqslant 5$ mA TO GUARANTEE

$V_{OL} \leqslant .6V$ AT I_{OL} = 13 mA

$I_{OH} \leqslant 250\ \mu A$

x ⩽ 25%

Figure 3.4.3.3   Worst Case Design of LSTTL to TTL Interface Using 5082-4360 Isolator that Allows for CTR Degradation.

The minimum I_F should be set approximately 30% above the minimum guaranteed I_{FH} for the isolator (5 mA for the 5082-4360, 5082-4364 and 10 mA for the 5082-4365) or as calculated in equation 3.4.3-5. The regulator will insure that any excess current is shunted through the external transistor. This has the added benefit of protecting the LED against harmful transients.

Isolators such as the 5082-4360 family require a bypass capacitor across V_{CC} and Gnd of each isolator (see section 3.3.5). If this bypass capacitor is mistakenly omitted or is located too far away from the isolator, the isolator amplifier can oscillate internally. The symptoms of these internal oscillations are the same as those of LED degradation of this type of isolator -- an increase in the switching threshold. However, this type of "degradation" can be cured by proper bypassing.

3.4.4   CTR "Degradation - Proof" Isolator Circuit Design Techniques

Some optically coupled isolator applications require overall system performance for several tens of thousands of hours. In designs of this type, a worst case circuit design such as shown in section 3.4.3 may not be adequate. No one knows exactly how much an LED will degrade in 10,000 hours, so how much CTR degradation should the design allow? The circuits illustrated in Figures 3.4.3-1 and 3.4.3-2 can not be designed for x = 80% because the isolator specified has insufficient initial CTR. Clearly an isolator that has degraded by 100% is a catastrophic failure but an isolator that has degraded by 99% is still marginally functional. For the purposes of this discussion, a "degradation-proof" design is one that will allow the isolator or isolators to degrade by 80% and still guarantee proper logic or analog signals to the output circuit. "Degradation-proof" circuits use either a very high CTR device or optical feedback that reduces the effects of CTR degradation.

One example of the first type of "degradation-proof" circuit is the use of an isolator with a high CTR in a circuit that was worst case designed to use a low CTR device. Then the high gain isolator can degrade until it reaches the minimum allowed CTR of the design. An example of this technique is shown in Figure 3.4.4-1a. In this example, a high gain isolator is simply retrofitted into a design for a low gain isolator. A worst case design that is optimized for a high gain isolator is shown in Figure 3.4.4-1b. Figure 3.4.4-1b is a better design than Figure 3.4.4-1a because in 3.4.4-1b, the isolator is operated at a lower I_F so its rate of degradation has been decreased. Especially in designs of this type, the designer should insure that the leakage current, I_{OH}, of the isolator be considerably lower than the "desired" output current, I_{OL}, if a low current application is being considered. Neither circuit has the switching

performance of Figures 3.4.3-1a, 3.4.3-1b, or 3.4.3-3, although an external resistor connected across pins 5 and 7 of the isolator will improve the switching performance of Figures 3.4.4-1a and 3.4.4-1b (see section 3.6.3 and Figure 3.6.3-5).

The analog amplifier explained in section 3.5.4 is an example of "degradation-proof" circuit that uses optical feedback to reduce the effects of CTR degradation. The circuit uses two isolators connected so that one isolator is forced to track the input current of a second isolator by servo action. The transfer function of this circuit is given by equation 3.5.4-1 and is reproduced below in simplified form:

$$I_{F2} = C \left[\sqrt[n_2]{\frac{R_2 K_1}{R_3 K_2}} \right] (I_{F1})^{n_1/n_2} \qquad (3.4.4-1)$$

where: each isolator can be described by equation 3.4.4-2:

$$I_o = K \left[\frac{I_F}{I_{F'}} \right]^n \qquad (3.4.4-2)$$

where: I_O is the collector current of the output transistor when biased in the active region, I_F is the input LED current, $I_{F'}$ is the input LED current where K is measured, K is the output current when $I_F = I_{F'}$, and n is the slope of I_O vs. I_F on logarithmic coordinates. Then C in equation 3.4.4-1 is equal to

$$\left[I_{F2'} \right] / \left[(I_{F1'})^{n_1/n_2} \right] \qquad (3.4.4-3)$$

This model applies particularly to the 5082-4350 type isolator, but can also be extended to the 5082-4370 series of isolators. Section 3.5.2 covers this model in more detail. At some value of I_{F1}, R_2 and R_3 can be adjusted so that $I_{F2} \equiv I_{F1}$. Then as I_{F1} is varied, I_{F2} will remain approximately equal to I_{F1} if $n_1 \approx n_2$. As the isolators degrade, K is reduced. If the isolators degrade at the same rate, then I_{F2} will remain approximately equal to I_{F1}. Since n_1 and n_2 are greater than one, the change in I_{F1}/I_{F2} will be less than the change in K_1/K_2. For instance, if isolator$_1$ degrades by 80% and isolator$_2$ does not degrade at all, $(R_2 K_1)/(R_3 K_2) = .2$ but if $n_2 = 1.8$, then $I_{F1}/I_{F2} = .41$ and the apparent effects of CTR degradation will be reduced. This servo feedback technique can also be applied to digital circuits.

3.5   Analog Applications of Optically Coupled Isolators

3.5.1   Introduction

Optically coupled isolators are useful for applications where analog or DC signals need to be transferred between two isolated systems in the presence of a large potential

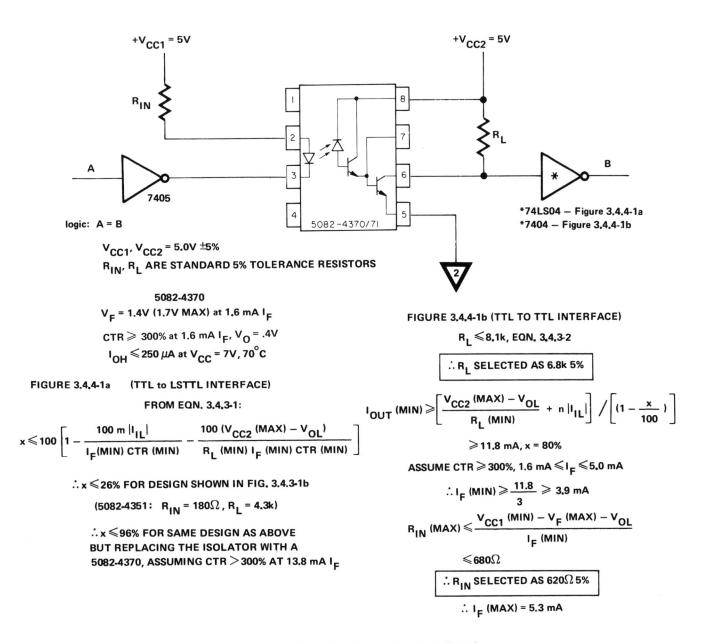

logic: A = B

$V_{CC1}, V_{CC2} = 5.0V \pm 5\%$

R_{IN}, R_L ARE STANDARD 5% TOLERANCE RESISTORS

5082-4370

$V_F = 1.4V$ (1.7V MAX) at 1.6 mA I_F

CTR $\geqslant$ 300% at 1.6 mA I_F, $V_O = .4V$

$I_{OH} \leqslant 250 \mu A$ at $V_{CC} = 7V$, 70°C

FIGURE 3.4.4-1a (TTL to LSTTL INTERFACE)

FROM EQN. 3.4.3-1:

$$x \leqslant 100 \left[1 - \frac{100\,m\,|I_{IL}|}{I_F (MIN)\,CTR\,(MIN)} - \frac{100\,(V_{CC2}\,(MAX) - V_{OL})}{R_L\,(MIN)\,I_F\,(MIN)\,CTR\,(MIN)} \right]$$

$\therefore x \leqslant 26\%$ FOR DESIGN SHOWN IN FIG. 3.4.3-1b

(5082-4351: $R_{IN} = 180\Omega$, $R_L = 4.3k$)

$\therefore x \leqslant 96\%$ FOR SAME DESIGN AS ABOVE
BUT REPLACING THE ISOLATOR WITH A
5082-4370, ASSUMING CTR > 300% AT 13.8 mA I_F

FIGURE 3.4.4-1b (TTL TO TTL INTERFACE)

$R_L \leqslant 8.1k$, EQN. 3.4.3-2

$\boxed{\therefore R_L \text{ SELECTED AS } 6.8k\ 5\%}$

$$I_{OUT}\,(MIN) \geqslant \left[\frac{V_{CC2}\,(MAX) - V_{OL}}{R_L\,(MIN)} + n\,|I_{IL}| \right] \bigg/ \left[\left(1 - \frac{x}{100}\right) \right]$$

$\geqslant 11.8$ mA, x = 80%

ASSUME CTR $\geqslant$ 300%, 1.6 mA $\leqslant I_F \leqslant 5.0$ mA

$\therefore I_F\,(MIN) \geqslant \dfrac{11.8}{3} \geqslant 3.9$ mA

$$R_{IN}\,(MAX) \leqslant \frac{V_{CC1}\,(MIN) - V_F\,(MAX) - V_{OL}}{I_F\,(MIN)}$$

$\leqslant 680\Omega$

$\boxed{\therefore R_{IN} \text{ SELECTED AS } 620\Omega\ 5\%}$

$\therefore I_F\,(MAX) = 5.3$ mA

Figure 3.4.4-1   Two Worst Case Designs of TTL Interfaces that are Optimized for Maximum Operating Life.

difference or induced noise. They are an inexpensive way to eliminate the shock hazard between an input transducer and an output circuit. Optically coupled isolators can also reduce the common mode noise generated between an isolated input circuit and an output circuit. Potential applications include those in which large transformers, expensive instrumentation amplifiers, or complicated A/D conversion schemes have been used. Examples include sensing circuits (thermocouples, transducers ...), patient monitoring equipment, power supply feedback, high voltage current monitoring, and audio or video amplifiers. In many applications, the isolator can transmit the analog signal directly. However, in applications where very high linearity and stability are critical, the analog signal can be converted into a digital form and then isolated. Overall circuit parameters like linearity bandwidth, and stability determine which approach is best.

3.5.2 Analog Model for an Optically Coupled Isolator

For an optically coupled isolator with a photodiode detector, or with an integrated photodiode and transistor detector such as the 5082-4350 family, the output current for a wide range of input currents can be expressed as:

$$I_O = K(I_F/I_{F'})^n \qquad (3.5.2\text{-}1)$$

where I_F is the input LED current, $I_{F'}$ is the input LED current where K is measured, K is the output current when $I_F = I_{F'}$, and n is the slope of I_O vs. I_F on logarithmic coordinates. For a photodiode opto isolator, I_O is the current flowing into the cathode of the zero or reverse biased photodiode. For an integrated photodiode and transistor isolator, I_O is the collector current of the output transistor when biased in the active mode.

If n is equal to one, then the input to output transfer function of the opto isolator is linear. For most optically coupled isolators, n is not equal to one. For the 5082-4350 isolator, n varies from approximately 2 at very low inputs currents to 1 or less at higher input currents. Typical I_O vs. I_F characteristics for the 5082-4350 type opto isolator are shown in Figure 3.5.2-1.

3.5.3 Types of Analog Circuits

There are several analog techniques that use optically coupled isolators to isolate an analog signal. The Servo and Differential techniques are dc coupled isolated amplifiers that use two opto isolators for improved linearity and temperature stability. The servo linearizer forces the input current of one optically coupled isolator to track the input current of a second opto isolator by servo action. If $n_1 \approx n_2$ over the excursion range of the input current of the isolators, then the non linearities will cancel and the overall transfer function will be linear. The differential linearizer causes the input current of one opto isolator to increase in response to an input signal while the input current of the second optically coupled isolator decreases by an equal amount. If $n_1 \approx n_2$ over the excursion range of the input currents of the isolators, then a gain increment in the first opto isolator will be approximately balanced by a gain decrement in the second opto isolator and the overall transfer function will be linear. For AC coupled applications, reasonable linearity can be obtained with a single optically coupled isolator. The opto isolator is biased at higher levels of input LED current where the ratio of incremental photodiode current to incremental LED current $(\partial I_P/\partial I_F)$ is more nearly constant. For the 5082-4350 series of opto isolators, this occurs at input currents greater than 15 mA, as shown by Figure 3.5.3-1. The input current of the optically coupled isolator is modulated by the input signal and the modulation of the photodiode current is detected and amplified. The linearity

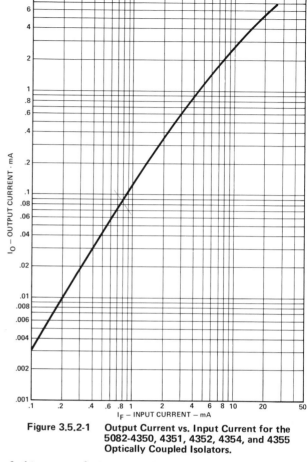

Figure 3.5.2-1 Output Current vs. Input Current for the 5082-4350, 4351, 4352, 4354, and 4355 Optically Coupled Isolators.

of this type of circuit is determined by the amount by which incremental CTR $(\partial I_P/\partial I_F)$ changes over the range over which the input LED current is modulated. For many applications, at least one of these three techniques can be used. Table 3.5.3-1 compares the advantages and disadvantages of each of these analog techniques.

3.5.4 Servo Isolation Amplifier

The servo isolation amplifier shown in Figure 3.5.4-1 operates on the principle that two optically coupled isolators will track each other if their gain changes by the same amount over some operating region. U_2 compares the outputs of each opto isolator and forces I_{F2} through the second isolator to be proportional to I_{F1} through the first isolator. The constant current sources bias each input LED at 3 mA quiescent current. R_1 has been selected so that I_{F1} varies from 2 mA to 4 mA as V_{IN} varies from -5V to +5V. R_1 can be selected to accomodate any desired input range. The zero adjustment potentiometer must have sufficient dynamic range to compensate for a worst case spread of opto isolator current transfer ratios (the values of $K/I_{F'}$ x 100%) at the input quiescent current of 3 mA. Then with V_{IN} at some value, R_4 can be adjusted for a gain of one. This potentiometer only requires sufficient dynamic range to compensate for I_{CC1} not being equal to to I_{CC2}. If

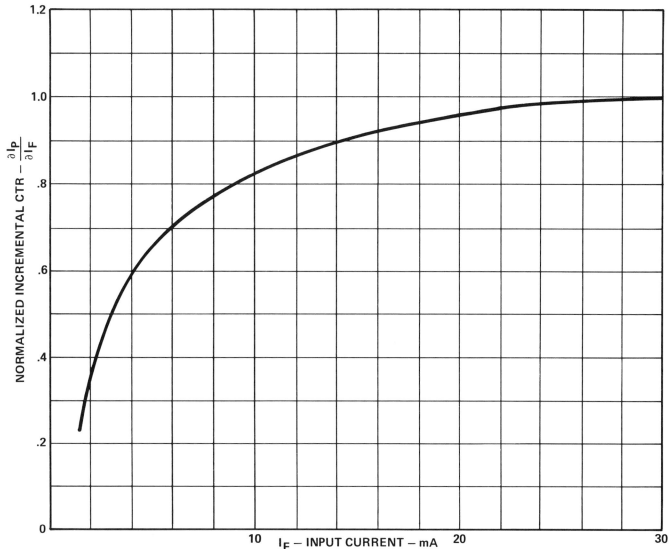

Figure 3.5.3-1   Normalized Incremental CTR vs. Input Current for 5082-4350, 4351, 4352 Optically Coupled Isolators.

	SERVO	DIFFERENTIAL	AC COUPLED
Number of Opto Isolators per Channel:	2: 5082-4350/1/2 OR 1: 5082-4354/5	2: 5082-4350/1/2 OR 1: 5082-4354/5	1: 5082-4350/1/2
Overall Linearity:	.5%	1%	1%
Input Current Range:	2–4 mA	2–4 mA	15–25 mA
Frequency Response:	DC	DC	AC
Bandwidth:	100 KHz	1 MHz	10 MHz
Temperature Stability — Offset:	±.01%/°C	±.04%/°C	Not Applicable
Temperature Stability — Gain:	−.03%/°C	−.4%/°C*	−.5%/°C*
Common Mode Rejection	>46 dB @ 1 KHz	>70 dB @ 1 KHz	>22 dB @ 1 MHz

*can be improved with additional thermistor

Table 3.5.3-1 Comparison of the Typical Characteristics of Servo, Differential, and ac Coupled Isolated Amplifiers.

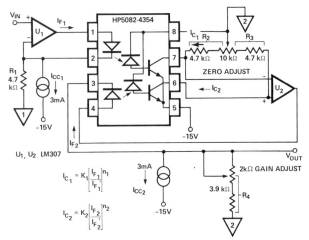

$$I_{C_1} = K_1 \left[\frac{I_{F_1}}{I_{F'_1}}\right]^{n_1}$$

$$I_{C_2} = K_2 \left[\frac{I_{F_2}}{I_{F'_2}}\right]^{n_2}$$

Figure 3.5.4-1   Servo Type DC Isolation Amplifier.

I_{CC1} is equal to I_{CC2}, R_4 can be replaced with a single resistor equal to R_1. The transfer function of the servo isolation amplifier is:

$$(3.5.4\text{-}1)$$

$$V_{OUT} = R_4 \left[(I_{F}'2) \left(\frac{K_1 R_2 (I_{CC1})^{n_1}}{K_2 R_3 (I_{F}'1)^{n_1}}\right)^{1/n_2} \right.$$

$$\left. \times \left(1 + \frac{V_{IN}}{R_1 I_{CC1}}\right)^{n_1/n_2} - I_{CC2} \right]$$

After zero adjustment, this transfer function reduces to:

$$(3.5.4\text{-}2)$$

$$V_{OUT} = R_4 I_{CC2} \left[(1 + x)^n - 1\right] \text{ where } x = \frac{V_{IN}}{R_1 I_{CC1}}, n = \frac{n_1}{n_2}$$

Using a binomial expansion, V_{OUT} can be written as:

$$(3.5.4\text{-}3)$$

$$V_{OUT} = R_4 I_{CC2} \left[nx + \frac{n(n-1) x^2}{2!} + \frac{n(n-1)(n-2)}{3!} x^3 + \ldots \right]$$

The linearity error in the transfer function when $n_1 \neq n_2$ can be written as:

$$\frac{\text{linearity error}}{\text{desired signal}} = \frac{(1+x)^n - nx - 1}{nx} \qquad (3.5.4\text{-}4)$$

For example, if $|x| \leq .35$, $n_1 = 1.9$ and $n_2 = 1.8$, then the error is 1% of the desired signal. Overall linearity can be improved by reducing $|x|$ or by matching the optically coupled isolators for n. The current transfer ratio ($K/I_{F'}$ x 100%) has no effect on overall linearity.

While temperature stability is also dependent on the stability of current sources and resistors, changes in current transfer ratio (the value of $K/I_{F'}$, x 100%) of the optically coupled isolators will have negligable effect on overall gain and offset as long as the ratio of K_1 to K_2 remains constant. Typical characteristics of the servo isolator amplifier (Figure 3.5.4-1) are given in Table 3.5.4-1.

| 1% linearity for 10V p-p dynamic range |
| Unity voltage gain |
| 25 KHz bandwidth (limited by U_1, U_2) |
| Gain drift: $-.03\%/^\circ C$ |
| Offset drift: ± 1 mV/$^\circ C$ |
| Common mode rejection: 46 dB at 1 KHz |
| 500V dc insulation (3000V if 2 single isolators are used) |

TABLE 3.5.4-1 Typical Performance for the Servo Linearized DC Amplifier.

3.5.5   Differential Isolation Amplifier

The differential isolation amplifier shown in Figure 3.5.5-1 operates on the principle that an operating region exists where a gain increment in one optically coupled isolator can be approximately balanced by a gain decrement in a second opto isolator. As I_{F1} increases in the first opto isolator due to changes in V_{IN}, I_{F2} in the second opto isolator decreases by an equal amount. If $n_1 \approx n_2$, then the gain increment caused by increases in I_{F1} will be approximately balanced by the gain decrement caused by decreases in I_{F2}. The constant current source biases each input LED at 3 mA quiescent current. R_1 and R_2 are selected so that I_F varies from 2 mA to 4 mA as V_{IN} varies from -5V to +5V. R_1 and R_2 can be selected to accomodate any desired input range. At the output, U_3 and U_4 force V_{OUT} to be proportional to the difference between I_{C1} and I_{C2}.

$$V_{OUT} = R_5 [(R_3/R_4) I_{C1} - I_{C2}] \qquad (3.5.5\text{-}1)$$

The zero adjustment potentiometer, R_3, allows V_{OUT} to be set to zero when V_{IN} is equal to zero. The gain adjustment potentiometer, R_5, allows the overall gain to be set to one. R_3 and R_5 must have sufficient dynamic range to compensate for a worst case spread of opto isolator current transfer ratios (the values of $K/I_{F'}$ x 100%) at the input quiescent current of 3 mA. The transfer function of the differential isolation amplifier is:

$$(3.5.5\text{-}2)$$

$$V_{OUT} = R_5 \left[\left(\frac{K_1 R_3}{R_4}\right) \left(\frac{I_{CC}}{2 I_{F}'1}\right)^{n_1} \left(1 + \frac{V_{IN}}{R I_{CC}}\right)^{n_1} \right.$$

$$\left. - K_2 \left(\frac{I_{CC}}{2 I_{F}'2}\right)^{n_2} \left(1 - \frac{V_{IN}}{R I_{CC}}\right)^{n_2} \right]$$

$$\text{if } R \equiv R_1 \equiv R_2$$

After zero adjustment, this transfer function reduces to:

$$(3.5.5\text{-}3)$$

$$V_{OUT} = R_5 K' \left[\left(1 + \frac{V_{IN}}{R I_{CC}}\right)^{n_1} - \left(1 - \frac{V_{IN}}{R I_{CC}}\right)^{n_2} \right]$$

$$\text{where } K' = \frac{K_1 R_3}{R_4} \left(\frac{I_{CC}}{2 I_{F}'1}\right)^{n_1} = K_2 \left(\frac{I_{CC}}{2 I_{F}'2}\right)^{n_2}$$

Using a binomial expansion, V_{OUT} can be written as:

$$(3.5.5-4)$$

$$V_{OUT} = R_5 K' \left[(n_1+n_2)x + \frac{[n_1(n_1-1)-n_2(n_2-1)]\, x^2}{2!} + \frac{n_1(n_1-1)(n_2-2)-n_2(n_2-1)(n_2-2)\, x^3}{3!} + \dots \right]$$

where $x = \dfrac{V_{IN}}{R\, I_{CC}}$

If both isolators have a square law response, i.e., $n_1 = n_2 = 2$, then all non linear terms will cancel and V_{OUT} will be proportional to V_{IN}. However, if $n_1 = n_2$, then all even order terms will cancel, and the total linearity error can be reduced below 1%. The linearity error in the transfer function for any values of n can be written as:

$$(3.5.5-5)$$

$$\frac{\text{linearity error}}{\text{desired signal}} = \frac{(1+x)^{n_1} - (1-x)^{n_2} - (n_1+n_2)\, x}{(n_1+n_2)\, x}$$

For example, if $|x| \leqslant .35$, $n_1 = 1.9$ and $n_2 = 1.8$, then the linearity error is 1.5% of the desired signal. Overall linearity can be improved by reducing $|x|$ or by matching the optically coupled isolators for n. The current transfer ratio has no effect on linearity.

While temperature stability is also dependent on the stability of current sources and resistors, changes in current transfer ratio of the optically coupled isolators over temperature will cause a change in gain of the circuit. This change in gain can be compensated with a thermistor in either the input or output circuit. For example, if R_5 is replaced by a positive TC thermistor, or R_1 and R_2 are replaced with negative TC thermistors, then the gain will tend to increase as temperature goes up. Zero offset over temperature will remain stable as long as the ratio of K_1 to K_2 remains constant. Typical characteristics of the differential isolation amplifier (Figure 3.5.5-1) are given in Table 3.5.5-1.

```
3% linearity for 10V p-p dynamic range
Unity voltage gain
25 kHz bandwidth (limited by U₁, U₂, U₃, U₄)
Gain drift: −.4%/°C
Offset drift: ±4 mV/°C
Common mode rejection: 70 dB at 1 kHz
3000V DC insulation
```

Table 3.5.5-1

Typical Performance of the Differential Linearized DC Amplifier

3.5.6 AC Coupled Isolation Amplifier

The AC coupled isolation amplifier shown in Figure 3.5.6-1 operates on the principle that a single optically coupled isolator can be biased in a region where incremental CTR $(\partial I_P/\partial I_F)$ is constant. Q_1 is biased by R_1, R_2 and R_3 for a quiescent collector current of 20 mA. R_3 allows I_F to vary from 15 mA to 25 mA for a 1V peak-to-peak input signal. Under these operating conditions, the 5082-4351 operates in a region of almost constant incremental CTR as shown by Figure 3.5.3-1. The varying photon flux is detected by the photodiode and amplified by Q_2 and Q_3. Q_2 and Q_3 form a cascade amplifier with feedback applied by R_4 and R_6. I_3 is selected to allow Q_3 to operate at its maximum gain bandwidth product. R_6 is selected as V_{be}/I_3 and R_7 is selected to allow maximum excursions of V_{OUT} without clipping: $R_7 = (V_{CC} - V_{be} - V_{ce\,sat})/(2I_3)$. R_5 provides dc bias for Q_3. Closed loop gain of the output amplifier can be adjusted with the gain adjustment potentiometer, R_4. Linearity can be improved at the expense of signal to noise ratio by reducing the excursions of I_F. This can be accomplished by increasing R_3 and adding a resistor from the collector of Q_1 to ground to obtain the desired quiescent collector current of 20 mA. This circuit has no feedback around the optically coupled isolator, so any parameter that causes the incremental CTR $(\partial I_P/\partial I_F)$ to vary will cause a change in gain of the circuit. Since the quantum efficiency of the input LED varies with temperature, the $\partial I_P/\partial I_F$ and thus the overall gain will also vary with temperature. However, a thermistor can be used in the output amplifier to compensate for this change in gain. For example, if R_7 is replaced by a positive TC thermistor, or R_6 is replaced by a negative TC thermistor, then the gain will tend to increase as temperature goes up. The transfer function of the AC coupled isolation amplifier is:

$$\frac{V_{OUT}}{V_{IN}} \cong \left[\frac{\partial I_P}{\partial I_F} \right] \left[\frac{1}{R_3} \right] \left[\frac{R_4\, R_7}{R_6} \right] \qquad (3.5.6-1)$$

Typical characteristics of the AC coupled isolation amplifier (Figure 3.5.6-1) are given in Table 3.5.6-1:

```
2% linearity over 1V p-p dynamic range
Unity voltage gain
10 MHz bandwidth
Gain drift: −.6%/°C
Common mode rejection: 22 dB at 1 MHz
3000V DC insulation
```

Table 3.5.6-1

Typical Performance of the Wide Bandwidth AC Amplifier

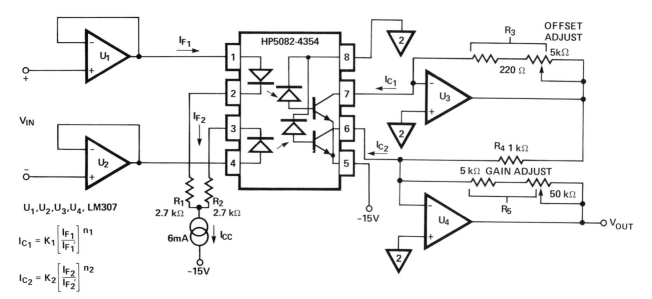

U_1, U_2, U_3, U_4, LM307

$$I_{C_1} = K_1 \left[\frac{I_{F_1}}{I_{F_1}'} \right]^{n_1}$$

$$I_{C_2} = K_2 \left[\frac{I_{F_2}}{I_{F_2}'} \right]^{n_2}$$

Figure 3.5.5-1   Differential Dc Isolation Amplifier.

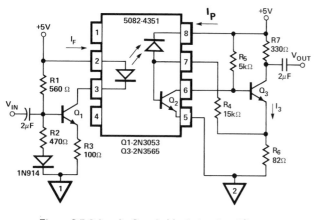

Figure 3.5.6-1   Ac Coupled Isolation Amplifier.

3.5.7   Digital Isolation Techniques

The Servo, Differential, and AC coupled analog circuits offer a simple and low cost means to optically isolate an analog signal. However, the smallest linearity error that can be achieved with these techniques is limited to about .5% to 1%. Moreover, stability of gain and offset over temperature may not be adequate for some applications. For applications that require higher linearity or better stability than can be achieved with analog techniques, isolation can be realized with digital techniques. A second reason for using digital isolation techniques is that in many applications, the analog signal must be converted into a digital format anyway before it can be processed by a microprocessor or by a digital output device. Optically coupled isolators can then provide isolation between the analog input device and the digital output device and at the same time provide all the advantages of digital isolation techniques.

With digital isolation techniques, the analog signal is converted into some type of digital code that can be transmitted through an optically coupled isolator. At the output, the digital code is converted back into analog form, or used directly by digital processing circuitry. Since the isolator is used only as a high speed switch, linearity, stability and bandwidth are determined by the conversion accuracy, temperature stability and conversion time of the A to D and D to A converters. Selection of the proper opto isolator is determined by such factors as maximum data rate, input current limitations, and proper logic interface at the output.

3.5.7.1 Isolated Analog to Digital Techniques

These techniques convert the analog signal into a digital signal through an isolated interface but leave the digital signal in a form that can be used by some type of digital processor. The A to D converter technique and variable pulse width monostable multivibrator technique illustrate two ways in which this analog to digital conversion can be accomplished. The circuits that are shown illustrate the use of optically coupled isolators to provide isolation between analog and digital circuitry and how high speed isolators can reduce systems cost.

The most commonly used digital technique to isolate an analog signal is the use of an A to D converter with optically coupled isolators providing isolation of the digital signals. This technique is outlined in Figure 3.5.7.1-1. The digital information can be transmitted through the isolators in either serial or parallel format, depending on the outputs available on the A to D converter. Serial transmission is especially practical for 8, 10 and 12 bit A to D converters because it replaces one isolator per bit with one or two high

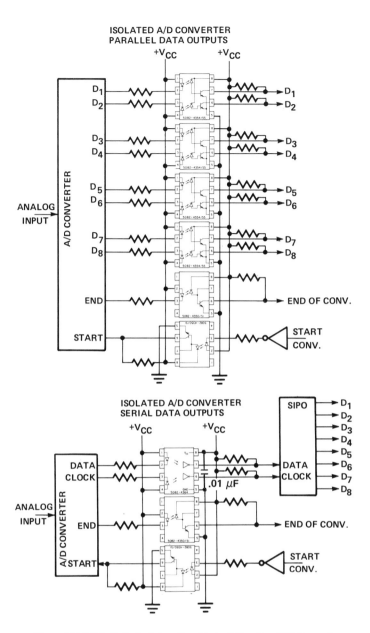

ISOLATED A/D CONVERTER
PARALLEL DATA OUTPUTS

ISOLATED A/D CONVERTER
SERIAL DATA OUTPUTS

Figure 3.5.7.1-1 Opto Isolators Provide Isolation for A/D Converters. High Speed Isolators can Reduce Cost by Transmitting Data in Serial Form.

speed isolators. If serial outputs are not available from the A to D converter, parallel outputs can be converted into serial format with a PISO shift register. The A to D converter digital isolation technique is especially useful where the digital information will be used as data for some type of digital system such as a microprocessor or LED display. A to D converters are available with straight binary, offset binary, two's complement binary, or binary coded decimal outputs.

A second technique that can be used to convert an analog signal into digital form while maintaining electrical isolation between the analog signal and the digital circuitry is the circuit shown in Figure 3.5.7.1-2. Whenever the monostable

multivibrator is triggered by a negative going pulse at TRIGGER, Q_0 goes high and stays high for a time proportional to the magnitude of V_{IN}. The time during which Q_0 is high is measured and used to provide a digital representation of V_{IN} to the output circuit. Q_0 can be used to gate an oscillator into a counter. After Q_0 goes low, the contents of the counter can be displayed by an LED display or read into a microprocessor. This concept is illustrated by Figure 3.5.7.1-2a. The counter can also be implemented in microprocessor software. With this approach, TRIGGER is attached to one bit of an output port of the microprocessor and Q_0 is attached to one bit of an input port. For this example, Q_0 is attached to D_7 to simplify the software implementation. With the program illustrated in Figure 3.5.7.1-2b, register HL will contain a binary number proportional to V_{IN} after the conversion is completed. The total time, t, that has elapsed will be equal to (29)(HL)(clock rate of microprocessor). While this approach uses less hardware than the first approach, the microprocessor must wait until the conversion is completed. Commercially available timers, such as the 555 timer, can be used to implement this technique at minimal cost.

3.5.7.2 Isolated Analog to Digital to Analog Techniques

These techniques perform the same type of transfer functions as the strictly analog servo, differential and AC coupled circuits described earlier. The difference between these techniques and the analog circuits discussed earlier is that the optically coupled isolators are used in a digital fashion. Thus, these circuits are more immune to CTR degradation and some types of common mode transients. These techniques are useful where some type of high stability isolated instrumentation amplifier is required. The pulse width modulator and the voltage to frequency converter illustrate two ways in which this analog to digital to analog conversion can be accomplished.

The circuit shown in Figure 3.5.7.2-1 shows a pulse width modulated scheme to isolate an analog signal. The oscillator operates at a fixed frequency, f, and provides a continuous trigger to the monostable multivibrator. Once triggered, the monostable multivibrator gives an output pulse proportional to the value of V_{IN}. The output of the monostable multivibrator is a square wave of frequency f but with a duty factor proportional to V_{IN}. The maximum frequency of the oscillator is determined by the required linearity of the circuit and the propagation delay of the optically coupled isolator.

$$(t_{max} - t_{min})(\text{required linearity}) \geqslant |t_{PLH} - t_{PHL}| \qquad (3.5.7.2-1)$$

For example, if f is 1 MHz, and the monostable multivibrator can vary the duty factor from 10% to 90%, then $(t_{max} - t_{min})$ = 800 nS. If the isolator is a 5082-4360, then $|t_{PLH} - t_{PHL}|$ varies from about 10 nS to 60 nS depending on drive currents, and temperature. Thus, the

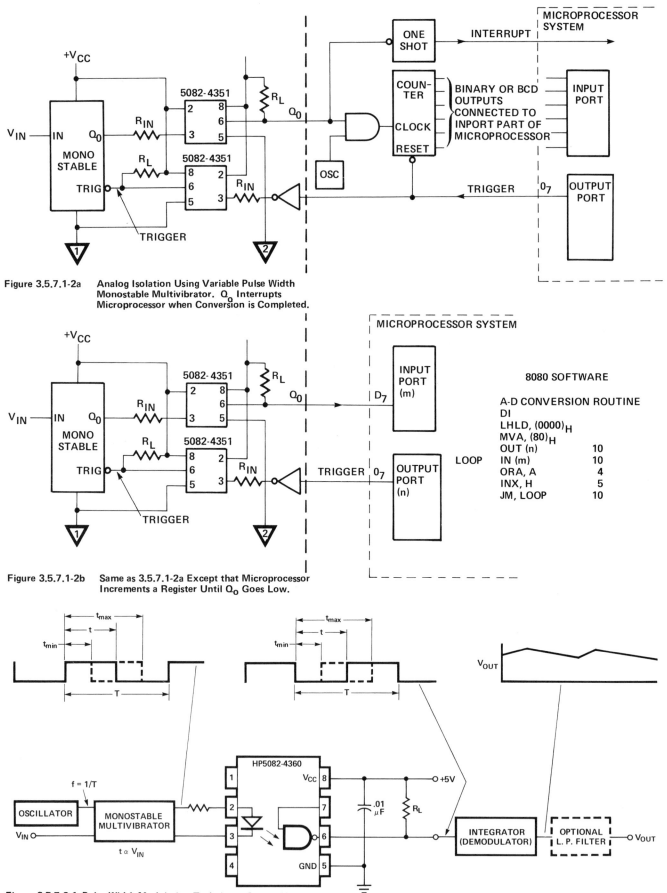

Figure 3.5.7.1-2a   Analog Isolation Using Variable Pulse Width
                     Monostable Multivibrator. Q_0 Interrupts
                     Microprocessor when Conversion is Completed.

Figure 3.5.7.1-2b   Same as 3.5.7.1-2a Except that Microprocessor
                     Increments a Register Until Q_0 Goes Low.

8080 SOFTWARE

A-D CONVERSION ROUTINE
DI
LHLD, $(0000)_H$
MVA, $(80)_H$

OUT (n)	10
IN (m)	10
ORA, A	4
INX, H	5
JM, LOOP	10

Figure 3.5.7.2-1 Pulse Width Modulation Techniques Can Be Used
                 to Isolate an Analog Signal.

3.38

worst case linearity error would be <7.5%. At 100 KHz, the worst case linearity error would be reduced to <.75%, assuming that the monostable multivibrator and output circuits had perfect linearity. At the output, the squarewave would be integrated or demodulated such that V_{OUT} would be proportional to the duty factor of the squarewave. An additional low pass filter would help to reduce the ripple at the output frequency f.

Figure 3.5.7.2-2 shows a voltage to frequency conversion scheme to isolate an analog signal. The voltage to frequency converter gives an output frequency that is proportional to V_{IN}. The maximum frequency that can be transmitted through the isolator is limited by the maximum data rate of the isolator. At the output, the frequency is converted back into a voltage. The overall circuit linearity is dependent only on the linearity of the voltage to frequency and frequency to voltage converters. Another modification on this technique is frequency modulation about a carrier frequency, f_0. Here, V_{IN} modulates f_0 by $\pm \Delta f$ depending on the amplitude of V_{IN}. At the output, V_{OUT} is reconstructed with a phase locked loop or similar circuit.

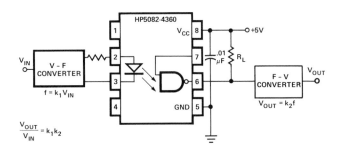

Figure 3.5.7.2-2 Voltage to Frequency Conversion Techniques Can Be Used to Isolate an Analog Signal.

3.6   Digital Applications

While there are many different digital applications of optoisolators, there are some input design considerations they all have in common. These considerations are:

a)   the "off" state: input diode forward current very small, zero, or negative (reverse biased)

b)   the "on" state: input diode forward current at a steady, adequate value.

c)   threshold: a transitional but definable level of input diode forward current.

For (a) and (c), the only consideration usually given is that they are unimportant, but this is not always true. For (b), the "worst case" limits are usually narrower than for (a) and (c), so (b) receives more attention.

In the "off" state, the voltage of the input diode is usually zero or negative. It may, however, be slightly positive. It is worth noting that at forward voltage as high as 1.2 volts, the forward current is often neglible ($<10\mu A$). The "off" state consideration may also have some relationship to the threshold condition that should be considered.

For the "on" state and threshold condition, the consideration is illustrated in Figure 3.6-1. In analog types of optoisolators, the output current is a relatively smooth function of input current, the ratio being the CTR (Current Transfer Ratio, see Section 3.2.5). As indicated in Figure 3.6-1(a) the required minimum level of forward current, I_F, for a proper "on" state is related to the output circuit. A good practice is to design the drive circuit so that I_F is at the level for which data sheet specs are given, then adjust the load accordingly, allowing an appropriate guardband for CTR degradation. Note also that there are maximum limits on I_F for both DC and pulsed operation.

The threshold condition can also be defined, with reference to Figure 3.6-1(a), as the threshold input current, I_{Fth}, at which the output voltage, V_O, is at the threshold level, V_{Oth}, for the particular logic type used in the output circuit. For example, with TTL logic, $V_{Oth} \approx 2V_{be} \approx 1.5V$. Because of the influence of temperature, CTR degradation, and unit-to-unit variability, I_{Fth} is not precise enough to describe the threshold for a comparator--type application unless the reference current is much greater than I_{Fth}. Nevertheless, I_{Fth} can have some influence on the propagation delay through the isolator, and should therefore be considered.

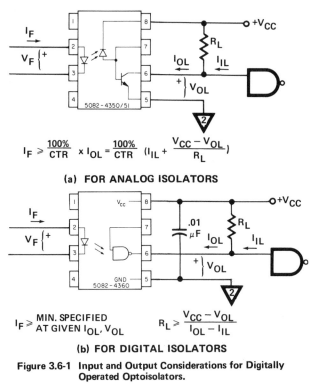

Figure 3.6-1 Input and Output Considerations for Digitally Operated Optoisolators.

In digital types of optoisolators, the threshold and "on" state input current levels are not subject to load adjustment. The output current is not a smooth function of input current. An example is shown in Figure 3.6-1(b). In the 5082-4360 type of optoisolator the threshold may be anywhere between the limits specified for I_{OH} and I_{OL}. There is, however, a specified level of I_F above which the output transistor will be capable of sinking a specified current. Guardbanding the load to allow for CTR degradation is not useful. Good practice is to design the drive circuit to provide a level of I_F slightly higher than the specified minimum. (This technique may also be applied to the circuit of Figure 3.6-1(a).) This form of guardbanding for CTR degradation narrows the spread between the minimum I_F required and the maximum data sheet rating. Techniques are given in Section 3.6.1.2 for dealing with a narrow spread.

3.6.1 Line Receivers

When digital data is transmitted over any appreciable length of transmission line (even less than a meter), there arises a possibility of ground shift, ground looping, etc. Optoisolators can reduce the amount of ground loop current and the effects of the resulting common mode voltage. They are, therefore, very useful as line receivers. This section discusses the fundamentals of designing optoisolators into line receivers.

Most line drivers are capable of sourcing a line voltage higher than the minimum ($\approx 1.5V$) needed to turn "on" an optoisolator. They usually also deliver a line current greater than the value of I_F as determined with reference to Figure 3.6-1. It is possible, of course, to design a line driver that will source the proper current and no more, but that is usually not good practice, as will be described later. Good practice is to drive the line with all the voltage and current available, then deal with the excess in the design of the termination of the line.

In the design of the termination, the "off" state is usually ignored unless pre-bias is used, and this is discussed in Section 3.6.3. There are usually **three objectives** in the design of the termination:

(A)   proper "on" state I_F
(B)   threshold level, I_{Fth}
(C)   reflections due to impedance mismatch.

3.6.1.1 Resistive Terminations

As seen earlier, objective (A) is mandatory; objectives (B) and (C) are discretionary. If discretion allows (B) and (C) to be neglected, the termination may be as simple as in Figure 3.6.1-1 requiring only a single resistor. In most cases, the

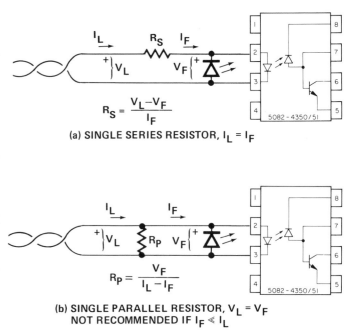

$$R_S = \frac{V_L - V_F}{I_F}$$

(a) SINGLE SERIES RESISTOR, $I_L = I_F$

$$R_P = \frac{V_F}{I_L - I_F}$$

(b) SINGLE PARALLEL RESISTOR, $V_L = V_F$
NOT RECOMMENDED IF $I_F \ll I_L$

Figure 3.6.1-1   Single-Resistor Terminations using Optoisolators as Line Receivers.

series resistor termination of Figure 3.6.1-1(a) would be used because it will accomodate a broader range of driver and line resistance variables. It is slower than the single parallel resistor termination of Figure 3.6.1-1(b) because the input diode is driven from a higher impedance. Slowness can be remedied with a peaking capacitor in parallel with R_S; if peaking capacitance is applied, the anti-parallel diode should be used (even if the driver is polarity *non*-reversing) to allow the peaking capacitor to charge and discharge the maximum amount.

The anti-parallel LED in Figure 3.6.1-1 is recommended also when polarity-reversing drive is used. It should be a GaAsP/GaAs device (e.g. HP type 5082-4850), or the input diode of another isolator of the same type to provide a balanced output.

When, in addition to proper I_F, consideration must be given to either threshold current, I_{Fth}, or to line reflections, an additional resistor provides one additional degree of freedom. A two-resistor termination can accomodate the additional objective of either setting the I_{Fth} level or of approximate impedance matching but not both. Approximate impedance matching will be described first because it is then easier to describe the condition for threshold setting.

First of all, the nature of the reflection problem must be understood. This can best be explained with reference to a "design" having large, oscillatory reflections, illustrated in Figure 3.6.1-2. (This "design", incidentally is not recommended; it is given here only as an example of what can take place.) When the driver makes a logic transition from "low" to "high", the source line is changed so rapidly

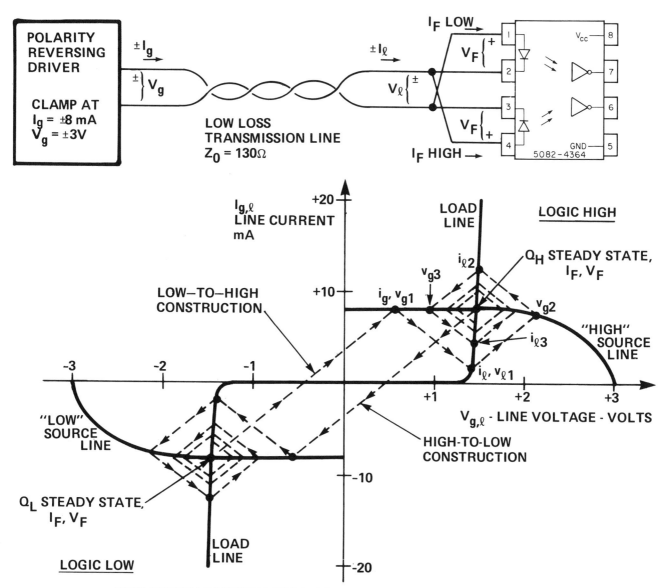

REFLECTION CONSTRUCTION FOR LOW-TO-HIGH TRANSITION
(ASSUME QUIESCENT LOW INITIAL CONDITION, Q_L)

1. FROM Q_L CONSTRUCT A LINE WITH SLOPE $\partial V/\partial I = \underline{+Z_0}$ TO WHERE IT INTERSECTS THE <u>SOURCE</u> LINE FOR LOGIC HIGH AT POINT i_g, v_{g1}. THIS IS THE INITIAL STEP CURRENT IN THE LINE.

2. FROM i_g, v_{g1}, CONSTRUCT A LINE WITH SLOPE $\partial V/\partial I = \underline{-Z_0}$ TO WHERE IT INTERSECTS THE <u>LOAD</u> LINE FOR LOGIC HIGH AT POINT i_ℓ, $v_{\ell 1}$. THIS IS THE INITIAL STEP CURRENT IN THE LOAD.

3. FROM i_ℓ, $v_{\ell 1}$ CONSTRUCT $+Z_0$ LINE TO v_{g2} ON SOURCE LINE.

4. FROM v_{g2} CONSTRUCT $-Z_0$ LINE TO $i_{\ell 2}$ ON LOAD LINE.

5. ETC., ETC., UNTIL CONSTRUCTION CONVERGES AT Q_H.

GENERAL RULE: $+Z_0$ LINES TO SOURCE LINE, $-Z_0$ LINES TO LOAD LINE.

Figure 3.6.1-2   Graphical Analysis of Reflections with Non-Linear
Line Driver & Load

as to be regarded as a step function. At this instant, the load on the driver is just the transmission line with dynamic resistance Z_0 and initial conditions of voltage and current corresponding to point Q_L. Thus the initial voltage and current in the line at the driver end are found as the intersection of the "high" source line with a line through Q_L and having a POSITIVE Z_0 slope. This point is labelled, i_g, v_{g1}, and this is the voltage and current that will propagate toward the load. Upon reaching the termination, the transmission line is now regarded as a source having Z_0 internal resistance with point i_g, v_{g1} as its initial condition. It is, therefore, represented as a line passing through i_g, v_{g1} with a NEGATIVE Z_0 slope, and its intersection with the load line gives the first voltage and current, i_ℓ, $v_{\ell 1}$, at the termination (load). This initial value, i_ℓ, $v_{\ell 1}$, will now travel back to the driver where it presents to the driver a load of dynamic resistance Z_0 and initial conditions i_ℓ, $v_{\ell 1}$. This is represented by a line through i_ℓ, $v_{\ell 1}$ with a POSITIVE Z_0 slope and, intersecting the source line gives a voltage v_{g2}, the second terminal voltage at the driver. The driver terminal voltage, v_{g2}, now travels back to the load, again along a NEGATIVE Z_0 slope, to intersect the load line at $i_{\ell 2}$, the second load terminal current. Then a POSITIVE Z_0 slope from $i_{\ell 2}$ intersects the source line at v_{g3} and a NEGATIVE Z_0 slope through v_{g3} intersects the load line at $i_{\ell 3}$. The construction is repeated until it converges at Q_H. Notice that the transmission line is alternately represented as a Z_0 load (POSITIVE Z_0 slope to intersect source line) and as a Z_0 source (NEGATIVE Z_0 slope to intersect load line).

This graphical method does not account for lossiness in the line, or for transient capacitive or inductive effects at the terminals. Nevertheless, it is a powerful aid in visualizing the effects of any load line or source line adjustments made by adding or changing resistances. Consider, for example, the addition of a series resistor, R_s, at the load in Figure 3.6.1-2. This would be represented as a positive slope of R_s starting from ≈ 1.5 volts at zero current. This would have brought $i_{\ell 2}$ much closer to v_{g2} without having changed the current at Q_H by very much. Consider also the case of a purely resistive Z_0 load; Q_L and Q_H would then be joined by a $+Z_0$ sloped line and there would be no reflections.

Such reflectionless design is not possible with optoisolator terminations because the impedance lowers as the voltage rises from below the turn-on voltage of the input diode to above turn-on. The best that can be done is to make sure that whatever reflections there are permit a monotonic approach to convergence at the quiescent point (Q_L or Q_H), rather than the oscillatory convergence seen in Figure 3.6.1-2. Monotonic convergence means that each successive step at the termination lies *between* the last step and the quiescent point -- not beyond the quiescent point. The problem with oscillatory convergence is that the first step may be high enough to initiate turn-on, while the next step

might be below turn-on, etc. Thus, in response to a single-edge logic transition there may be one or more extraneous pulses, i.e., one or more extra edges resulting from one edge.

In working out a design, there are three cases to consider:

CASE 1: quiescent voltage-to-current ratio is equal to Z_0

CASE 2: dynamic load resistance never greater than Z_0

CASE 3: dynamic load resistance never less than Z_0

The optimal termination lies somewhere between CASE 2 and CASE 3, with CASE 1 being a good starting point. Graphical construction of two-resistor terminations are shown in Figures 3.6.1-3 and 3.6.1-4. They are called the low-threshold (LT) and high threshold (HT) circuits with reference to the line voltage at which turn-on begins. They have the same relative merits, respectively, as the series and parallel one-resistor terminations of Figure 3.6.1-1.

Applying CASE 1, then doing a reflection construction by the method outlined in Figure 3.6.1-2 suggests that CASE 1 would be an optimal design, but while this is nearly true, it neglects the fact that turn-on of the input diode does not happen instantly. The first load voltage point must lie on an extension of the dynamic resistance below turn-on. This is why CASE 1 is a good first approximation, but upward or downward adjustment of the dynamic resistance above and below turn-on might be necessary. Direction of adjustment depends on the dynamic resistance of the driver in the neighborhood of the intersection of its source line with the Z_0 line. If its dynamic resistance here exceeds Z_0 adjust toward CASE 3, but if less than Z_0 adjust toward CASE 2. Going all the way to the CASE 2 or CASE 3 limits might be too far because they may place the quiescent point in a region where the dynamic resistance of the source line is radically different from what it is at its intersection with the Z_0 line.

If the source is very non-linear, a good approach is to select as the quiescent condition that point on the source line where the dynamic resistance is Z_0 ($-Z_0$ slope along source line). Then use general construction to find the resistor values for either the LT or the HT termination.

For line lengths greater than 50 meters, reflections are unlikely to be a problem, so attention can be focused on making a threshold selection.

There are THREE SITUATIONS for which threshold adjustment makes any sense:

(a)   to establish the threshold for a comparator-type application with slowly varying line drive.

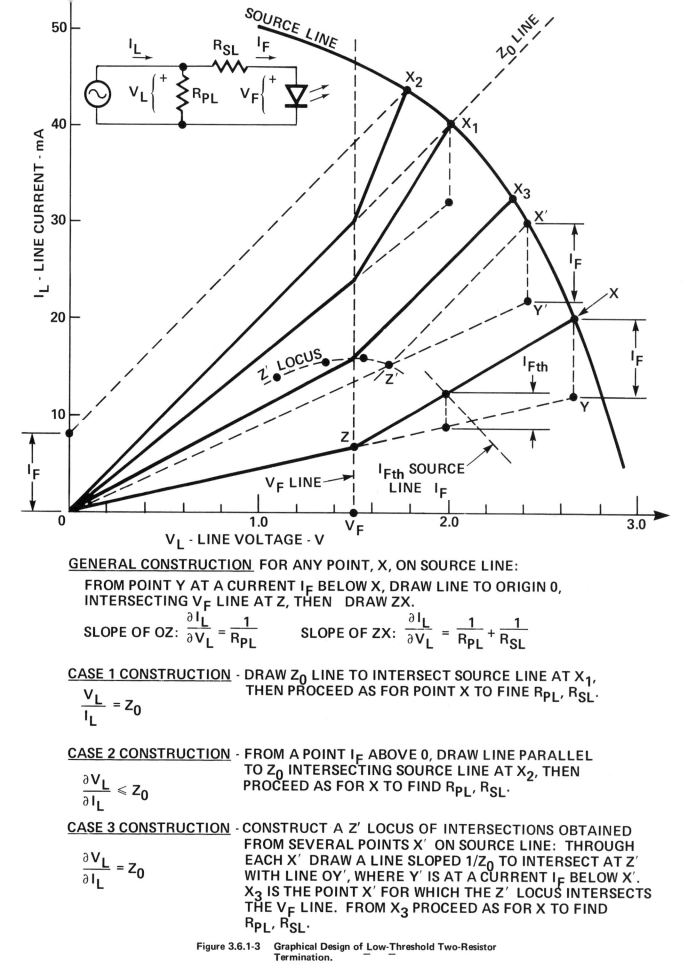

GENERAL CONSTRUCTION FOR ANY POINT, X, ON SOURCE LINE:

FROM POINT Y AT A CURRENT I_F BELOW X, DRAW LINE TO ORIGIN 0, INTERSECTING V_F LINE AT Z, THEN DRAW ZX.

SLOPE OF OZ: $\dfrac{\partial I_L}{\partial V_L} = \dfrac{1}{R_{PL}}$ SLOPE OF ZX: $\dfrac{\partial I_L}{\partial V_L} = \dfrac{1}{R_{PL}} + \dfrac{1}{R_{SL}}$

CASE 1 CONSTRUCTION - DRAW Z_0 LINE TO INTERSECT SOURCE LINE AT X_1,
$\dfrac{V_L}{I_L} = Z_0$ THEN PROCEED AS FOR POINT X TO FINE R_{PL}, R_{SL}.

CASE 2 CONSTRUCTION - FROM A POINT I_F ABOVE 0, DRAW LINE PARALLEL
$\dfrac{\partial V_L}{\partial I_L} \leqslant Z_0$ TO Z_0 INTERSECTING SOURCE LINE AT X_2, THEN
PROCEED AS FOR X TO FIND R_{PL}, R_{SL}.

CASE 3 CONSTRUCTION - CONSTRUCT A Z' LOCUS OF INTERSECTIONS OBTAINED
$\dfrac{\partial V_L}{\partial I_L} = Z_0$ FROM SEVERAL POINTS X' ON SOURCE LINE: THROUGH
EACH X' DRAW A LINE SLOPED $1/Z_0$ TO INTERSECT AT Z'
WITH LINE OY', WHERE Y' IS AT A CURRENT I_F BELOW X'.
X_3 IS THE POINT X' FOR WHICH THE Z' LOCUS INTERSECTS
THE V_F LINE. FROM X_3 PROCEED AS FOR X TO FIND
R_{PL}, R_{SL}.

Figure 3.6.1-3 Graphical Design of Low-Threshold Two-Resistor
Termination.

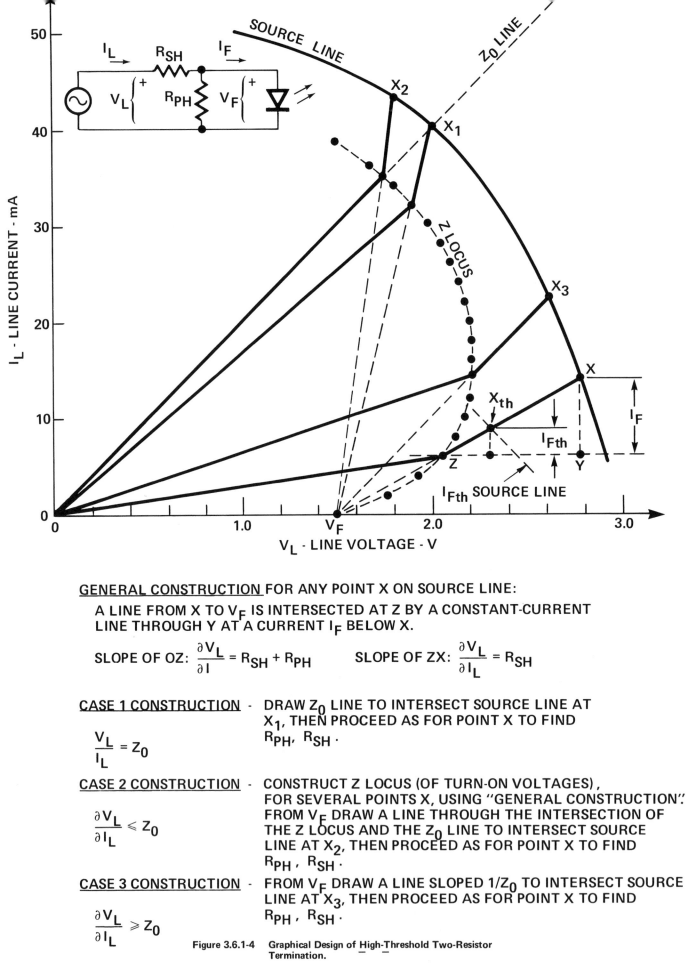

GENERAL CONSTRUCTION FOR ANY POINT X ON SOURCE LINE:

A LINE FROM X TO V_F IS INTERSECTED AT Z BY A CONSTANT-CURRENT LINE THROUGH Y AT A CURRENT I_F BELOW X.

SLOPE OF OZ: $\dfrac{\partial V_L}{\partial I} = R_{SH} + R_{PH}$ SLOPE OF ZX: $\dfrac{\partial V_L}{\partial I_L} = R_{SH}$

CASE 1 CONSTRUCTION -   DRAW Z_0 LINE TO INTERSECT SOURCE LINE AT X_1, THEN PROCEED AS FOR POINT X TO FIND R_{PH}, R_{SH}.

$\dfrac{V_L}{I_L} = Z_0$

CASE 2 CONSTRUCTION -   CONSTRUCT Z LOCUS (OF TURN-ON VOLTAGES), FOR SEVERAL POINTS X, USING "GENERAL CONSTRUCTION". FROM V_F DRAW A LINE THROUGH THE INTERSECTION OF THE Z LOCUS AND THE Z_0 LINE TO INTERSECT SOURCE LINE AT X_2, THEN PROCEED AS FOR POINT X TO FIND R_{PH}, R_{SH}.

$\dfrac{\partial V_L}{\partial I_L} \leqslant Z_0$

CASE 3 CONSTRUCTION -   FROM V_F DRAW A LINE SLOPED $1/Z_0$ TO INTERSECT SOURCE LINE AT X_3, THEN PROCEED AS FOR POINT X TO FIND R_{PH}, R_{SH}.

$\dfrac{\partial V_L}{\partial I_L} \geqslant Z_0$

Figure 3.6.1-4   Graphical Design of High-Threshold Two-Resistor Termination.

(b)   to balance delays when a single isolator is used with polarity *non*-reversing drive.

(c)   to optimize data rate with either single or dual isolators and polarity reversing drive.

The last one (c) is easy: make the threshold as low as possible. This is best done with an active voltage-clamp termination, as will be seen later. In designing for (a) or (b), it is first necessary to determine what the SOURCE LINE is for I_{Fth}.

The most precise determination of the I_{Fth} source line is by transient measurement of the V-I characteristics at the termination for the instant at which threshold current is desired. This is usually more labor than is justifiable by the results because any solution will yield only approximate resistor values and subsequent tweaking will be needed for either (a) or (b) above.

The next best is to make shrewd assumptions based on known system characteristics. Some good possibilities:

1. If the driver is back-matched (impedance = Z_0) construct the I_{Fth} SOURCE LINE parallel to the steady-state source line -- spaced halfway between zero and steady-state for situation (b), and spaced at whatever is appropriate for (a).

2. If the line is long enough that the round trip travel time of a pulse exceeds t_{PHL} or t_{PLH}, then, for (b) construct the first two legs of the reflection diagram (Figure 3.6.1-2) starting from $V_L, I_L = 0$. If the second leg (NEGATIVE Z_0 slope through i_g, v_{g1}) falls below the steady-state source line, take this as the I_{Fth} source line. If it falls above the steady-state source line, forget it.

3. For either (a) or (b), assume that the driver voltage and current rise proportionately and construct the I_{Fth} source line as the locus of points obtained by taking points on the steady state source line and reducing the voltage and current by the same proportion -- by 0.5 for (b), and by whatever is appropriate for (a).

Next, for points on the steady-state source line perform the general construction (Figure 3.6.1-3 or 3.6.1-4) to find the point at which the desired values of I_F and I_{Fth} are obtained. With a backmatched source and assumption No. 1, the solution can be found analytically by substituting I_{Fth} and E_{th} for I_F and E in the circuit equation (E and E_{th} are intercepts of the steady-state and I_{Fth} source lines at $I_L = 0$). These substitutions yield a pair of simultaneous equations to be solved for the resistor values. For the LT circuit, use:

$$E = I_F Z_0 + (V_F + I_F R_{SL})\left(1 + \frac{Z_0}{R_{PL}}\right) \qquad (3.6.1\text{-}1)$$

For the HT circuit use:

$$E = (R_{SH} + Z_0)\left(I_F + \frac{V_F}{R_{PH}}\right) \qquad (3.6.1\text{-}2)$$

To satisfy all three of the design objectives requires a three-resistor termination in either a "π" or "T" configuration. This is a rather complicated procedure, and for some driver/load combinations, solutions do not exist. The subject is discussed in EDN, in a five-part article: Feb. 5, 20, Mar. 5, 20, and April 5, 1976.

3.6.1.2 Active Terminations

Unlike resistive terminations, active terminations can respond to a variety of driver and line conditions to deliver the proper I_F to the optoisolator input, provide a low threshold, and absorb re-reflections from a mismatched driver. They should, therefore, be considered for the following situations:

*   Busing or current looping where changing the number of stations affects the current available to each station.
*   Power supply fluctuation at the driver.
*   Design flexibility requirement; same termination to be used with any of several types of drivers.
*   Accomodation of variation in line characteristics or length, where line resistance is significant.
*   Temperature change on long lines ($\Delta R/R \approx 0.4\%/°C$ for copper wire).
*   Data rate enhancement where a long or lossy line degrades the rise time at the termination.
*   Driver mismatch causing reflections.

Active terminations are simply current regulators for the optoisolator input diode. There are two classes: current-clamp regulators allowing terminal voltage to rise, and voltage-clamp regulators allowing terminal current to rise. Whereas for resistive terminations, there are three design objectives to consider, with active terminations there is really only one: proper steady-state I_F. So far as threshold is concerned, a voltage-clamp regulator allows it to be lower than any resistive termination can permit without allowing I_F to rise excessively. Reflections are of no concern unless there are other stations on the line that might suffer -- the active termination absorbs the current/voltage fluctuations caused by reflections. If the active termination causes reflection problems for other stations, these can be reduced by addition of series or shunt resistance.

Current-clamp circuits are shown in Figure 3.6.1-5, and the operating principles for polarity non-reversing drive are seen in the V-I characteristics. Note that isolator input current does not flow until $V_L > (V_F + V_{be})$, and beyond that, there is some time delay in turning on Q2. Lowering R2 to speed up the Q2 turn-on might cause unsatisfactory regulation of I_F in the one-port circuit, but is no problem with the two-port. The main advantage of the one-port over the two-port is the simplicity of converting it to polarity-reversing drive if $|V_L| > (4 V_{be} + V_F)$ *(if this were not so, a current-clamp regulator would be a poor choice anyway)* -- a diode bridge allows the same regulator to limit current in both directions.

Voltage Clamp circuits are shown in Figure 3.6.1-6 for polarity non-reversing drive, along with V-I characteristics. Circuits for polarity-reversing drive are shown in Figure 3.6.1-7. Note that turn-on begins at V_F, and fully regulated current is obtained with $V_L = (V_F + V_{be})$ for either polarity-reversing or polarity non-reversing drive. Compare this with the current-clamp circuits requiring $V_L = (V_F + 2V_{be})$ for polarity non-reversing and $V_L = (V_F + 3V_{be})$ or $(V_F + 4V_{be})$ for polarity-reversing drive.

Note also that the regulator does not become active until Q1 turns on. That is, there is no transistor turn-on delay of isolator input current (as for Q2 in the current-clamp circuits). Moreover, since the turn-on of Q1 lags the flow of line current, there will be a momentary surge of line current into the isolator input diode. A similar lag in turn-off of Q1 hastens the turn-off of the optoisolator. *This peaking effect enhances the data rate especially where long or lossy lines cause long rise-time at the termination.*

The terminal voltage compatibility makes the voltage-clamp regulator ideal for busing. Although slight differences in V_F, V_{be}, or β might cause one regulator to operate at higher current than others in parallel with it, the voltage will be the same on all, so each isolator input diode will have its own proper I_F. That is, isolators requiring high I_F can be terminated in parallel with low-I_F types, providing, of course, the line driver can supply the total current needed.

In current looping, the build-up of terminal voltages around the loop limits the number of stations that can be operated. Also, as stations are added or removed, there may be a considerable change in loop current. For these reasons, the low operating voltage and broad range of regulated operation makes the voltage-clamp regulator a good choice.

3.6.2 Common Mode Rejection (CMR) Enhancement

Common mode interference is so named because it appears as a voltage which is, as referred to output ground, *common* to both input terminals of the optoisolator. It is labelled e_{CM} in Figure 3.6.2-1. The desired signal appears *differentially* and is therefore called the differential mode signal, e_{DM}.

Characterization of the basic optoisolator CMR properties are discussed in Section 3.2.1. Attention here is given to various circuit techniques for improving CMR over that of the basic isolator.

e_{CM} comes about in either of two ways: "induced", represented by e_1 in Figure 3.6.2-1, or "conducted", represented by e_2. In some situations, e_2 is inherent to the system; for example, if the isolator is used to control a module floating at some large voltage with respect to driver ground, this large voltage is represented by e_2. Another example of conducted interference is that which comes through the interwinding capacitance of a power transformer. e_1 represents the voltage between the transmission line and ground or between the line and some parallel conductor. Although shown as capacitively coupled in Figure 3.6.2-1, e_1 could also be magnetically induced, as might occur in industrial machine control applications in which control and power lines are close together over appreciable distances.

If e_1 is more tightly coupled to one side of the line than to the other, the difference can cause a substantial voltage difference between the two lines. Twisted pair lines are used to balance the common mode coupling, and a shield over the twisted pair aids in balancing induced interference even if the shield floats; that is, no shield connection at either end. If the shield is to be connected anywhere, it should be to a point at the junction of two resistors connected in series across the receiving end of the transmission line. The only exception to this rule occurs when the shield is used as a third conductor to pre-bias a balanced split-phase receiver, as in Figure 3.6.3-3. A further aid in balancing out e_1 is to have the internal driver impedance balanced to driver ground. The effect of unbalanced induced interference can be somewhat compensated by adjustment of the resistors from each side of the line to the shield at the receiving end.

Even with e_1 balanced, it is still present on both terminals at the receiver, and, along with e_2, comprises e_{CM}, which can be capacitively coupled to the amplifier on the output side of the isolator, as discussed in Section 3.2.1. There are a number of techniques for dealing with e_{CM}:

- Neutralization
- Balanced differential amplification
- Amplifier de-sensitization
- Selective flip-flop output
- Exclusive-OR flip-flop
- Use of high-CMR devices

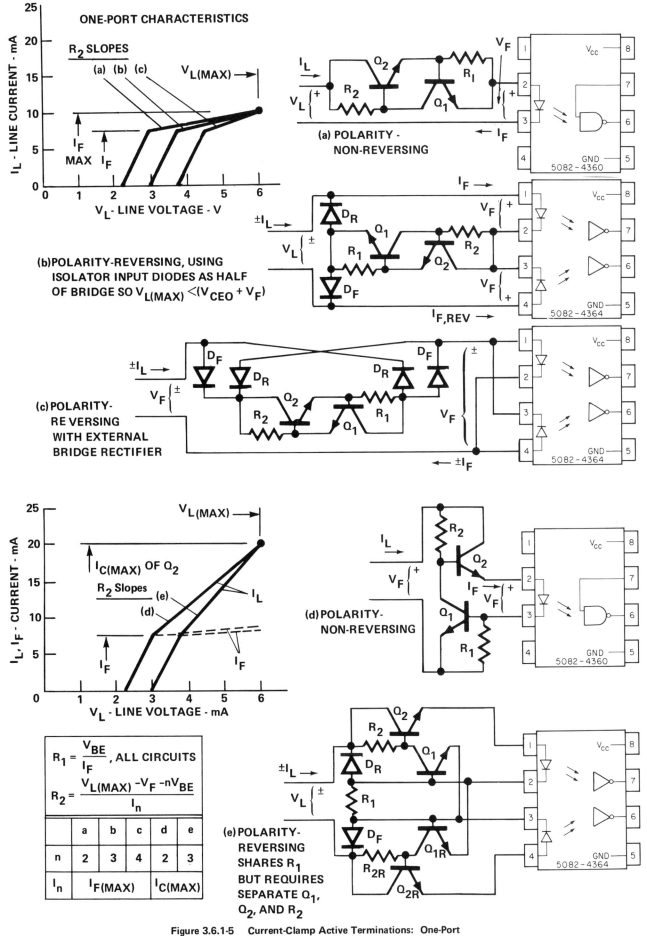

Figure 3.6.1-5   Current-Clamp Active Terminations: One-Port
(a, b, c) and Two-Port (d, e).

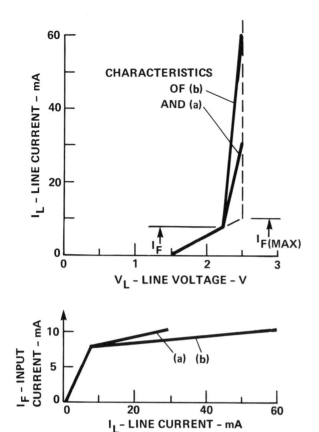

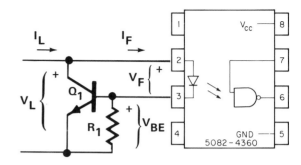

CHARACTERISTICS
OF (b)
AND (a)

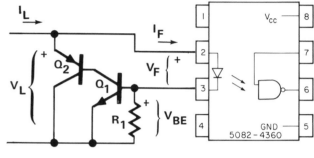

(a) LOW-GAIN

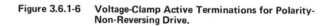

IN BOTH (a) AND (b) $R_1 = V_{BE}/I_F$

(b) HIGH-GAIN

Figure 3.6.1-6   Voltage-Clamp Active Terminations for Polarity-Non-Reversing Drive.

(a) LOW-GAIN ——
DIODES D_F AND D_R
ARE REQUIRED TO
PREVENT REVERSE
TURN-ON OF THE
TRANSISTORS

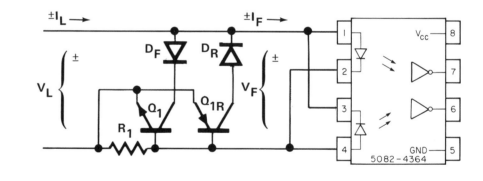

(b) HIGH-GAIN ——
THE EMITTER-BASE
JUNCTIONS OF Q_2
AND Q_{2R} SERVE
AS D_F AND D_R IN (a).
REVERSE BREAKDOWN
IS NO CONCERN
BECAUSE THE LINE
VOLTAGE IS CLAMPED.

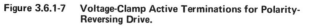

Figure 3.6.1-7   Voltage-Clamp Active Terminations for Polarity-Reversing Drive.

3.48

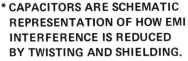

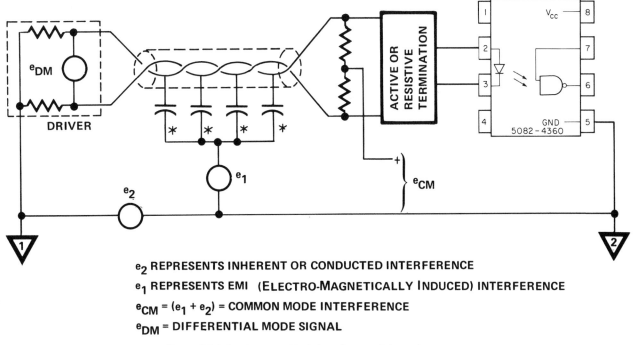

e₂ REPRESENTS INHERENT OR CONDUCTED INTERFERENCE

e₁ REPRESENTS EMI (ELECTRO-MAGNETICALLY INDUCED) INTERFERENCE

$e_{CM} = (e_1 + e_2)$ = COMMON MODE INTERFERENCE

e_{DM} = DIFFERENTIAL MODE SIGNAL

Figure 3.6.2-1   Common Mode Interference; Schematic Representation.

Neutralization can be used with optoisolators having a single-transistor amplifier operated common-emitter. Also, it is effective only while the transistor is active -- not saturated or cut-off. However, since it is while the transistor is active that e_{CM} is most troublesome, it is a technique worth considering. It consists simply of a neutralizing capacitor, C_N in Figure 3.6.2-2(a) that sources a current $C_N (de_{CM}/dt)$ which is opposite in polarity to the collector current $\beta C_{CM} (de_{CM}/dt)$ caused by e_{CM}. If $C_N = \beta C_{CM}$, these opposing currents will be balanced and the collector voltage will not be affected by e_{CM}. Even if C_N does not precisely neutralize e_{CM}, its effect is beneficial. *CAUTION: C_N must have a voltage rating high enough to accomodate the maximum voltage that e_{CM} might attain.*

Balanced differential amplification also works only if the isolator output amplifiers are active. It is clear in Figure 3.6.2-3 that if both isolators are driven to cutoff or saturation by an e_{CM} transient, they cannot maintain a differential response to e_{DM}.

Amplifier de-sensitization is perhaps the simplest defense to raise against e_{CM}. The base bypass resistor, R_{BE} in Figure 3.6.2-4 reduces the impedance in which the current flows that is coupled via C_{CM} by any e_{CM} transient. Thus a larger e_{CM} transient can be tolerated. It also reduces the amplifier's sensitivity to photocurrent resulting from e_{DM}, and therefore, requires a higher current in the input diode

for a proper "on" state. The base bypass can be used with either single-transistor or split-darlington types of amplifiers and offers the additional benefit of higher data rate capability (see Figure 3.6.3-5). R_{SK} is recommended where there is a risk that a body charged with static electricity may be discharged with all or part of the surge current passing through C_{CM}. Being amplified by the first transistor, such surge current can then be destructive.

A selective flip-flop output circuit can take advantage of a situation in which the e_{CM} transients have a higher rate of change in one direction than in the other, such as a sawtooth. It can also take advantage of a circuit with $CM_H > CM_L$ or vice versa (see Section 3.2.1). Since a NAND flip-flop can tolerate having both inputs high, it should be used where the likelihood is greater for both isolator outputs to be high (due to e_{CM} transient) than is the likelihood that both outputs will be low. Conversely, a NOR flip-flop can tolerate having both inputs low, so the rule is reversed. These rules are summarized in Figure 3.6.2-5.

Exclusive-OR flip-flop, whether of NOR or NAND construction can tolerate either both inputs high or both inputs low without either of its outputs changing state. The outputs can change state only in response to a change in differential input, so it has **infinite** common mode rejection for a static condition of e_{DM} in either logic state. Note in

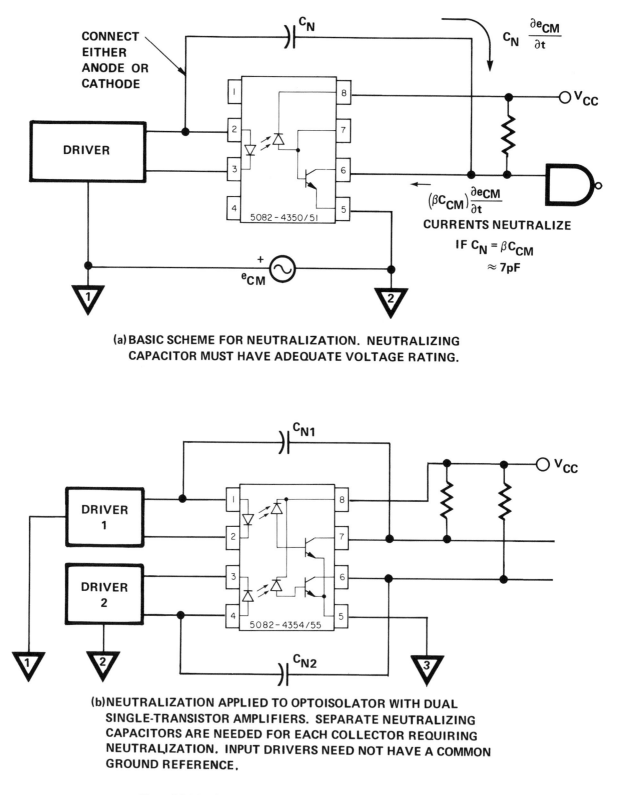

$$C_N \frac{\partial e_{CM}}{\partial t}$$

$$(\beta C_{CM}) \frac{\partial e_{CM}}{\partial t}$$

CURRENTS NEUTRALIZE

IF $C_N = \beta C_{CM}$

$\approx 7pF$

(a) BASIC SCHEME FOR NEUTRALIZATION. NEUTRALIZING
CAPACITOR MUST HAVE ADEQUATE VOLTAGE RATING.

(b) NEUTRALIZATION APPLIED TO OPTOISOLATOR WITH DUAL
SINGLE-TRANSISTOR AMPLIFIERS. SEPARATE NEUTRALIZING
CAPACITORS ARE NEEDED FOR EACH COLLECTOR REQUIRING
NEUTRALIZATION. INPUT DRIVERS NEED NOT HAVE A COMMON
GROUND REFERENCE.

Figure 3.6.2-2   Neutralization of Common-Mode Interference with
External Capacitor.

3.50

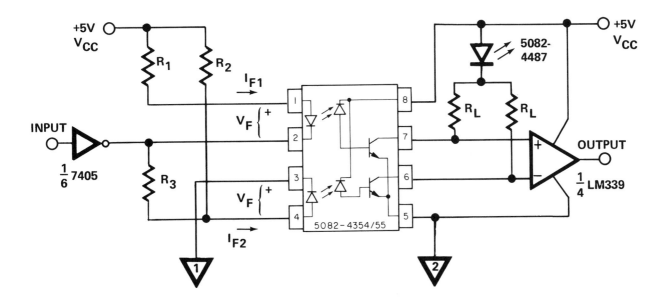

INPUT	I_{F1}	I_{F2}	OUTPUT
0	I_L	I_H	0
1	I_H	I_L	1

INPUT	I_{F1}	I_{F2}
0	$\dfrac{V_{CC} - 2V_F}{R_1 + R_3}$	$\dfrac{V_{CC} - 2V_F}{R_1 + R_3} + \dfrac{V_{CC} - V_F}{R_2}$
1	$\dfrac{V_{CC} - V_F}{R_1}$	$\dfrac{V_{CC} - V_F}{R_2} - \dfrac{V_F}{R_3}$

RESULTS OBTAINED WITH
$R_1 = R_3 = 180$, $R_2 = 270$

INPUT	I_{F1} (mA)	I_{F2} (mA)
0	5.6	18.5
1	19.4	4.6

REQUIRED RATIO OF RESISTORS : $\quad R_1 = R_3 = \dfrac{R_2}{2\left(1 - \dfrac{V_F}{V_{CC}}\right)}$

$$\frac{I_H}{I_L} = 2\left(\frac{V_{CC} - V_F}{V_{CC} - 2V_F}\right) \qquad R_1 \geqslant \frac{V_{CC} - V_F}{I_H}$$

$$R_L \geqslant \frac{V_{CC} - V_F}{I_{C(MAX)}} \approx 470$$

IT IS NOT NECESSARY TO HAVE PRECISE BALANCE OF ON- AND OFF-STATE INPUT CURRENTS AS LONG AS THE RATIO OF THE LOWER HIGH TO THE HIGHER LOW EXCEEDS THE RATIO OF THE HIGHER CTR TO THE LOWER CTR. SUCH UNBALANCE CAN ALSO BE COMPENSATED BY ADJUSTMENT OF R_L.

Figure 3.6.2-3   Active Differential Drive for CMR Enhancement
and High Data Rate.

REDUCE R_{BE}, R_L FOR:
$\left\{\begin{array}{l}\text{LOWER AMPLIFIER GAIN}\\\text{HIGHER CMR}\\\text{LOWER CTR}\end{array}\right.$

R_{SK} RESTRICTS SURGE; DOES NOT HELP OR HURT

CMR OR CTR. RECOMMENDED VALUE: $R_{SK} = \dfrac{1\ \text{VOLT}}{.15 \times I_F}$

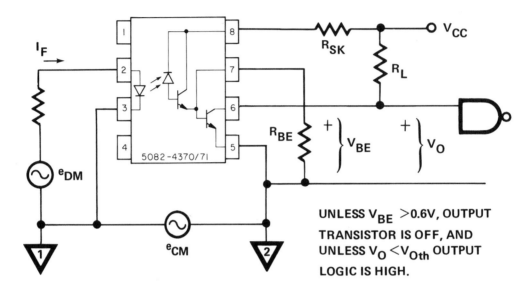

UNLESS $V_{BE} > 0.6V$, OUTPUT
TRANSISTOR IS OFF, AND
UNLESS $V_O < V_{Oth}$ OUTPUT
LOGIC IS HIGH.

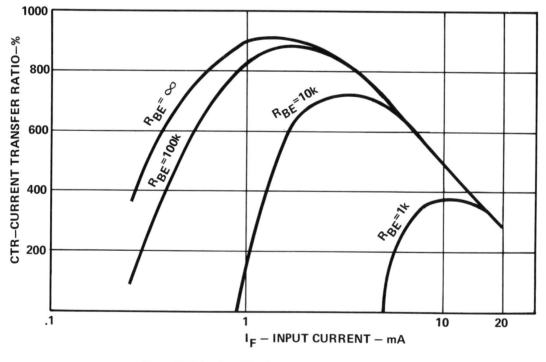

Figure 3.6.2-4   Amplifier Desensitization for CMR Enhancement
by Lowering CTR.

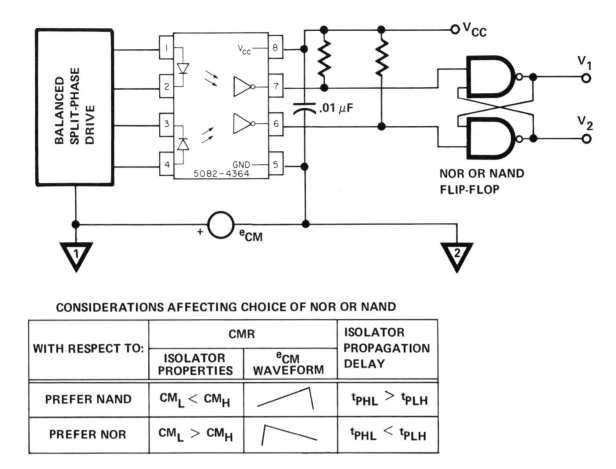

CONSIDERATIONS AFFECTING CHOICE OF NOR OR NAND

WITH RESPECT TO:	CMR		ISOLATOR PROPAGATION DELAY
	ISOLATOR PROPERTIES	e_{CM} WAVEFORM	
PREFER NAND	$CM_L < CM_H$		$t_{PHL} > t_{PLH}$
PREFER NOR	$CM_L > CM_H$		$t_{PHL} < t_{PLH}$

Figure 3.6.2-5   Selective Flip-Flop Output for CMR Enhancement and Edge Sharpening.

Figure 3.6.2-6 that the exclusive -OR flip-flop requires the output of each isolator to drive two gate inputs. If the isolator cannot handle two inputs, a buffer inverter can be inserted at no loss of CMR.

CAUTION: although the exclusive-OR flip-flop has infinite static immunity to e_{CM} its *dynamic immunity* is **not** *infinite*. If a common mode transient capable of holding both isolator outputs either high or low should persist throughout the duration of a differential mode pulse, the pulse will escape detection. Other than this, the worst a common mode transient can do to the exclusive-OR flip-flop is to advance or retard the timing of the flip-flop output transition, i.e., cause jitter of the edge.

High CMR devices are those in which the ratio of optical coupling to capacitive coupling has been made higher than that of the *usual* "sandwich"-type optoisolators. One such technique is the use of a conductive transparent screen over the detector (see Section 3.2.1). Another is the use of lenses or fiber-optics to couple the optical signal efficiently over longer distances, thereby physically separating the detector from the input and reducing capacitive coupling without a severe penalty in optical coupling. Such arrangements are discussed further in Section 2.5.5.

3.6.3   Data Rate Enhancement

In addition to the t_{PHL} and t_{PLH} data rate restrictions imposed by the optoisolator there are system limitations. Some of these can be optimized and others can be compensated. In *compensating,* the usual tradeoffs are reduction in differential-mode noise margin or common-mode rejection. This is especially true when single-ended (rather than balanced split-phase) optoisolators are used because the compensation reduces the amount of differential-mode voltage change needed to cause the isolator output circuit to change logic state. The circuit is therefore more vulnerable to differential-mode noise and to common-mode noise which has been partly differentialized by unbalanced impedance.

System optimization consists of reducing the degradation in the rise time of step changes in the differential mode signal, e_{DM}. If such degradation is due to stray wiring capacitance or inductance the remedies include reducing the impedance of the e_{DM} signal source, enlarging the wires or circuit board traces, and dressing leads properly.

The potential data rate devastation of a transmission line is often underestimated. This is because its transient response

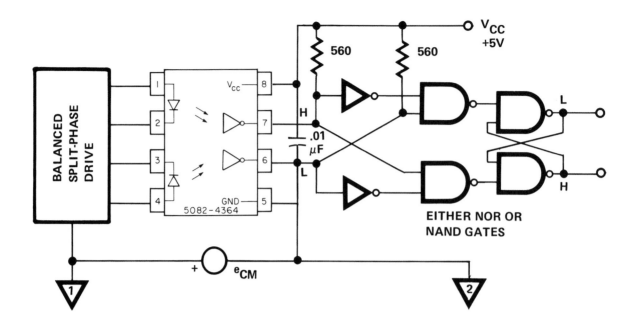

Figure 3.6.2-6 Exclusive-Or Flip-Flop for Infinite CMR Under Steady State Conditions.

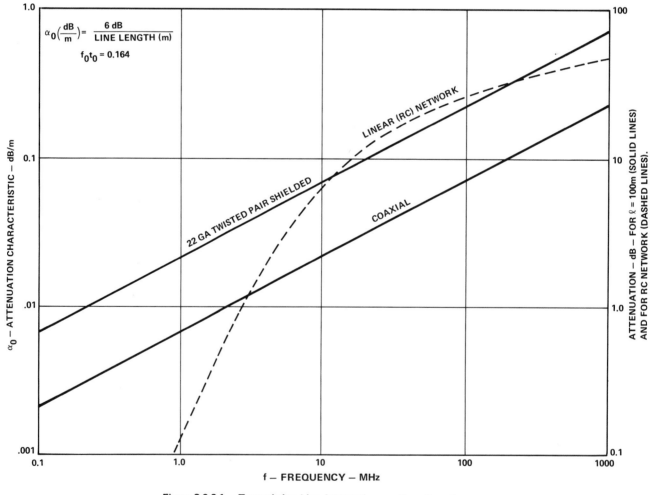

Figure 3.6.3-1 Transmission Line Attenuation as a Function of Frequency.

is not the same as the familiar RC-type transient. A good estimate of transmission line performance, and appropriate design choices, can be made using the relationships described in Figures 3.6.3-1 and 3.6.3-2.

Due to skin effect, most transmission lines have an attenuation vs. frequency that varies as the square root of frequency over the frequency range of the solid lines in Figure 3.6.3-1. If a curve is not available, it is simple enough to generate one by measuring the relative output voltage, V_1 and V_2, from two lengths, ℓ_1 and ℓ_2, of cable at some particular frequency, f. Then on coordinates of log α_0 vs. log f, plot the point obtained by measurement at f:

$$\alpha_0 \ (f) = \frac{[20 \log (V_2/V_1)]}{[\ell_1 - \ell_2]} \quad \left(\frac{dB}{m}\right) \qquad (3.6.3\text{-}1)$$

Through this point, draw a line sloping upward at half a decade of α_0 per decade of f. The lengths of line used in the measurement need not be the same as required for the system -- sample lengths will do, but the longer the samples, the greater the precision. If the generator and load impedances match the characteristic impedance of the line, a single length, ℓ_1, can be used; then substitute $\ell_2 = 0$ and V_2 = generator voltage (with matched load connected) in equation 3.6.3-1. The frequency used should be in the range of 1 to 10 MHz.

Response of a transmission line of length, ℓ, to a step input is found as follows:

1. Compute α_0 by dividing the line length, ℓ, into 6 dB:

$$\alpha_0\left(\frac{dB}{m}\right) = \frac{6 \ dB}{\ell(m)} \qquad (3.6.3\text{-}2)$$

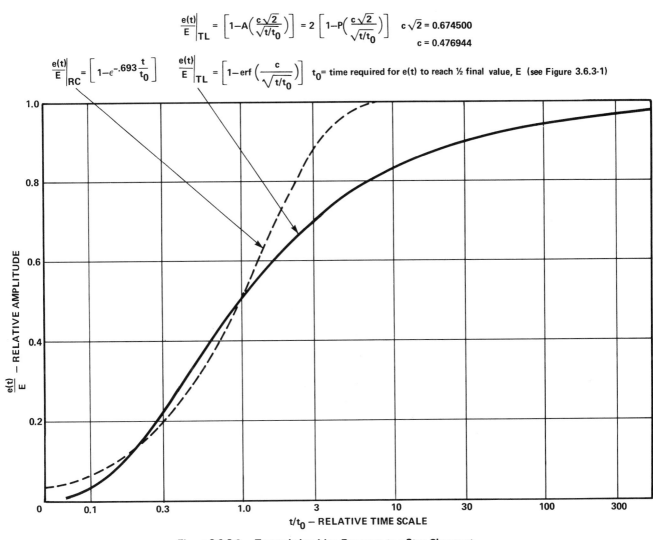

Figure 3.6.3-2   Transmission Line Response to a Step Change at the Input.

2. With α_O from Step 1, enter Figure 3.6.3-1 and find f_O at the intersection of α_O with the curve for the selected cable. (f_O is the 6-dB frequency for the length, ℓ, and could be obtained by the measurement procedure described above.)

3. Compute t_O from the relationship:

$$f_O t_O = 0.164 \tag{3.6.3-3}$$

4. Step response, excluding travel time, is found from the normalized curve (solid line) in Figure 3.6.3-2, where t_O is the time required for the transient to change through half its asymptotic value. The logic delay, t_{TL}, imposed by the transmission line is then:

$$t_{TL} = t_O \times (t_{th}/t_O) \tag{3.6.3-4}$$

where (t_{th}/t_O) is the value of (t/t_O) at which

$$\frac{e(t)}{E} = \frac{E_{th} - E_O}{E_{as} - E_O} = \frac{\Delta E}{E} \tag{3.6.3-5}$$

where: E_O = initial line voltage
E_{as} = asymptotic voltage
E_{th} = voltage where switching occurs

EXAMPLE: For ℓ = 100m of twisted pair shielded cable,

Step 1 gives $\alpha_O = 0.06$

Step 2 gives f_O = 8.5 MHz (typical curve, Figure 3.6.3-1)

Step 4 if $\Delta E/E = 0.7$, $(t_{th}/t_O) = 3$
t_{TL} = 19.3 ns x 3 = 57.9 ns

Note that if $\Delta E/E$ had been 0.8, t_{TL} = 19.3 ns $\times$ 7 = 135.1 ns, which is more than twice the time delay for only a small change of threshold. The importance of threshold in dealing with transmission line effects can hardly be given enough emphasis.

Shown for comparison in Figure 3.6.3-1 is a linear (RC) transient which also rises to half its final value at $t = t_O$, but note that it attains 90% in just 3.3 $\times$ t_O while the transmission line transient requires $\approx$ 30 $\times$ t_O. The attenuation vs. frequency of the same RC network is shown in Figure 3.6.3-1 for comparison with the performance of 100m of twisted pair shielded cable -- scales at right side. The normalized transmission line transient can be plotted either from tables of $erf(x)$ or from tables of the area under the "normal curve of error", A(x):

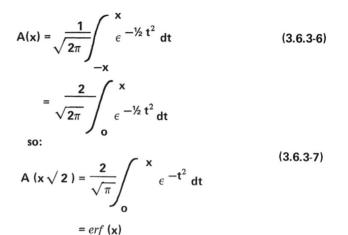

$$A(x) = \frac{1}{\sqrt{2\pi}} \int_{-x}^{x} \epsilon^{-\frac{1}{2}t^2} dt \tag{3.6.3-6}$$

$$= \frac{2}{\sqrt{2\pi}} \int_{0}^{x} \epsilon^{-\frac{1}{2}t^2} dt$$

so:

$$A(x\sqrt{2}) = \frac{2}{\sqrt{\pi}} \int_{0}^{x} \epsilon^{-t^2} dt \tag{3.6.3-7}$$

$$= erf(x)$$

Some math tables give the area A(x) between the limits -x and +x; others the area P(x) for -∞ to +x; still others give the area between +x and +∞. Since the normal curve is symmetrical and the total area, -∞ to +∞, is just 1.0, any of these tables can be used. Jahnke and Emde (4th Edition, p.24) gives $erf(x)$ with adequate precision. A great deal of precision is not worthwhile because the results of the procedure given here are only estimates to be used for avoiding serious problems with data rate limitation.

The procedure can also be used in reverse:

4. Given threshold, initial, and asymptotic voltages, use equations 3.6.3-4 and 3.6.3-5 and Figure 3.6.3-2 to find a value of t_O for tolerable delay, t_{TL}.

3. Find f_O from t_O using equation 3.6.3-3.

2. Enter Figure 3.6.3-1 with f_O to where it intersects the cable curve at α_O.

1. From equation 3.6.3-2, compute the maximum length of cable that can be used. If that is too short, calculate α_O from equation 3.6.3-2 for the desired length; then the intersection of this value of α_O with f_O from Step 3 gives a point on Figure 3.6.3-1 below which the cable curve must lie. A proper cable may thus be defined.

A handy relationship to bear in mind is that t_O increases as the square of the length of the cable.

The curve of Figure 3.6.3-2 can be used even if the line is mismatched at both ends. During the transient, the transmission line, as seen at the termination, is a source with impedance Z_0 and open-circuit (Thevenin) voltage changing in the manner described by Figure 3.6.3-2. The voltages, E_O, E_{as}, E_{th} to be used in equation 3.6.3-5 are found most easily by graphical construction, as shown in Figure 3.6.3-3. Worst-case delays occur under initial conditions as determined by the quiescent point (Q_L or Q_H). For the transient going from Q_L toward Q_H, a line with NEGATIVE Z_0 slope through Q_L intersects the $I_L = 0$ line at E_0. (NOTE: Because $I_L = 0$ at Q_L, the NEGATIVE Z_0 line has zero length in this example. For a non-zero example, see Figure 3.6.3-4b.) To find E_{as} begin with a POSITIVE Z_0 sloped line through Q_L to intersect the SOURCE LINE for Q_H (not necessarily at Q_H); from this intersection, a line with NEGATIVE Z_0 slope intersects the $I_L = 0$ line at E_{as}. E_{th} is a property of the load characteristic upon which there is a point, X_{th}, defining threshold conditions for the termination. Thus a line with NEGATIVE Z_0 slope through X_{th} intersects the $I_L = 0$ line at E_{th}. For a transient going from Q_H to Q_L, construction is symmetrical with a balanced polarity-reversing system; with polarity non-reversing, construction is not symmetrical but the rules above are the same for finding E_0, E_{as}, and E_{th}.

Compensation for either linear or transmission line transients can be done by peaking or by threshold-lowering, or both.

Without peaking or threshold-lowering the lowest possible threshold voltage at the terminals of an isolator input circuit are:

$$V_{Lth} = V_F + I_{Fth} R_S \qquad (3.6.3-8)$$

where the series resistance may be either R_S (Figure 3.6.1-1) or R_{SL} (Figure 3.6.1-3) or R_1 (Figures 3.6.1-6,7). V_{Lth} is the voltage at the point X_{th} in Figure 3.6.3-3. By prebiasing the input diode, this threshold can be lowered. As long as the prebias voltage is below 1.2V, the forward current in the isolator input diode will be less than $10\,\mu$A.

The effects of such prebias on the logic delay, where the isolator terminates a transmission line, are shown in Figure 3.6.3-3. The technique is useful also where no transmission line is involved, but transient delays are present.

With a single-ended system, the threshold may already be near optimum, as seen in Figure 3.6.3-3(a). The net effect of threshold adjustment is determined with reference to Figure 3.6.3-2 for a transmission line transient -- in other situations, the effects can be found with reference to the particular kind of transient involved, e.g. RC. In general, threshold adjustment in a single-ended circuit shortens delay in one direction while lengthening it in the other, so it can be used to compensate for unbalanced t_{PLH} and t_{PHL} in the isolator.

With a balanced, polarity reversing system, the optimum threshold is at $V_L = 0$, so the objective is to make it as low as possible. The effect on data rate can be profound. In

Figure 3.6.3-3(b), the one-volt prebias changes the delay from $\approx 12 \times t_0$ to $\approx 2.4 \times t_0$. If t_0 is only 20 ns, this changes the delay from 240 ns down to 48 ns and raises the NRZ data rate from less than 3.5 Mb/s to approximately 10 Mb/s.

Peaking is another form of compensation. The usual technique is to place a capacitor across the series current-limiting resistor. Then when the driver changes state, the voltage on the capacitor, for as long as it persists, augments the source. With respect to the circuit in Figure 3.6.3-4(a), the peaking can be analyzed with or without transmission line effect. The construction at the left shows the "temporary" load line caused by the voltage, V_P, on the peaking capacitor. As the source line changes from Q_L to Q_H, turn-off begins immediately in one input diode and requires only a $2 \times V_F$ change of source voltage for turn-on to begin in the other. If V_P persists while the source line changes all the way from Q_L to Q_H, there will be a surge of forward current in the diode being turned on. As described in Section 3.2.3, this will enhance the data rate. **Caution:** Peaking must not allow I_{SURGE} to exceed the amplitude or duration permitted by maximum ratings. If the source or transmission line impedance is too low, peaking within limits permitted can be done with bypassing around only part of R_S. In general, the rule for construction of the "temporary" load line is to represent only that portion of the load which is NOT bypassed, and shift it by the voltage on the peaking capacitor.

If peaking is applied to a single-ended polarity-non-reversing termination, the construction rule is the same as above. It should be apparent, in construction, that an anti-parallel diode is needed across the isolator input diode to allow C_P to charge and discharge, even though the drive polarity does not reverse.

A safe, but slightly less effective peaking technique for polarity-reversing balanced drive is shown in Figure 3.6.3-4(b). With respect to steady state conditions (Q_L, Q_H) the series resistor limits input current. In the transient state, the diode being turned off has its charge transferred to the diode being turned on. Until the initially-on diode is discharged, the voltage on it augments the line-to-line drive. Threshold is very near zero, being offset only by the series resistance, R_D, of the initially-on diode. Discharge time is very short (≈ 20 ns), and if the drive transient does not rise through $\Delta E/E$ within this time, the threshold shifts over to the X_{th} point on the steady state load line. This form of peaking is, therefore, effective only for short lines or very low-loss lines, having $t_0 < 20$ ns (see Figure 3.6.3-1).

Safe peaking is also obtained with the voltage-clamp regulators of Figure 3.6.1-6,7. Until the regulator transistor is turned on, current through the input diode of the isolator is limited only by r_1 and Z_0. The effect can be enhanced

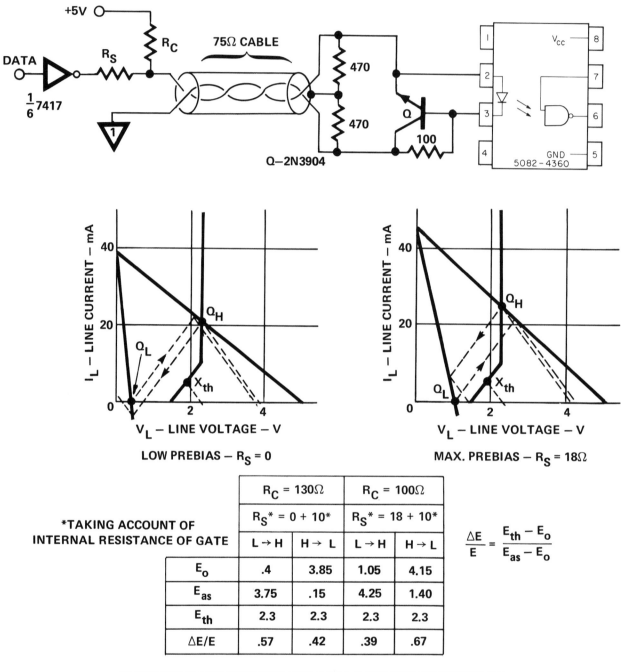

	$R_C = 130\Omega$		$R_C = 100\Omega$	
	$R_S^* = 0 + 10^*$		$R_S^* = 18 + 10^*$	
*TAKING ACCOUNT OF INTERNAL RESISTANCE OF GATE	$L \rightarrow H$	$H \rightarrow L$	$L \rightarrow H$	$H \rightarrow L$
E_o	.4	3.85	1.05	4.15
E_{as}	3.75	.15	4.25	1.40
E_{th}	2.3	2.3	2.3	2.3
$\Delta E/E$	.57	.42	.39	.67

$$\frac{\Delta E}{E} = \frac{E_{th} - E_o}{E_{as} - E_o}$$

THRESHOLD ADJUSTMENTS FOR DATA RATE ENHANCEMENT OR DELAY BALANCING WITH SINGLE-ENDED RECEIVERS — POLARITY NON-REVERSING.

Figure 3.6.3-3a Threshold Adjustment (PRE-BIAS) for Data Rate Enhancement; Polarity Non-Reversing.

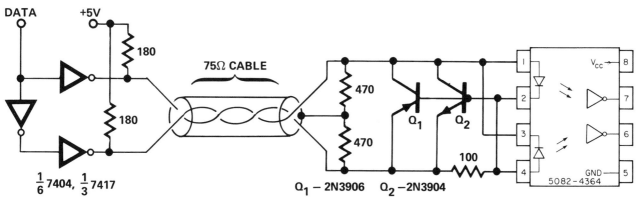

NO PREBIAS – SHIELD BALANCED FOR OPTIMAL CMR AT THE EXPENSE OF DATA RATE.

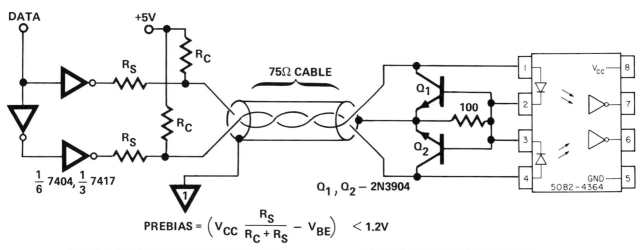

$$\text{PREBIAS} = \left(V_{CC}\,\frac{R_S}{R_C + R_S} - V_{BE}\right) < 1.2V$$

SHIELD RETURNS CURRENT OF "ON" SIDE, ALLOWS "OFF" SIDE TO BE PREBIASED.

ALTERNATIVE USE OF SHIELD WITH SPLIT-PHASE POLARITY-REVERSING DRIVE AND BALANCED DUAL-ISOLATOR LOAD – EITHER CMR OPTIMIZATION OR DATA RATE ENHANCEMENT.

Figure 3.6.3-3b Threshold Adjustment (PRE-BIAS) for Data Rate Enhancement; Polarity Reversing

with a capacitor bypass across all or part of R_1. This also gives more consistent peaking because it is less dependent on the transistor turn-on time.

CAUTION: Both peaking and threshold-lowering tend to reduce CMR. If common-mode interference includes large, rapidly changing transients, the CMR enhancement techniques of Section 3.6.2 should be applied. For balanced terminations, the exclusive-OR flip-flop of Figure 3.6.2-6 is recommended. Single-ended terminations may require internally shielded optoisolators.

Output circuitry may also require attention in data rate optimization. With very high speed optoisolators the pullup resistor should be made as small as possible, consistent with current sinking capability of the optoisolator output and current sinking requirements of the logic inputs being driven.

Slower types of optoisolators can be made faster by "swamping" the base of the output transistor. With a resistor, R_{BE}, connected from the base to the emitter, the logic delay, especially t_{PLH}, is greatly enhanced by the discharge path that R_{BE} provides for the base. The gain (CTR), however, is reduced, especially for low input currents. It is therefore necessary to raise the input current to obtain proper operation. Lowering R_{BE}, at any particular level of input current, reduces t_{PLH} but raises t_{PHL}, as seen in Figure 3.6.3-5. Thus at any level of input current there is a value of R_{BE} at which t_{PLH} and t_{PHL} are approximately balanced. The optimum value of R_{BE} depends also on the value of the pullup resistor. Figure 3.6.3-5 shows recommended values of R_{BE} and the results to be expected as a function of input current.

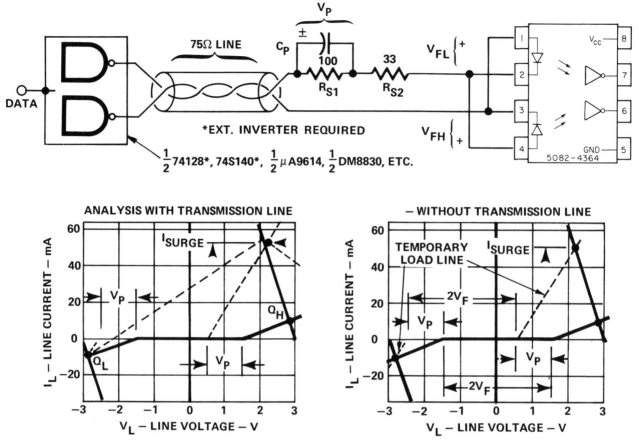

(a) USE OF PEAKING CAPACITOR — $R_{S1} \times C_p > t_{LH}$ OR t_{HL}

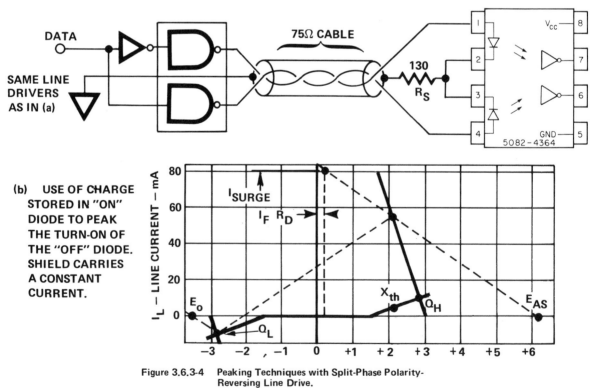

(b) USE OF CHARGE STORED IN "ON" DIODE TO PEAK THE TURN-ON OF THE "OFF" DIODE. SHIELD CARRIES A CONSTANT CURRENT.

Figure 3.6.3-4 Peaking Techniques with Split-Phase Polarity-Reversing Line Drive.

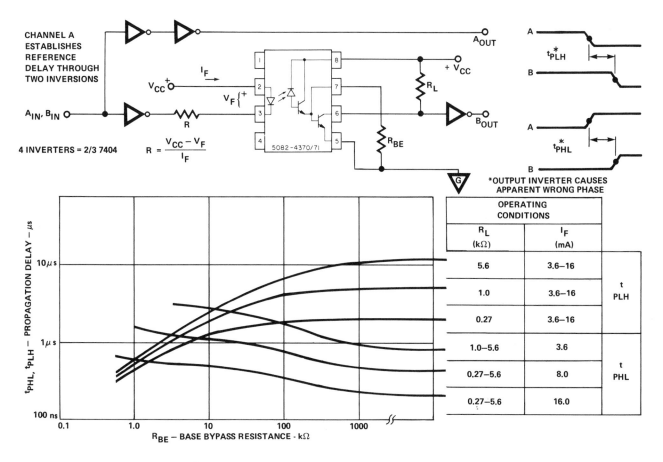

Figure 3.6.3-5   Base Bypass Resistor, R_{BE}, Effect on
Propagation Delay.

3.6.4   Party Line Operation (Bussing, Current Looping)

Some situations require two or more stations to operate from a common transmission line. There may be only one station capable of transmitting while all others on the line are receiving; this is called a simplex system. If two stations are each capable of transmitting to the other, it is called a duplex system. The more general case, where any one of several stations is capable of transmitting to all others, as well as receiving, is called a multiplex system.

RS-232 is a popular form of simplex bussing, documented by EIA as RS-232-C. The EIA specifications do not require ground separation, however, optoisolators are often used as extra protection with lines longer than specified in RS-232C. The EIA receiver input specifications are completely met using the termination in Figure 3.6.1-1(a) with $R_s \geqslant 3,000$ ohms as per RS-232C. Since RS-232-C allows a line voltage as low as 3 volts, the isolator input is required to operate with I_F as low as 0.5 mA. Also, RS-232-C limits the maximum line voltage excursions to ±25V, so the power into the input diode cannot exceed $25^2/(4 \times 3k) = 52$ mW, and the antiparallel LED, while recommended, is not essential.

RS-422 and **RS-423** are EIA documents for simplex data transmission at data rates higher than that specified in RS-232-C. RS-422 is for line voltage balanced with respect to driver ground and RS-423 is for unbalanced (single-ended), but both specify polarity-reversing line drive. Here, also, ground separation is not required, but optoisolators of adequate speed of response can be used. The RS-422/423 receiver specifications include a zero threshold voltage, so the specification cannot be completely met by any optoisolator due to the V_F threshold for turn-on. However, the RS-422/423 driver specifications are compatible with the termination in Figure 3.6.1-7 if R_1 is 90 to 100 ohms.

Wire cost in long simplex runs can be reduced using the scheme in Figure 3.6.4-1 to transmit both clock and data along the same line. To avoid distortion of the reconstructed clock pulse, the data transitions at the input should be permitted only while the clock is low. By using a pair of high-gain optoisolators at the output, the buffer inverters can be omitted. Omitting the buffers will invert the data but the clock output will be the same.

3.61

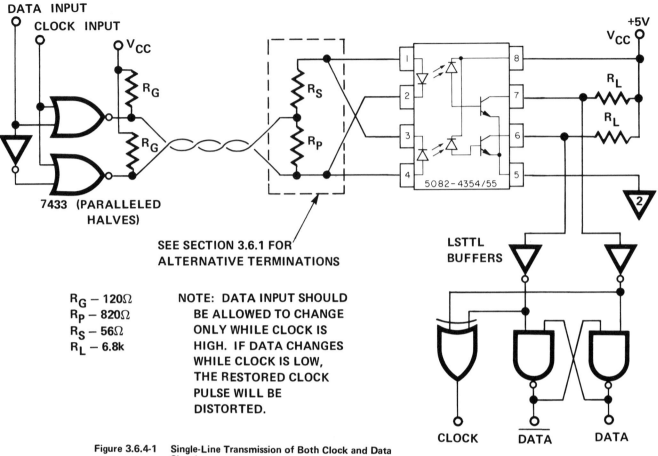

DATA INPUT

CLOCK INPUT

V_{CC}

R_G

R_G

7433 (PARALLELED HALVES)

R_S

R_P

5082-4354/55

+5V

V_{CC}

R_L

R_L

LSTTL BUFFERS

CLOCK   DATA   DATA

SEE SECTION 3.6.1 FOR ALTERNATIVE TERMINATIONS

R_G — 120Ω
R_P — 820Ω
R_S — 56Ω
R_L — 6.8k

NOTE: DATA INPUT SHOULD BE ALLOWED TO CHANGE ONLY WHILE CLOCK IS HIGH. IF DATA CHANGES WHILE CLOCK IS LOW, THE RESTORED CLOCK PULSE WILL BE DISTORTED.

Figure 3.6.4-1   Single-Line Transmission of Both Clock and Data Signals.

Duplex data transmission between two modules is usually done along a single line, with each module normally in the "receive" mode. If ground separation between the modules is required, a pair of optoisolators can be connected as shown in Figure 3.6.4-2(a) to establish two-way data transmission. In the quiescent state, i.e., if there are no externally applied gates pulling P_1 or P_2 low, they will be pulled up by pullup resistors and both isolators will be off. Applying a zero at P_1 turns on U_{12} causing a zero to appear at P_2; the output of N_2 remains low, and U_{21} remains off, so when the zero is removed from P_1 the circuit returns to its quiescent state. Similarly, if a zero is applied at P_2, U_{21} goes on and U_{12} remains off. With both isolators "off" in the quiescent state, loss of power in either module does not affect the other. Use of optoisolators with output current sinking capability high enough to accomodate more than one gate loading permits dropping the buffer inverters (B_1, B_2) and leads to the modification in Figure 3.6.4-2(b). Operation is similar, except that both isolators are "on" in the quiescent state so loss of power in one module causes a zero at P in the other. By driving the cathodes of the input diodes, rather than the anodes, NOR gates may be used in place of OR gates.

Multiplexing is commonly done by a "wired-or" bus as in Figure 3.6.4-3. Any station can drop a zero on the line, causing a zero to appear at all other stations. Obviously, such a system requires some protocol to establish transmission priorities. "Wired-or" bussing requires the logic family in any station to be compatible with that of other stations. Moreover, the common ground, shared by all stations, may allow ground loop interference and here optoisolators offer a solution by permitting ground separation at any or all of the stations. Optoisolators also allow use of any logic family in any station, with the isolator's "line" side performing the interface.

Each station to be ground-separated requires one optoisolator to receive and one to transmit. The "receive" isolator must be capable of operating with low enough input current that the required number of stations can be served, while the "transmit" isolator must be capable of sinking enough current to cause a low at all the other stations.

3.62

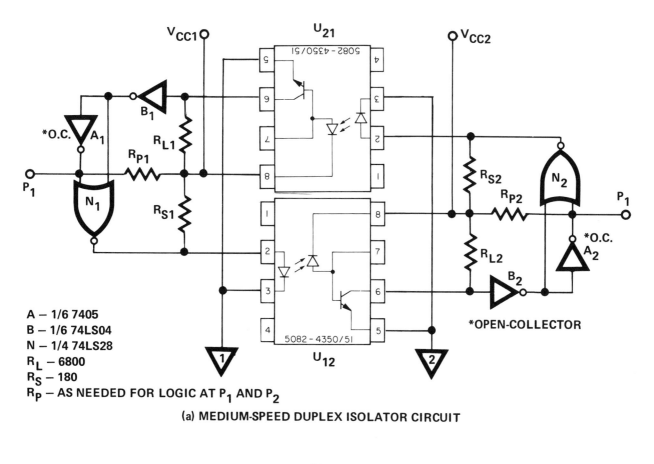

A — 1/6 7405
B — 1/6 74LS04
N — 1/4 74LS28
R_L — 6800
R_S — 180
R_P — AS NEEDED FOR LOGIC AT P_1 AND P_2

(a) MEDIUM-SPEED DUPLEX ISOLATOR CIRCUIT

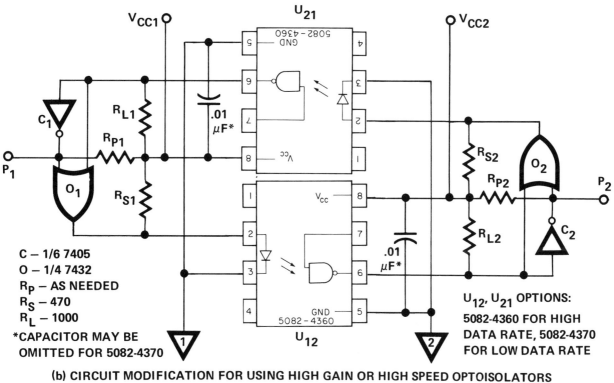

C — 1/6 7405
O — 1/4 7432
R_P — AS NEEDED
R_S — 470
R_L — 1000
*CAPACITOR MAY BE
OMITTED FOR 5082-4370

U_{12}, U_{21} OPTIONS:
5082-4360 FOR HIGH
DATA RATE, 5082-4370
FOR LOW DATA RATE

(b) CIRCUIT MODIFICATION FOR USING HIGH GAIN OR HIGH SPEED OPTOISOLATORS

Figure 3.6.4-2 Duplex Data Transmission with Optoisolators for
Ground Separation.

3.63

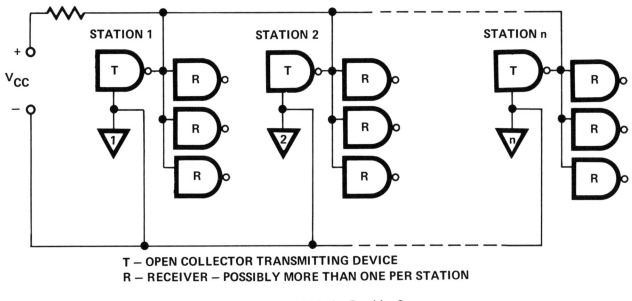

T – OPEN COLLECTOR TRANSMITTING DEVICE
R – RECEIVER – POSSIBLY MORE THAN ONE PER STATION

Figure 3.6.4-3   Wired-OR Bus Multiplexing Requiring Common
Ground for all Stations.

In the circuit of Figure 3.6.4-4(a), the buffer inverters match the isolator input/output characteristics to whatever logic family is used in the module. A low applied to P_T turns on the "transmit" isolator, causing the line voltage to drop. Whenever the line voltage drops low enough to approach the threshold voltage of the "receive" isolator, it will turn off and cause P_R to go low. Also, a low would appear at the P_R terminal of each optoisolator receiver on the line. The optional diodes D_3 and D_2 enhance the data rate by keeping the line voltage from dropping very far below the "receive" threshold. An additional advantage is that D_3 will glow to indicate which, of the several stations on the bus, is the one from which the "zero" is originating.

To have in each station a single terminal at which data can be received as well as transmitted requires some kind of anti-lock logic. For example, if in Figure 3.6.4-4(a), the terminals P_R and P_T were to be connected together, the entire system would lock low. Figure 3.6.4-4(b) shows one form of anti-lock logic; in quiescent state the line is high so V_O is low and P_{TR} is high. In the receiving mode, as the line goes from high to low, V_O goes from low to high, and P_{TR} goes from high to low, but the output of the NOR gate *remains low* because V_O is applied directly to one of its inputs, and inverted to the other. In transmitting, when a zero is *applied* at P_{TR}, since V_O is already a zero, the output of the NOR gate will rise and allow the "transmit" isolator to turn on, and its output will drop the line voltage. This will turn off the "receive" isolator, but V_O cannot rise because the Open-Collector (O.C.) anti-lock inverter, B, is holding V_O low. With the input to inverter A held low, P_{TR} will go to a high when the zero ceases to be applied.

In general, the principle of anti-lock logic is to have within the station a means to:

(a)   hold the "transmit" isolator off during reception

(b)   allow the "transmit" isolator to turn on when a zero is applied at P_{TR}

(c)   make P_{TR} return to a high when the zero ceases to be applied.

Requirements (a) and (b) are met with O.C. inverters A followed by OR-type logic (NOR or OR), however, implied in (b) is a means to hold the input of inverter A at a low when a low is applied at P_{TR}, and this would then also take care of requirement (c). Inverter B does this; note that without inverter B, applying a low at P_{TR} would initiate a low line which would raise V_O and cause the line to rise again -- oscillation would result as long as P_{TR} is held low. The O.C. inverter B works because the isolator output is also O.C. If it were not, i.e., if the isolator had active pullup, the anti-lock loop would require a gate, rather than the "wired-or" situation at the output of the isolator.

Alternatives to the use of inverter B are shown in Figure 3.6.4-4(c) and (d). In (c) the "receive" isolator's output transistor is held "on" with base current supplied from the cathode of the "transmit" isolator's input diode. This reduces component count, but the data rate capability is reduced because of the capacitance of the input diode being added at the base. In (d) the "receive" isolator is held "on" by running the entire line source current through the input diode. This reduces component count still further (eliminates D3) and has a higher data rate capability than (c), but the large current forced into the transmit isolator's input will cause a higher rate of CTR degradation, especially if the "transmit" mode occurs frequently. Both (c) and (d) will have a longer delay than (b) in recovering

3.64

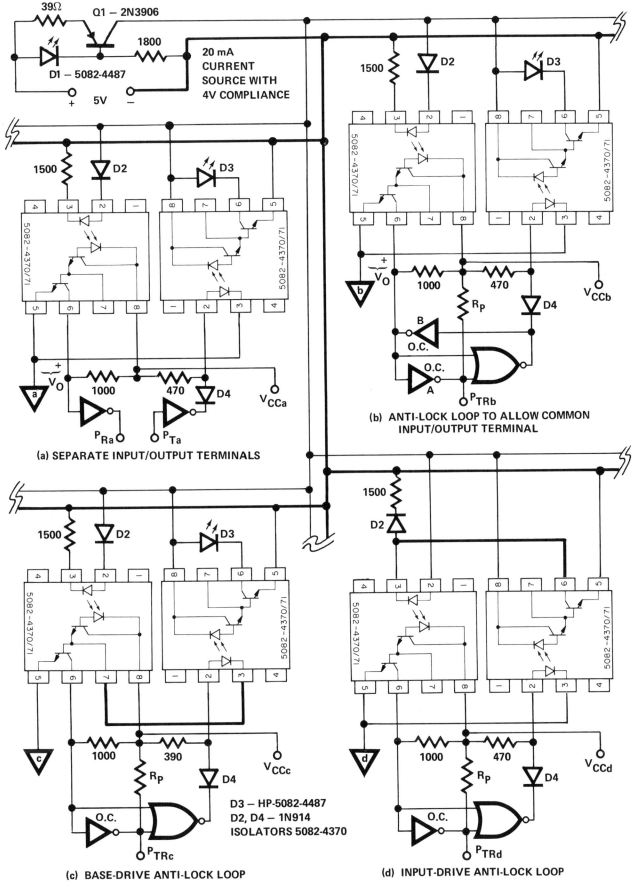

Figure 3.6.4-4 Bus Multiplexing with Optoisolators for Ground Separation of Stations.

from the "transmit" mode to the "receive" mode, due to the higher base current in the isolator output transistor during the "transmit" mode.

Other means of changing from "receive" mode to "transmit" mode are discussed in Section 3.6.6 describing the use of optoisolators with microprocessors.

Current-looping is another form of data transmission which also allows multiplex as well as simplex operation. The basic difference between current-loop multiplexing and bus multiplexing is that in a current loop each transmitting station transmits by interrupting, rather than shunting, the line current. Thus, as seen in Figure 3.6.4-5(a) and (b), the "transmit" isolator is turned "off", rather than "on" when a zero is applied at P_T (or P_{TR}).

When the "transmit" isolator is "off" the terminal voltage, V_T, rises to a value, V_{TL}, which depends on how many stations are in the loop. Worst case would be with only one station in the loop, in which case the full supply voltage would appear across the terminals of the "transmit" isolator's output. If the loop supply voltage exceeds the isolator output voltage rating, a buffer switch can be used. By using the isolator to bypass the buffer switch base current, the voltage across the isolator output never exceeds the base-to-emitter voltage of the buffer switch. Also, when the "transmit" isolator is "on", loop current may exceed the rating of the "receive" isolator input diode. A shunt regulator, such as that of Figure 3.6.1-6, can be used to bypass line current in excess of that required by the isolator.

Terminal voltage for the high state, V_{TH} limits the number of stations the loop can accomodate. It is, therefore, desirable to make it as low as possible. If separate buffer switch and input-current regulator are used, the least possible terminal voltage would be the sum of $(V_F + V_{be})$ for the regulator and $(2V_{be})$ for the buffer, and would require four transistors, two of which must carry full line current. By combining the buffer/regulator functions as in Figure 3.6.4-5(c), two transistors can be eliminated.

V_{TH} may be as low as $(V_F + V_{be} + V_{SAT,Q2}) \approx 2.5V$ but may also be somewhat higher, depending on the β of Q2, as seen in Figure 3.6.4-5(c). Such dependence on β can be eliminated by using a darlington in place of Q2, making $V_{TH} \approx 3.0V$, but not dependent on β.

Notice that with the shunt-operated buffer in Figure 3.6.4-5(c), the logic relationship between the input of the "transmit" isolator and loop is inverted from what it is in (a) or (b). Thus, the anti-lock loop of Figure 3.6.4-4(b) or (c), but not (d), can be used.

3.6.5 Telephone Circuit Applications

Any apparatus connected to telephone lines must meet two requirements. The primary requirement is that it must respond to the desired signal only. The second is that it must not interfere with the normal function of the line. Optoisolators are a nearly ideal solution to the second requirement, and proper circuit design can satisfy the first.

Ring detection requires the circuit to respond to ring signals only. These can occur over a fairly broad range of frequencies. Simple peak detection of the ac ring signal is unsatisfactory because there may be other high-amplitude voltage variations on the line, such as dialing "spikes". When a ring occurs, there are several cycles in succession at a higher repetition rate than dialing spikes. In Figure 3.6.5-1, the input diode of the optoisolator and the antiparallel diode allow ac current to flow in the coupling capacitor, C1, with each half cycle causing a peak current of about 0.5 mA in the input diode. The resulting current spikes are integrated by C2 in the output circuit. After a few cycles of the ring signal, the charge on C2 has changed far enough to turn on the output transistor. Dial spikes are ignored by this circuit. Loading of the telephone line cannot be worse than the 200k established by the series resistors. At 20 Hz, it is more nearly 450k, and at dc it is limited only by capacitor leakage.

On/off hook detection requires a means to sense current in the line, flowing in a particular direction. With no interference a single optoisolator will do, as in Figure 3.6.5-2(a). If transient common mode current is present, a pair of optoisolators with OR (or NOR) logic can be used, as in Figure 3.6.5-2(b). Common mode current in either direction can only turn on one of the two isolators -- a differential current is required to turn on both and get an output. If the common mode current is not transient, but a persistent dc level, it may prevent response of one side or the other to differential mode current. In such a situation, the circuit of Figure 3.6.5-2(c)can be used. Here the common mode current, I_{CM}, decrements the input current in one isolator as it increments the input current in the other. The isolators should be a type that is approximately linear and the threshold set at a level requiring the combined collector current resulting from differential current, I_{DM}, to obtain an output. This circuit can respond only to I_{DM} as long as $|I_{CM}| < I_{DM}$. In the LED/resistor network, the purpose of the resistor is to adjust each isolator channel to have the same ratio of collector current to line current. The LED is there to make this ratio nearly constant, down to very low current levels, without requiring a large voltage drop across the resistors.

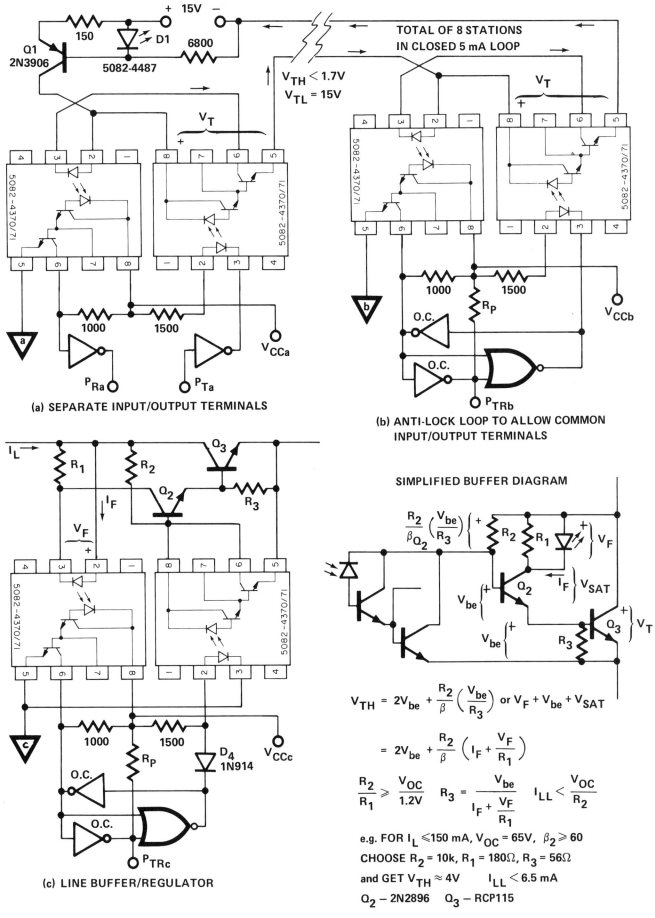

(a) SEPARATE INPUT/OUTPUT TERMINALS

(b) ANTI-LOCK LOOP TO ALLOW COMMON INPUT/OUTPUT TERMINALS

(c) LINE BUFFER/REGULATOR

SIMPLIFIED BUFFER DIAGRAM

$$V_{TH} = 2V_{be} + \frac{R_2}{\beta}\left(\frac{V_{be}}{R_3}\right) \text{ or } V_F + V_{be} + V_{SAT}$$

$$= 2V_{be} + \frac{R_2}{\beta}\left(I_F + \frac{V_F}{R_1}\right)$$

$$\frac{R_2}{R_1} \geqslant \frac{V_{OC}}{1.2V} \quad R_3 = \frac{V_{be}}{I_F + \frac{V_F}{R_1}} \quad I_{LL} < \frac{V_{OC}}{R_2}$$

e.g. FOR $I_L \leqslant 150$ mA, $V_{OC} = 65V$, $\beta_2 \geqslant 60$
CHOOSE $R_2 = 10k$, $R_1 = 180\Omega$, $R_3 = 56\Omega$
and GET $V_{TH} \approx 4V$ $I_{LL} < 6.5$ mA
$Q_2 - 2N2896$ $Q_3 - RCP115$

Figure 3.6.4-5 Current-Loop with Optoisolators to Accommodate Loop Voltage Drops.

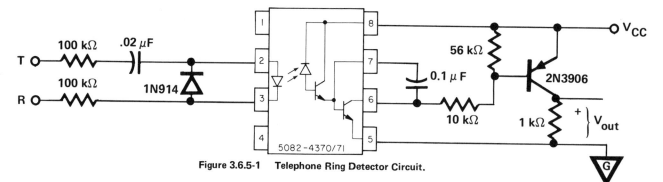

Figure 3.6.5-1   Telephone Ring Detector Circuit.

If $|I_{CM}| > I_{DM}$, four isolators are required, and all four must be adjusted to the *same ratio* of *collector current* to *line current,* as shown in Figure 3.6.5-2(d). In this scheme, I_{CM} of either polarity causes current both into and out of the collector current summing node, leaving its voltage changed only by the amount of unbalance in the collector-current-to-line-current ratio. I_{DM} of either polarity unbalances the output causing the summing node voltage to move up or down according to the polarity of I_{DM}. For the on/off hook application, only R_{L1} and the "$-I_{DM}$" comparator are needed, but for general application, this scheme, by using two load resistors and two comparators can not only detect I_{DM}, but also its polarity.

3.6.6 Microprocessor Applications

Ground looping in microprocessors has been blamed for everything from giving free "games" to blowing the entire circuit. Certainly there are static electricity hazards to circuits. These can be relieved by using optoisolators to open the ground loops, with little or no impairment of the system function. The slight additional expense is minuscule compared with the cost of troubleshooting.

The hazards usually arise where peripheral hardware (memory, I/O units, etc.) are connected with long bus runs, but can also exist in some intra-modular situations. The standard bus interface devices do offer some CMR protection, but are themselves vulnerable to large ground loop surges.

An arrangement whereby a peripheral unit can exchange information with the data bus and yet be isolated is shown symbolically in Figure 3.6.6-1 and schematic details are in Figure 3.6.6-2. As the truth table shows, the direction of data flow is basically controlled by the Receive Enable (RE) and Transmit Enable (TE) functions. With RE low, the R_n outputs remain high regardless of what B_n is; with TE low, the B_n outputs are "open collector" so the T_n inputs cannot control B_n. With both RE and TE low, the bus side of the interface unit is entirely "open collector"; not only are the outputs "open collector", but the inputs also present only "open collector" loading, thus making the current sinking efforts of other bus drivers more effective. The schematic of Figure 3.6.6-2 shows why.

When RE is high, the Q_{RE} transistors are switched on and are capable of sourcing 2 mA to any of the four isolator inputs facing the bus side; so whenever B_n is pulled low (by some other transmitter on the bus) the isolator at B_n is turned on and R_n goes low. Thus $R_n = B_n$.

When TE is high, four of the 20 mA current sources at Q_{TE} source current to the bus lines, while the fifth energizes the V_{CC} terminals of the four "transmit" isolators; so whenever T_n goes low, the corresponding isolator is turned on and B_n goes low. Thus $B_n = T_n$.

RE and TE are permitted to be simultaneously high, but if R_n is connected to T_n, the line will lock low. If it is desirable to have R_n and T_n connected so as to have a common receive/transmit terminal on the I/O side, the lock-low situation can be prevented by using an anti-lock loop, such as described in Section 3.6.4. A more common way is to use the logic driving RE and TE, shown in Figures 3.6.6-1 and 3.6.6-2. When CS (Chip Select) is high, both RE and TE are low. With CS low, either RE **or** TE is high, depending on whether R/T (Receive/Transmit) is high or low. If R/T is high, only RE can go high, and since TE is low, a low at T_n cannot be transmitted to B_n, so the line will not lock low with R_n connected to T_n. If R/T is low, only TE can go high; B_n will then go low when T_n goes low, but with RE low, Q_{RE} is off and the output transistor of the "receive" isolator remains off, so again the line will not lock low if R_n is connected to T_n.

The electrical ports on the bus side of the optoisolator interface are all compatible with the usual logic levels required at inputs of those modules not requiring isolation. The bus-side inputs require that non-isolated (as well as isolated) drivers sink no more than 2 mA, because that is all the Q_{RE} sources make available, and as little as 1 mA will do, depending on what logic current the R_n output must sink. At the 2 mA level of input current, the R_n output can handle TTL.

It is usually not necessary to install optoisolator interfacing as extensively as Figure 3.6.6-3 suggests, but it can be done. If only one or two of the modules is likely to suffer (or cause) difficulty, it is usually necessary only to isolate those

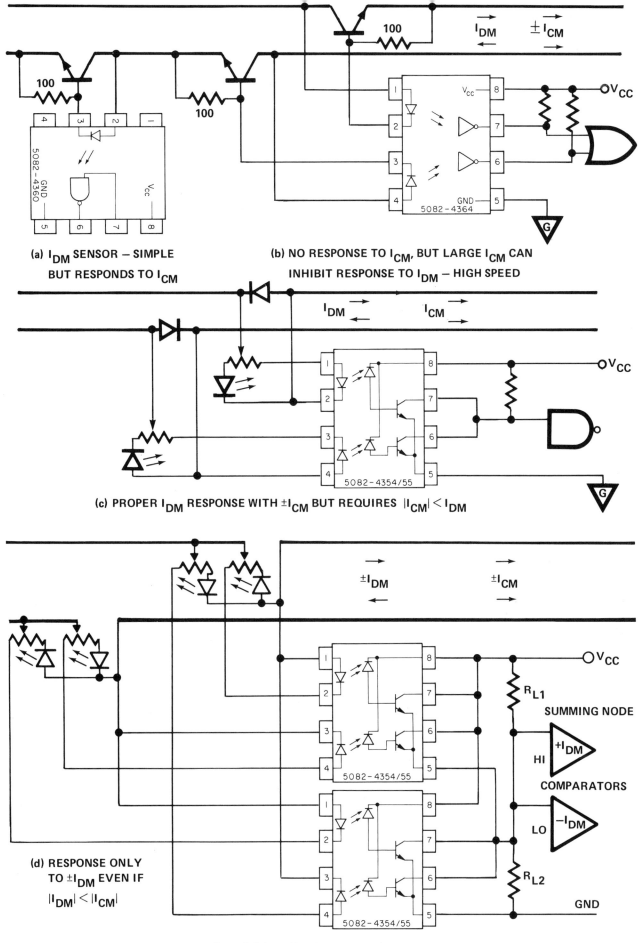

(a) I_{DM} SENSOR – SIMPLE
BUT RESPONDS TO I_{CM}

(b) NO RESPONSE TO I_{CM}, BUT LARGE I_{CM} CAN
INHIBIT RESPONSE TO I_{DM} – HIGH SPEED

(c) PROPER I_{DM} RESPONSE WITH $\pm I_{CM}$ BUT REQUIRES $|I_{CM}| < I_{DM}$

(d) RESPONSE ONLY
TO $\pm I_{DM}$ EVEN IF
$|I_{DM}| < |I_{CM}|$

Figure 3.6.5-2   Telephone Line On/Off Hook Detectors.

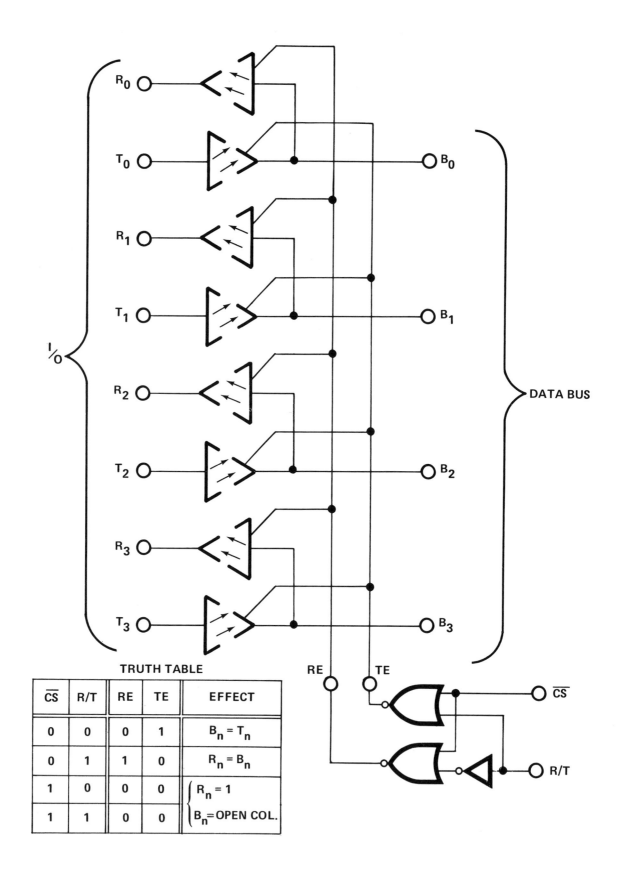

$\overline{CS}$	R/T	RE	TE	EFFECT
0	0	0	1	$B_n = T_n$
0	1	1	0	$R_n = B_n$
1	0	0	0	$R_n = 1$
1	1	0	0	$B_n = $ OPEN COL.

TRUTH TABLE

Figure 3.6.6-1   Symbolic Representation of Isolated Bus-I/O
Interface Circuit.

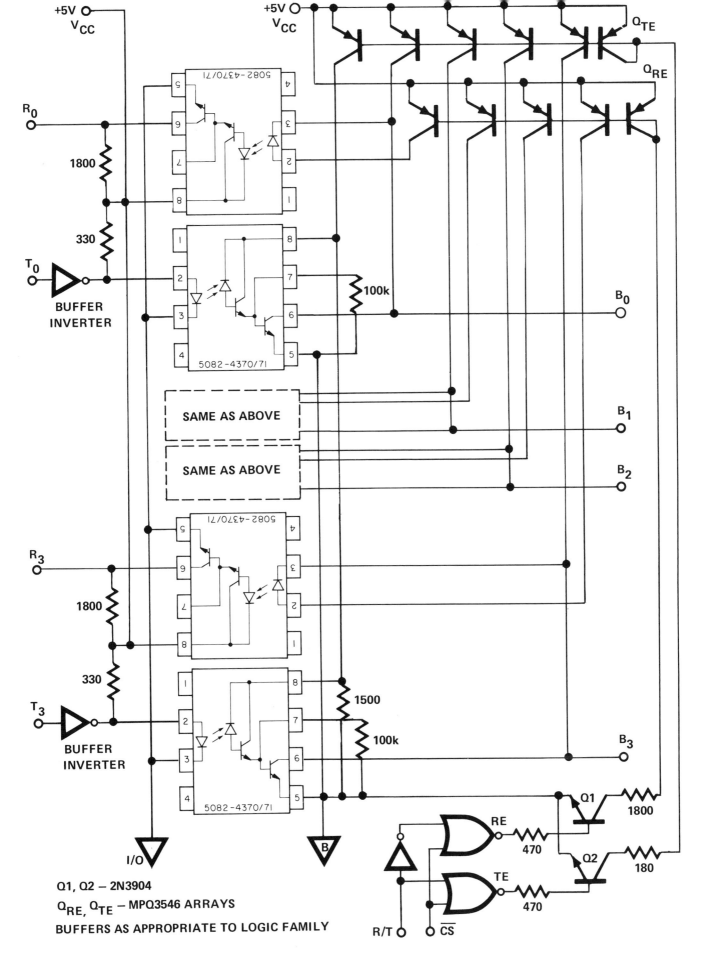

Figure 3.6.6-2 Schematic Diagram of Optoisolators wired for Bus-I/O Interface Circuit.

Q1, Q2 – 2N3904

Q_{RE}, Q_{TE} – MPQ3546 ARRAYS

BUFFERS AS APPROPRIATE TO LOGIC FAMILY

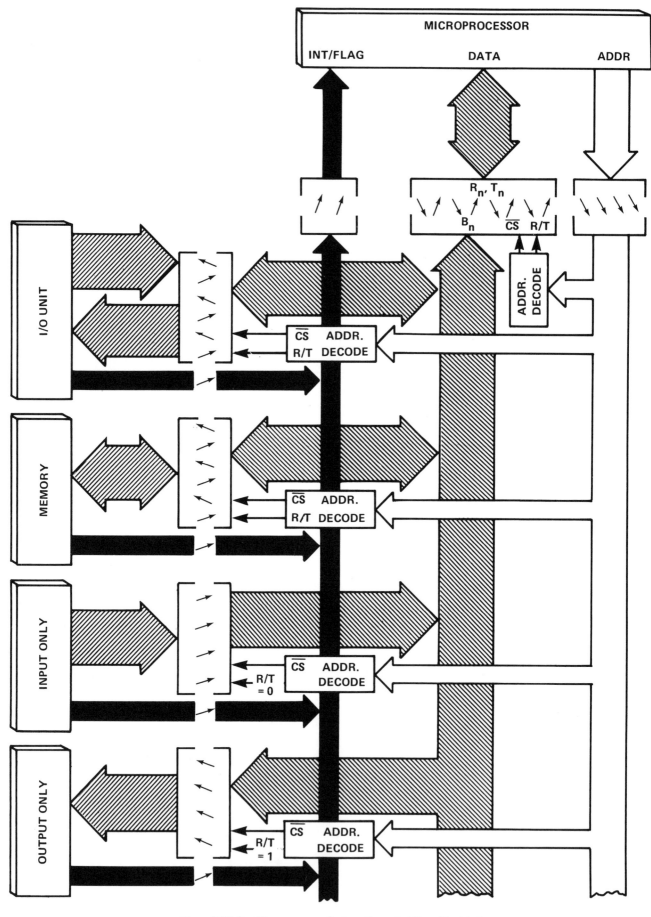

Figure 3.6.6-3   Microprocessor System; Suggested Use of Isolated
Bus-I/O Interface Circuit.

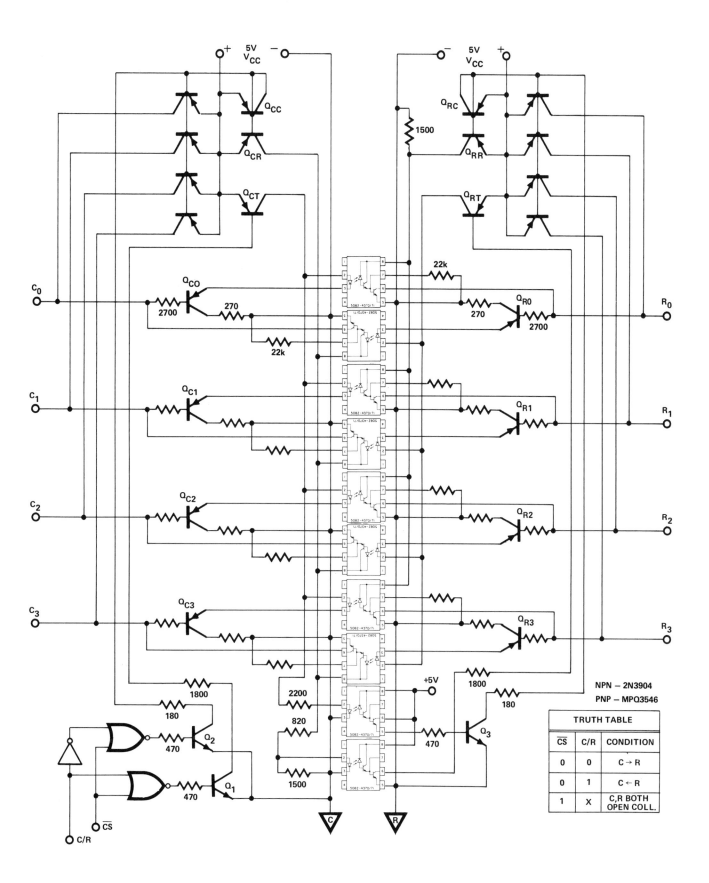

Figure 3.6.6-4 Symmetrical Four-Bit Data Bus Isolator.

3.73

particular units. If isolation is used on *any* of the wires connecting such a particular unit, then, for best isolation, *all* wires to that unit should be isolated. Any un-isolated wires allow a ground loop closure that can affect those which are isolated.

The microprocessor can also be isolated from the data bus it is controlling, as in Figure 3.6.6-3, top right. The reason the microprocessor appears to be addressing itself is that directional control of the flow of data (to or from the bus) is through inputs on the bus side of the optoisolator interface. In most cases, the $\overline{CS}$ input of the microprocessor isolator would be left low and its "address decoder" would only decode direction information from appropriate lines in the address bus. If for any reason, modules are connected on the bus lines between the microprocessor and the microprocessor isolator, care should be taken to make sure their electrical requirements are compatible with the R_n, T_n terminal characteristics. Such caution is especially required if the realization in Figure 3.6.6-2 is used because it is not symmetrical. That is, the R_n outputs do not have as large a current sinking capability as the B_n outputs.

A symmetrical interface isolator circuit is shown in Figure 3.6.6-4. Achieving symmetry requires the buffer transistor (Q_{Cn}, Q_{Rn}) to obtain 10mA of isolator input current from less than 1mA of current being sunk at the input (base). If there is assurance of at least a 10mA current sinking capability from any other drivers on the line, then Q_{Cn} and/or Q_{Rn} can be omitted, and the cathode of the isolator input diode would be connected through 270 ohms to the collector of the output isolator. This change would still leave an open-collector condition whenever $\overline{CS}$ is high.

The symmetrical interface isolator is especially useful when the data bus must serve several units that do not require isolation as well as some that do require isolation from the microprocessor but not from each other. The alternative to using the symmetrical interface isolator would be to use an interface isolator, such as the one in Figure 3.6.6-2 at each of the units requiring isolation, and this would require many more parts than symmetrical isolation.

NOTES

Section 4
Photodiodes

4.0   PHOTODIODES

This section deals with the bare fundamentals of photodiode design and construction and with the basic characteristics of PIN photodiodes. Amplifier configurations are described for linear and logarithmic response to optical signals. Also given are circuits and suggested applications for utilizing the performance features of PIN photodiodes.

4.1   Theory and Characterization

4.1.1   Photodiode Design and Construction

When a photon is absorbed in a semiconductor, an electron-hole pair is formed. Photocurrent results when the photon-generated electron-hole pairs are separated, electrons going to the N side, holes to the P side.

Separation of a photon-generated electron-hole pair is more likely to occur when the pair is formed in a region of the semiconductor where there is an electric field (see Figure 4.1.1-1). The alternative to separation is for the electron-hole pair to simply recombine, thereby causing no charge displacement and thus no contribution to photocurrent.

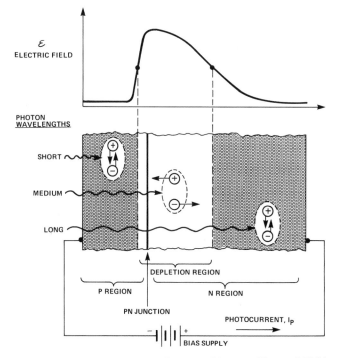

Figure 4.1.1-1   P-N Photodiode Junction; Diagram of Internal Field Effect on Detection.

Electric field distribution in a semiconductor diode is not uniform. In the regions of the P-type diffusion (front) and N-type diffusion (back) the field is much weaker than it is in the region between, known as the depletion region. For best performance, a photodiode should be made to allow the largest possible number of photons to be absorbed in the depletion region. That is, the photons should not be absorbed until they have penetrated as far as the depletion region, and should be absorbed before penetrating beyond the depletion region.

The depth to which a photon penetrates before it is absorbed is a function of the photon wavelength. Short wavelength photons are absorbed near the surface while those of longer wavelength may penetrate the entire thickness of the crystal. For this reason, if a photodiode is to have a broad spectral response (respond well to a broad spectrum of wavelengths) it should have a very thin P-layer to allow penetration of short wavelength photons, as well as a thick depletion region to maximize photocurrent from long wavelength photons.

The thickness of the depletion region depends on the resistivity of the region to be depleted and on the reverse bias.

A depletion region exists even if no reverse bias is applied. This is due to the "built-in" field produced by diffusion of minority carriers across the junction. Reverse bias aids the built-in field and expands the depletion region.

The extent of the depletion region at any voltage is larger in devices made with higher resistivity at the junction; but low resistivity is required at both surfaces for making ohmic contact to the device. P-N photodiodes, such as solar cells, are made with P diffusion into N-type material of low resistivity. In P-N photodiodes, a thin P diffusion allows good short-wavelength response, but a relatively high reverse bias is required to extend the depletion to the depth required for good long-wavelength response. A deep P diffusion degrades the short-wavelength response but lowers the bias required for good response at longer wavelengths.

Optimization of both short- and long-wavelength response at low reverse bias requires a P-I-N, rather than a P-N diode structure. A PIN diode has a thin P-type diffusion in the front and an N-type diffusion into the back of a wafer of very high resistivity silicon (see Figure 4.1.1-2). The high resistivity material between the P-type and N-type diffusions is called the intrinsic region, or I-layer.

In Hewlett-Packard PIN photodiodes, the I-layer has a resistivity so high that even at zero bias, the depletion region extends from the P-layer to approximately halfway through the I-layer. With as little as 5 volts reverse bias, depletion is extended all the way to the N-layer; this is called the "punch-through" voltage. Since breakdown voltage is over 200V, it is often desirable to operate at reverse voltages well above punch-through so as to keep the I-layer fully depleted even at high flux levels. This insures best linearity and speed of response.

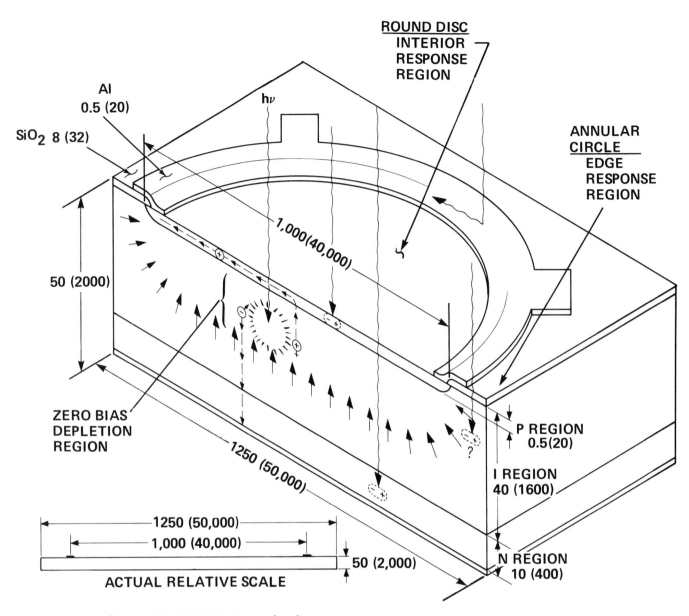

AI
0.5 (20)

SiO$_2$ 8 (32)

hν

ROUND DISC
INTERIOR
RESPONSE
REGION

ANNULAR
CIRCLE
EDGE
RESPONSE
REGION

50 (2000)

1,000(40,000)

ZERO BIAS
DEPLETION
REGION

1250 (50,000)

P REGION
0.5(20)

I REGION
40 (1600)

N REGION
10 (400)

1250 (50,000)

1,000 (40,000)

50 (2,000)

ACTUAL RELATIVE SCALE

ALL DIMENSIONS IN μm (μin)

Figure 4.1.1-2   P-I-N Photodiode; Isometric Cutaway Distorted to
Clarify Main Features.

As reverse voltage is increased, the depletion region expands sideways, extending beyond the contact rim with as little as 20 volts applied. This permits photons incident at the edge outside the ring to enter and be absorbed in the depletion region without having to penetrate the P-layer. Since the silicon dioxide surface passivation is very transparent, even to very short (UV) wavelengths, the quantum efficiency of edge response can actually exceed unity (see Figure 4.1.2-1). Quantum efficiency exceeding unity is possible because the shorter wavelength photons have energies greater than twice the bandgap of silicon.

Virtually any kind of semiconductor junction exhibits photoresponse, but most devices are designed and packaged

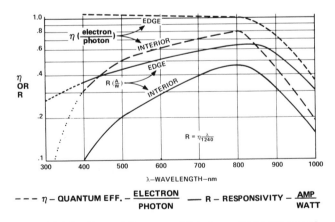

Figure 4.1.2-1   Spectral Response of Interior and Edge Regions of
HP P-I-N Photodiode.

4.2

to prevent radiant flux from interfering with their intended function. LEDs are packaged to permit radiation of flux from their junctions, so their junctions are easily exposed to radiation. Of course, they are not designed for optimal performance as photodiodes; nevertheless, they perform well enough to be useful in many applications. Their spectral response peak is at a wavelength much shorter than a silicon photodiode spectral peak, so, in the absence of filtering, they give a closer approximation to photopic (human visual) response.

4.1.2 Photodiode Characterization

A figure of merit for photodiode response can be described by the quantum efficiency. Ideally, each photon (or quantum of energy) should cause a contribution of one electron to the stream of photocurrent. *Quantum efficiency*, η_q, is therefore dimensioned as "electrons per photon".

For most engineering, a more familiar performance parameter is the *flux responsivity*, R_ϕ, which takes account of the photon energy. It is the ratio of photocurrent to spot flux:

$$R_\phi = \eta_q \frac{\lambda}{1240} = \frac{I_P}{\phi_e} \qquad (4.1.2\text{-}1)$$

where: R_ϕ = flux responsivity in amps per watt

η_q = quantum efficiency in electrons per photon

λ = photon wavelength in nanometers

I_P = photocurrent in amperes

ϕ_e = radiant flux in watts

In Figure 4.1.2-1 the spectral quantum efficiency is shown with dashed lines and the responsivity with solid lines. These same values apply to all Hewlett-Packard PIN photodiodes, regardless of their size or lens magnification. This is because responsivity is area-independent, being defined for an incrementally small spot. Note that there is a substantial difference between the responsivity of the interior region and that of the edge region (see also Figure 4.1.1-2). Actually, the edge response is obtained only when reverse bias is applied. Interior response is nearly independent of reverse bias at wavelengths shorter than peak; at longer wavelengths interior responsivity does increase slightly when reverse bias is applied — depending on how much reverse bias is applied and how high the flux level is. At reverse bias greater than punch through, the variation of interior responsivity with reverse bias is nearly zero for moderate flux level.

Another handy performance parameter is the incidance response, R_E, which takes account of the photosensitive area (or apparent area, for photodiodes with magnifying lenses). It is the ratio of photocurrent to incidance:

$$R_E = \frac{I_P}{E_e} = \int [R_\phi (A_D)] \, d A_D \approx R_\phi \, A_D \qquad (4.1.2\text{-}2)$$

where: R_E = incidance response in amps per watt per square millimeter*

E_e = radiant incidance in watts per square millimeter*

A_D = effective photosensitive area in square millimeters*

I_P, R_ϕ are defined in equation 4.1.2-1

*The SI units of area are square millimeters and square meters but square centimeters are still frequently used in describing incidance.

Incidance response describes the photodiode performance when it is floodlighted (uniform incidance over the entire device) so that edge effects are included along with interior response. The incidance response varies with reverse bias as the depletion region spreads out from the P-diffusion area. Consequently, the description of incidance response as the product of responsivity times area is a useful approximation only. The perimeter-to-area ratio decreases as photodiode area is increased, so the approximation improves as the area is increased. With very large reverse bias (>100V) the entire surface of the chip can be photosensitive — that is, the depletion region can be extended all the way from the edge of the P-diffusion to the edge of the chip.

Speed of response also depends on what areas of the photodiode are irradiated and how much reverse bias is applied, as seen in Figure 4.1.2-2. The speed observed also depends on the load resistance. Photocurrent begins to flow just a few picoseconds after flux is applied, but there is junction capacitance, package capacitance, and stray wiring capacitance to be charged. The rise/fall time constant therefore depends largely on load resistance unless the load resistance is so small that the internal resistance of the photodiode limits the speed. The internal resistance is mainly the sheet resistance of the very thin P-diffusion. With flux applied at the very center of the interior region, the resulting photocurrent encounters maximum sheet resistance and this may be as much as 50 ohms. Flux applied to the interior at portions closer to the aluminum contact ring produces photocurrent that encounters a lower sheet resistance.

Low noise is another benefit resulting from the extremely high resistivity of the I-layer in Hewlett-Packard PIN photodiodes. Diode-noise-limited operation can be achieved over a modulation frequency range extending from dc to more than 10 KHz, as seen in Figure 4.1.2-3. The horizontal dashed line is the noise calculated from the shot noise formula:

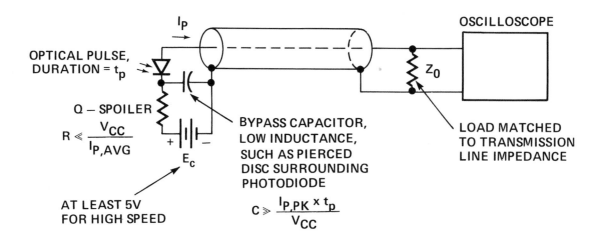

OPTICAL PULSE, DURATION = t_p

I_P

OSCILLOSCOPE

Q – SPOILER

$R \ll \dfrac{V_{CC}}{I_{P,AVG}}$

E_c

AT LEAST 5V FOR HIGH SPEED

BYPASS CAPACITOR, LOW INDUCTANCE, SUCH AS PIERCED DISC SURROUNDING PHOTODIODE

$C \gg \dfrac{I_{P,PK} \times t_p}{V_{CC}}$

Z_0

LOAD MATCHED TO TRANSMISSION LINE IMPEDANCE

(a) CIRCUIT ARRANGEMENT FOR SPEED OF RESPONSE OBSERVATION

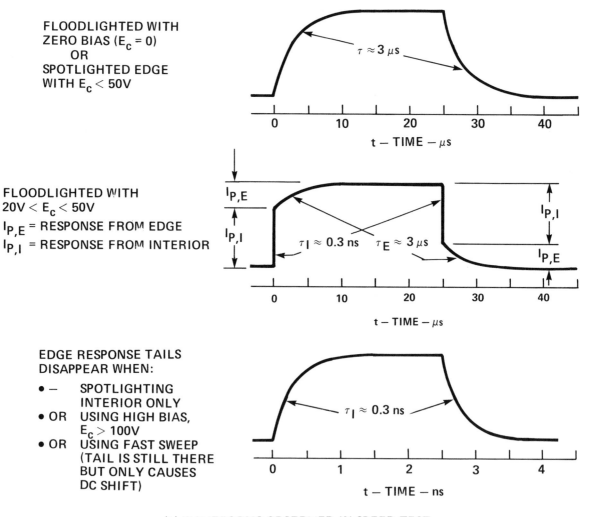

FLOODLIGHTED WITH ZERO BIAS ($E_c = 0$) OR SPOTLIGHTED EDGE WITH $E_c < 50V$

$\tau \approx 3\ \mu s$

t – TIME – μs

FLOODLIGHTED WITH $20V < E_c < 50V$

$I_{P,E}$ = RESPONSE FROM EDGE

$I_{P,I}$ = RESPONSE FROM INTERIOR

$I_{P,E}$

$I_{P,I}$

$\tau_I \approx 0.3\ ns$ $\tau_E \approx 3\ \mu s$

$I_{P,I}$

$I_{P,E}$

t – TIME – μs

EDGE RESPONSE TAILS DISAPPEAR WHEN:

- – SPOTLIGHTING INTERIOR ONLY
- OR USING HIGH BIAS, $E_c > 100V$
- OR USING FAST SWEEP (TAIL IS STILL THERE BUT ONLY CAUSES DC SHIFT)

$\tau_I \approx 0.3\ ns$

t – TIME – ns

(b) WAVEFORMS OBSERVED IN SPEED TEST

Figure 4.1.2-2   Speed of Response Observation Apparatus and Waveforms.

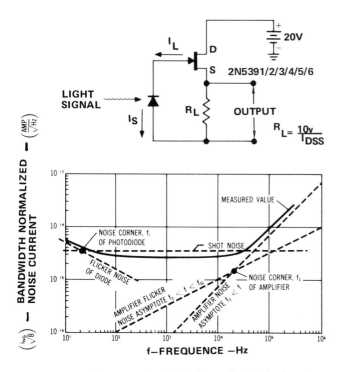

Figure 4.1.2-3   Modulation-Bandwidth-Normalized Noise Spectrum Referred to Amplifier Input.

$$\frac{I_{N,SHOT}}{\sqrt{B}} = \sqrt{2\,q\,I_{dc}} = 17.9\sqrt{I_{dc}\,(nA)}\;\left(\frac{fA}{\sqrt{Hz}}\right) \qquad (4.1.2\text{-}3)$$

where: $I_N/\sqrt{B}$ = bandwidth normalized noise current in femtoamps per root hertz

q = electron charge = 1.602×10^{-19} coulombs

I_{dc} = dc current flowing in the photodiode in nanoamps -- usually taken as total dark current.

The noise current calculated from equation 4.1.2-3 is usually greater than the measured noise at frequencies above the flicker noise corner, f_N. This is because the dc current used in the formula actually consists of two components: leakage current and junction current. Only junction current causes full shot noise; the only noise arising from leakage current is the thermal noise of the leakage resistance and the flicker noise. The two components are difficult to distinguish, so a worst case value is obtained by applying the shot noise formula (equation 4.1.2-3) to the entire dark current.

In a frequency band extending from f_1 at the low end to f_2 at the high end, the total noise current is:

$$i_N\,(f_2, f_1) = i_{NO}\sqrt{(f_2 - f_1) + f_n\,\ell n\,(f_2/f_1)} \qquad (4.1.2\text{-}4)$$

where: $i_N\,(f_2, f_1)$ = total noise current in amps

i_{NO} = bandwidth normalized noise current in amps per root hertz from equation 4.1.2-3

f_2, f_1 = upper and lower 3 dB frequency in hertz

f_N = flicker noise corner in hertz, typically less than 20 Hz

Zero bias noise is just the thermal noise in the dynamic resistance of the photodiode at zero bias. According to the thermal noise formula:

$$I_{N,\,THERM} = \sqrt{\frac{4\,k\,T}{R_{DO}}} \qquad (4.1.2\text{-}5)$$

$$= \frac{4}{\sqrt{R_{DO}\,(G\Omega)}}\;\left(\frac{fA}{\sqrt{Hz}}\right)$$

where: $I_N/\sqrt{B}$ = bandwidth normalized noise current in femtoamps per root hertz

k = Boltzmann's constant 1.38×10^{-23} joules per degree Kelvin

T = absolute temperature, taken as $290°$ K

R_{DO} = photodiode dynamic resistance at zero bias in gigaohms

The zero bias dynamic resistance is difficult to measure directly, but can be calculated from:

$$R_{DO}\,(G\,\Omega) = \left[\frac{k\,T}{q}\,(V)\right]\frac{1}{I_S\,(nA)} \qquad (4.1.2\text{-}6)$$

where: I_S = "reverse saturation current", measured as in Figure 4.2.2-2, in nanoamps

$\dfrac{kT}{q} \approx 0.025V$, defined in equations 4.1.2-3 and 4.1.2-5

Inserting R_{DO} from equation 4.2.1-6 into equation 4.2.1-5 leads to the interesting result:

$$I_{N,\,THERM} = \sqrt{4\,q\,I_S} \qquad (4.1.2\text{-}7$$

$$= 25.3\sqrt{I_S\,(nA)}\;(fA/\sqrt{Hz})$$

Comparing this with equation 4.1.2-3 suggests that a low leakage diode would be noisier at zero bias — and this is true. However, at zero bias there is no flicker noise, so this mode is usually preferred for low noise operation. Because there is no flicker noise, the noise in a bandwidth, $B = (f_2 - f_1)$ is that found from equation 4.1.2-4 with $f_N = 0$.

When signal flux is applied to a photodiode, the resulting photocurrent produces full shot noise. (This sometimes comes as a surprise when the photodiode is used with signals of low modulation depth.) Nevertheless, signal-to-noise ratio is defined as the ratio of the photocurrent when signal is applied to the noise current when there is no signal (i.e., dark):

$$\frac{S}{N} \triangleq \frac{I_P}{I_N} = \frac{\varphi \times R_\phi}{I_N} \qquad (4.1.2\text{-}8)$$

The **Noise Equivalent Power NEP** is defined as the signal flux level for which S/N = 1.0 for B = 1 Hz:

$$NEP \triangleq \frac{I_N/\sqrt{B}}{R_\phi} \left(\frac{fW}{\sqrt{Hz}}\right) \qquad (4.1.2\text{-}9)$$

Because NEP varies inversely as the responsivity, reversal of the log scale of NEP in Figure 4.1.2-4 gives the log of relative spectral response.

Under floodlight conditions, the modulation-bandwidth-normalized signal/noise ratio is:

$$\frac{S}{N} = \frac{E_e R_\phi A_D}{I_N/\sqrt{B}} \qquad (4.1.2\text{-}10)$$

The S/N ratio increases as the square root of the photodiode area, A_D, because the signal rises linearly as the area while noise current varies only as the square root of the area. A figure of merit called **Detectivity, D*** (DEE-STAR) characterizes the area normalized quality of a photodiode surface by normalizing the S/N ratio with respect to the square root of the area and the incidance (flux per unit area). Thus from equation 4.1.2-10:

$$D^* \triangleq \frac{S/N}{E_e\sqrt{A}} = \frac{R_\phi \sqrt{A_D}}{I_N/\sqrt{B}} \qquad (4.1.2\text{-}11)$$

Then from equation 4.1.2-9, the relationship between D* and NEP is derived:

$$D^* = \frac{\sqrt{A_D}}{NEP} \qquad (4.1.2\text{-}12)$$

The usual units for D* are centimeters root hertz per watt.

Because they both depend on responsivity, which varies with wavelength; and on noise, which varies with modulation frequency and bandwidth, NEP and D* are both usually given with parenthetical conditions, e.g.: D* (λ, f, Δf) or NEP (λ, f, Δf).

4.2   Photodiode Operation

4.2.1   Circuit Model

The simplest model of a photodiode is just an ordinary diode having in parallel with it a current source as in Figure 4.2.1-1. The magnitude of this current source is proportional to the radiant flux being detected by the photodiode. The polarity of this photocurrent is from cathode to anode. It is thus apparent that if zero external

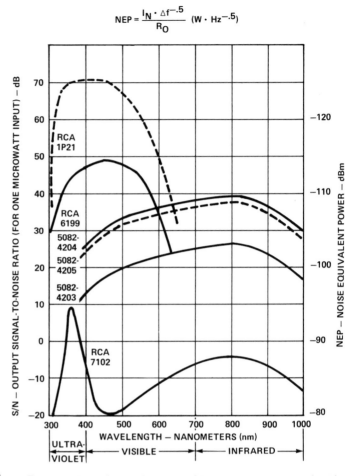

$$NEP = \frac{I_N \cdot \Delta f^{-.5}}{R_O} \quad (W \cdot Hz^{-.5})$$

Figure 4.1.2-4   Optical Spectrum of Noise Equivalent Power (NEP) for HP P-I-N Photodiodes.

bias is applied, the photocurrent will cause the anode to become positive with respect to the cathode. Part of the photocurrent will flow back through the photodiode, and part will flow in the load resistance. If the load resistance is open or extremely high, most of the photocurrent flows in the forward direction through the diode. This may seem paradoxical, but such a model does describe the results obtained in an open circuit.

Operation with zero bias is called the photovoltaic mode because the photodiode is actually generating the load voltage. Photovoltaic operation can be either linear or logarithmic depending upon the value of the load resistance. Logarithmic operation is obtained if the load resistance is very high ($>10^{11}$ ohms). Linear operation is obtained if the load resistance is very low with respect to the dynamic resistance of the photodiode. The upper limit of linear zero-bias operation is at $V_L \approx 100$ mV depending on the precision of the linearity requirement. With higher values of R_L, sensitivity can be increased for detecting very low level signals, but the dynamic range of linear response is decreased. The maximum practical value of R_L ranges from 25MΩ for large area photodiodes to 550MΩ for the smaller devices.

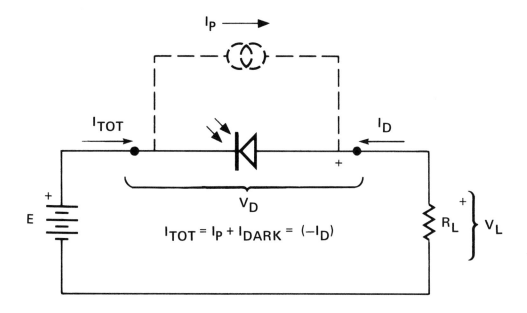

(a) CIRCUIT MODEL OF A PIN PHOTODIODE

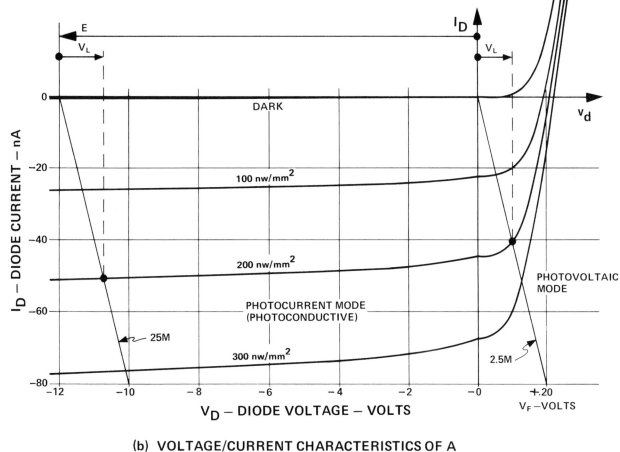

(b) VOLTAGE/CURRENT CHARACTERISTICS OF A
    PIN PHOTODIODE (NOTE CHANGE OF VOLTAGE
    SCALE THROUGH ZERO)

Figure 4.2.1-1   Electrical Characteristics of HP P-I-N Photodiode,
5082-4207.

4.7

Operation with reverse bias is called the photocurrent or the photoconductive mode. As compared with photovoltaic mode, the photocurrent mode offers:

(1) higher speed
(2) better stability
(3) larger dynamic range
(4) lower temperature coefficient
(5) improved long-wavelength response over the interior region
(6) short-wavelength (ultraviolet) response in the edge region.

The main drawback in photocurrent-mode operation is the flow of dark current, due to the reverse bias. Dark current is that which flows when no radiant flux is applied to the photodiode. Dark current flowing in R_L produces an offset voltage that varies exponentially with temperature. There is also likely to be some flicker noise due to reverse bias. Since offset and flicker noise disappear when zero bias is used, the photovoltaic mode is usually preferred unless the application requires one of the six advantages enumerated above.

4.2.2 Basic Amplifier Arrangements

For linear operation, the photodiode should be operated with as small a load resistance as possible. Figure 4.2.2-1 shows the recommended amplifier arrangement. The negative-going input is at virtual ground; the dynamic resistance seen there by the photodiode is R_1 divided by loop gain. If the op-amp has extremely high input resistance, loop gain is very nearly the forward gain of the op-amp. R_2 can be omitted if the photocurrent is reasonably high — its purpose is only to balance off the effect of offset current. As shown, the output voltage will rise in response to the optical signal. If it is preferable to have the output drop in response to optical input, then *both* the photodiode *and* E_c should be reversed. E_c may, of course, be zero. Speed of response is usually limited by the time constant of R_1 with its own capacitance, so it is improved by using a string of two or more resistors in place of a single R_1.

Logarithmic operation requires the highest possible load resistance — at least $10G\Omega$. With an FET-input op-amp, this is easily achieved as in Figure 4.2.2-2. If the offset current of the amplifier poses a problem, a resistor can be added between the positive- and negative-going inputs. Its value should not be less than $10G\Omega$ divided by loop gain. If the amplifier has a very high input resistance, loop gain is equal to the forward gain of the amplifier divided by $(1 + R_2/R_1)$ so making $R_2 = 0$ allows the smallest possible resistance between the inputs. The speed of response of this amplifier will be very low, with a time constant $\tau \approx 0.1s$. If high

speed logarithmic operation is required, it is best to use the linear amplifier of Figure 4.2.2-1 followed by a logarithmic converter.

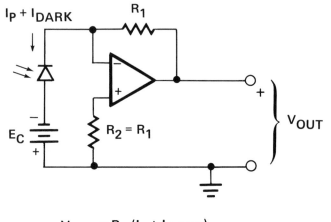

$$V_{OUT} = R_1 (I_P + I_{DARK})$$

Figure 4.2.2-1   Linear Response; Photodiode and Amplifier Circuit Arrangement.

4.2.3 Suggested Applications

PIN photodiodes are extremely stable, have a zero temperature coefficient for $\lambda < 800nm$, and operate linearly over 100 dB with <1% distortion. The success achieved with a servo system such as that in Figure 4.2.3-1 is limited more by mechanical stability of the components than by the photodiodes. The loop consisting of amplifier A_1 with optical feedback from D_1 to D_2 stabilizes the intensity I_{e2} and makes it linearly proprotional to V_{REF}. If the beam splitter is stable, then the ratio of I_{e3} to I_{e2} is

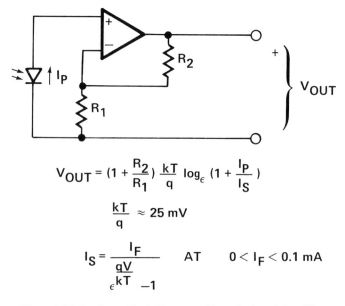

$$V_{OUT} = (1 + \frac{R_2}{R_1}) \frac{kT}{q} \log_\epsilon (1 + \frac{I_P}{I_S})$$

$$\frac{kT}{q} \approx 25 \text{ mV}$$

$$I_S = \frac{I_F}{\epsilon^{\frac{qV}{kT}} - 1} \qquad AT \qquad 0 < I_F < 0.1 \text{ mA}$$

Figure 4.2.2-2   Logarithmic Response; Photodiode and Amplifier Circuit Arrangement.

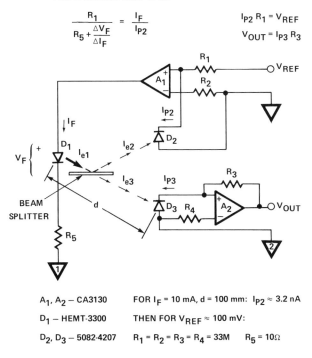

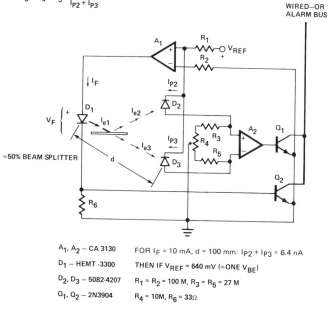

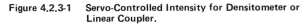

FOR BALANCED LOOP GAIN

$$\frac{R_1}{R_5 + \frac{\Delta V_F}{\Delta I_F}} = \frac{I_F}{I_{P2}}$$

$I_{P2} R_1 = V_{REF}$

$V_{OUT} = I_{P3} R_3$

$$R_1 = R_2 = \frac{V_{REF}}{I_{P2} + I_{P3}}$$

$$R_3 + R_4 + R_5 = \frac{400 \text{ mV}}{I_{P2} + I_{P3}}$$

WIRED–OR ALARM BUS

≈50% BEAM SPLITTER

A₁, A₂ – CA3130   FOR I_F = 10 mA, d = 100 mm: $I_{P2} \approx 3.2$ nA

D₁ – HEMT-3300   THEN FOR $V_{REF} \approx 100$ mV:

D₂, D₃ – 5082-4207 $R_1 = R_2 = R_3 = R_4 = 33M$ $R_5 = 10\Omega$

Figure 4.2.3-1   Servo-Controlled Intensity for Densitometer or Linear Coupler.

A₁, A₂ – CA 3130   FOR I_F = 10 mA, d = 100 mm: $I_{P2} + I_{P3} = 6.4$ nA

D₁ – HEMT -3300   THEN IF V_{REF} = 640 mV (≈ONE V_{BE})

D₂, D₃ – 5082-4207 $R_1 = R_2 = 100$ M, $R_3 = R_5 = 27$ M

Q₁, Q₂ – 2N3904 $R_4 = 10M, R_6 = 33\Omega$

Figure 4.2.3-2   Sum-&-Difference Amplifiers Improve Sensitivity in Obscuration Alarm.

constant and V_{OUT} is linearly proportional to V_{REF}. This remains true even if D_1 degrades — the A_1, D_1, D_2 servo would simply force more current to D_1 to compensate the degradation. In this mode, since ground 2 need not be the same as ground 1, it is a linear optical coupler (optoisolator). With I_{e3} held stable, the incidence at D_3 would have a linear relationship to the transmittance of any material inserted between the beam splitter and D_3, making a stable optical transmissometer. If material is instead inserted between the beam splitter and D_2, I_{P3} would rise in proportion to the attenuation. Then if, instead of a linear amplifier, A_2 were a log amplifier (Figure 4.2.2-2) V_{OUT} would be logarithmically related to the attenuation of I_{e2}, thus making an optical densitometer.

The very high dynamic resistance of PIN photodiodes raises some interesting amplifier possibilities. Sum-and-difference amplification, for example, is vastly simplified because the diode impedance is high enough to isolate the "sum" side from the "difference" side as in Figure 4.2.3.2. At the input of amplifier A_1, the photocurrents I_{P2} and I_{P3} are summed to make the loop gain that stabilizes I_{e1} greater by 6 dB than it is in Figure 4.2.2-1. The expression for the loop gain for A_1 is:

$$A_{LOOP\,1} = \left(\frac{I_{p2} + I_{p3}}{I_F}\right)\left(\frac{R_1}{R_6 + \frac{\Delta V_F}{\Delta I_F}}\right)(A_{V1})$$

$$= 1.80 \times 10^5 \text{ or } 105.12 \text{ dB}$$

assuming that A_1 has a forward gain $A_{V1} = 10^5$. The extra 6dB loop gain is probably not needed with $A_{V1} = 10^5$, but as long as loop gain is high, the cathodes of D_2 and D_3 are at virtual ground. Consequently, the voltages at the anodes of D_2 and D_3 may be as high as 100mV and still provide linear operation. Thus the resistance from either anode to ground may be as high as 100 mV/3.2 nA = 31.3MΩ.

If A_2 is connected as a linear differential amplifier, it would be wise to keep the anode voltages below 50 mV to allow 0 to 100% excursion of I_{P2} or I_{P3}. The system can then be used to give linearly the relative transmission of the I_{e2} and I_{e3} flux paths for the full range of 0 to 100% transmittance.

As shown, however, A_2 is connected as a very sensitive obscuration alarm, with R_4 adjusting the threshold. If the obscuration should increase in the I_{e3} flux path, or decrease in the I_{e2} flux path, A_2 would turn on Q_1. Should obscuration of both flux paths increase, LOOP 1 reacts to raise I_F, and at approximately 20 mA Q_2 will turn on. Also, if the efficiency of D_1 should degrade, Q_2 will turn on, indicating that something is wrong. Notice that if obscuration increases in flux path I_{e3} there will be proportional increase of the flux in path I_{e2} to keep $I_{P2} + I_{P3}$ constant. Thus the use of sum-and-difference amplifier provides differential sensing with 6dB higher gain than would be obtained by simply stabilizing flux path I_{e2} and sensing a change in I_{e3}.

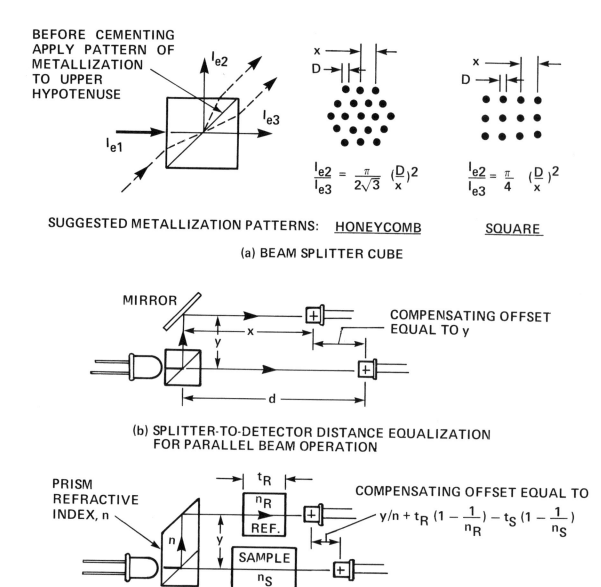

BEFORE CEMENTING APPLY PATTERN OF METALLIZATION TO UPPER HYPOTENUSE

$$\frac{I_{e2}}{I_{e3}} = \frac{\pi}{2\sqrt{3}} \left(\frac{D}{x}\right)^2$$

$$\frac{I_{e2}}{I_{e3}} = \frac{\pi}{4} \left(\frac{D}{x}\right)^2$$

SUGGESTED METALLIZATION PATTERNS: HONEYCOMB   SQUARE

(a) BEAM SPLITTER CUBE

MIRROR

COMPENSATING OFFSET EQUAL TO y

(b) SPLITTER-TO-DETECTOR DISTANCE EQUALIZATION FOR PARALLEL BEAM OPERATION

PRISM REFRACTIVE INDEX, n

COMPENSATING OFFSET EQUAL TO

$$y/n + t_R \left(1 - \frac{1}{n_R}\right) - t_S \left(1 - \frac{1}{n_S}\right)$$

(c) DISTANCE EQUALIZATION ACCOUNTING FOR VARIATION OF REFRACTIVE INDEX THROUGH BOTH OPTICAL PATHS

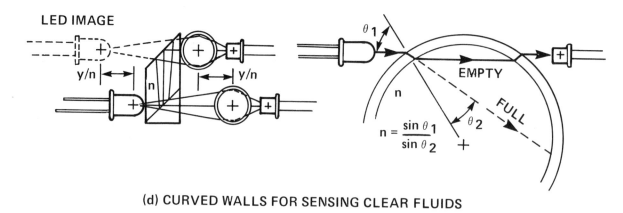

LED IMAGE

EMPTY

FULL

$$n = \frac{\sin\theta_1}{\sin\theta_2}$$

(d) CURVED WALLS FOR SENSING CLEAR FLUIDS

Figure 4.2.3-3   Optical Configurations for Use In Densitometer and Obscuration Alarm.

4.10

Some suggested optical path considerations are shown in Figure 4.2.3-3 to be used with the circuits of Figures 4.2.3-1,-2. Beam splitting to a very high precision can be done by applying partial metallization to the hypotenuse face of one prism, then cementing it to the hypotenuse face of another, as in Figure 4.2.3-3(a). The repetition interval of the metallization pattern should be very small to avoid interference patterns with LED details. To obtain parallel output beams, the incident beam can be applied parallel to the hypotenuse face, thus causing the split beams to emerge parallel (similar to a Dove prism), as shown by the dashed lines. Separation between the two beams may not be adequate, however, and the off-axis angle of incidence introduces "tilted-plate" astigmatism. Alternative means for obtaining parallel beams are shown in Figure 4.2.3-3(b,c) — these require distance compensation for beam offset, which (a) does not. All three, however, require compensation for sample length, as in (c).

Figure 4.2.3-3(d) shows how sample cells with curved walls can be applied with the densitometer or the obscuration detector circuits to respond to changes of index of refraction of clear fluids, as well as relative opacity of diffuse fluids.

NOTES

4.11

NOTES

Section 5

Displays

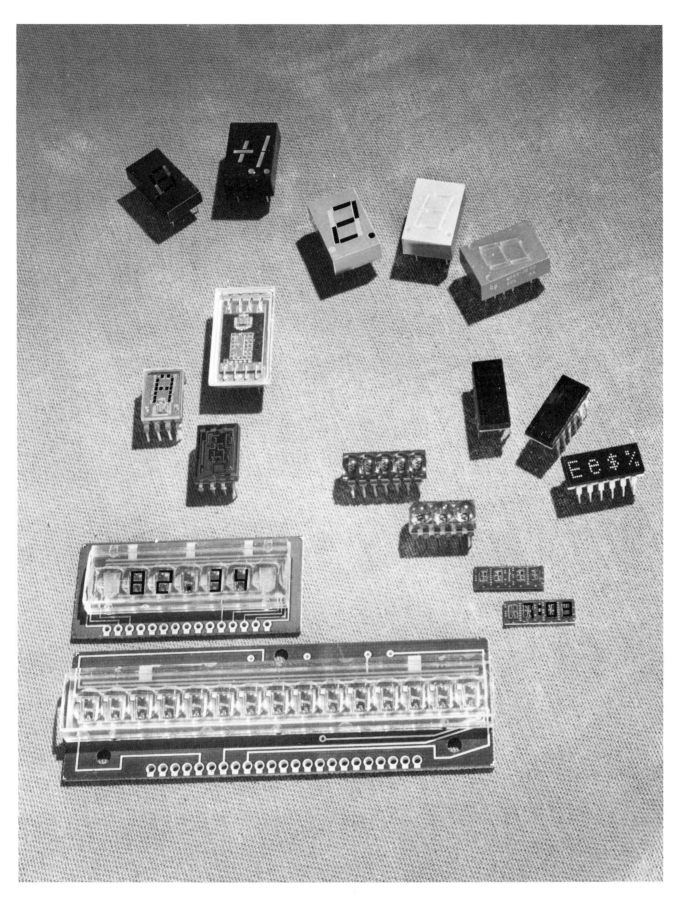

Figure 5.0.1-1 A Collection of LED Display Devices

5.0   DISPLAYS

The function of any display is to attract the attention of an observer by some recognizable arrangement of objects, illustrations, images or symbols. An electronic display has the advantage of easy control of the various light emitting or light modulating elements permitting substantial information transfer across the man-machine interface. Electronic displays may be implemented using any one of a large number of different technologies. A few of the more commonly known are:

- Incandescent
- Cathode ray tube
- Liquid crystal
- Fluorescent (operation is analogous to that of the CRT)
- Electrochromic
- Electrofluoridic
- Light emitting diode

Each of these technologies is characterized by distinct capabilities and requirements with respect to element density, color, power dissipation, packaging, unit size, drive voltages, environmental capabilities, cost, availability and other properties. The choice of any one display technology is generally a matter of optimizing some of these properties at the expense of others.

5.0.1   Types of LED Displays

The LED display technology is relatively new compared to most of the other technologies listed above. The first commercial LED products entered the marketplace in 1968. Since that time, the scope of LED display devices has expanded rapidly so that it now includes a wide variety of distinctly different products. Based on use, size, and drive requirements, these display products may be divided into four generalized categories:

1)   Displays with On-Board Integrated Circuits
2)   Strobable-Seven-Stretched Segment Displays
3)   Magnified Monolithic Displays
4)   Dot Matrix Alphanumeric Displays

Figure 5.0.1-1 depicts some of the different display products available. All of these display types have been originally developed using 655 nm red GaAsP type diode technology. Technologically, it is feasible to manufacture most of these different display categories utilizing the GaP transparent substrate colors high-efficiency red, yellow and green. However, at the present time, the availability of displays utilizing these higher technology LEDs has, for the most part, been limited to the stretched segment type displays. The GaP transparent substrate LEDs will become available in more and more products as this technology advances.

5.0.2   Display Fonts

The most prominent feature of any display product is the physical arrangement of the display elements. This arrangement, or "font" as it is commonly termed, is important not only from the standpoint of the type information which can be transmitted but also important in that it dictates the nature and complexity of the support electronics required by the display.

Figure 5.0.2-1 depicts some of the common display fonts. Fonts A and E are by far the most common in use in LED technology. Both arrangements have some positive and negative attributes. The seven segment technique is easy to utilize from an electrical standpoint, however, it is limited to displaying numeric and a small range of alphabetic information. The 5x7 dot matrix can display a wide range of numeric, alphabetic and other characters but involves some rather expensive electronic circuitry to implement the other portions of the display subsystem. The sixteen segment approach shown in Figure C has a full alphanumeric capability but has had rather limited acceptance. This font is not capable of displaying some of the specialized characters required by sophisticated systems. Font F illustrates a 9-segment display which has a somewhat more pleasing font than a 7-segment display. It can display the same numeric information and has more alphabetic capability. The dot matrix fonts in Figures B and D are abbreviated versions of the 35 dot matrix of Figure E and are used primarily to display only numeric and hexidecimal information.

5.0.3   The Display Subsystem

The visible portion of a display (be it LED or some other technology) is only a minor portion of the electronic system necessary to convert electrically coded data into a form readily comprehended by a human viewer. Figure 5.0.3-1 is a block diagram of what may be generally termed a display subsystem. A display user can purchase one, some, or all of the functions in the block diagram as a single product. It is necessary, however, that all of the functions depicted be present in some part of any information display.

By way of example, a seven segment decoder/driver may be combined with a data latch and a seven segment diode matrix to form a display subsystem. Similarly, all four of the above functions may be purchased in a single package to perform the same end result. Often, the data latch depicted in the block diagram will be present as a part of

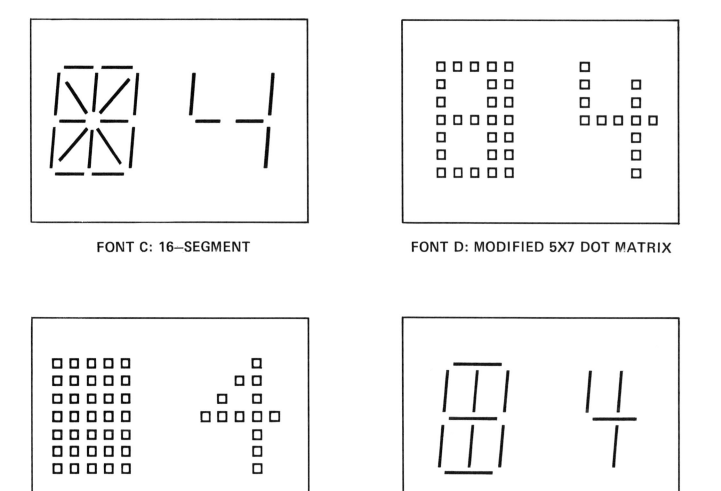

FONT A: 7–SEGMENT

FONT B: MODIFIED 4X7 DOT MATRIX

FONT C: 16–SEGMENT

FONT D: MODIFIED 5X7 DOT MATRIX

FONT E: 5X7 DOT MATRIX FONT

FONT F: 9–SEGMENT

Figure 5.0.2-1 Display Fonts Used in LED Displays.

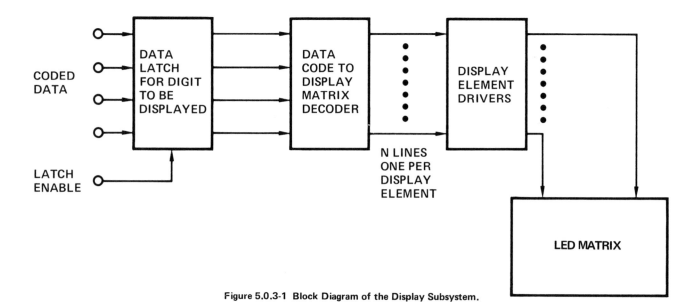

Figure 5.0.3-1 Block Diagram of the Display Subsystem.

the data source and does not have to be supplied as a portion of the display. Price, digit size, viewing conditions, power limitations, system architecture, and other considerations must be taken into account when deciding how to best design and partition a display subsystem.

5.0.4 Data Handling in Display Systems

The natural-partitioning of systems utilzing displays often leads to a situation where the display subsystem is physically separated from the data source. In these situations, it becomes necessary to transmit coded data from the data source to the display subsystem. Figure 5.0.4-1 depicts two of the most common techniques for transmitting four line BCD data plus decimal point status. The choice of technique usually depends on data source and display system requirements. For long distances, the cost of the full parallel (character parallel/bit parallel) approach can become prohibitive. If full parallel data must be converted to a character serial/bit parallel data format, the circuit depicted in Figure 5.0.4-2 may be implemented using the 3 state data bus buffers such as the National DM8095. A low true signal at the enable input of each buffer will apply the input data to the data bus. Timing of the enable signals will permit either RZ (return to zero) or NRZ (non return to zero) data formatting. In the RZ system, clocking information can be extracted at the receiving end and hence no additional transmission lines need be utilized for clocking.

5.1 Numeric Displays with an On-Board Integrated Circuit (OBIC)

For many applications, it is desirable to have the display subsystem assembled into a compact configuration and packaged separately at some distance from the data source.

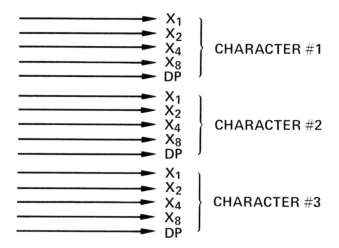

CHARACTER PARALLEL/BIT PARALLEL DATA TRANSMISSION (FULL PARALLEL)

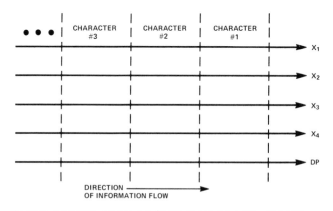

CHARACTER SERIAL/BIT PARALLEL DATA TRANSMISSION.

Figure 5.0.4-1 Two Common Techniques for Transmitting 4-Line BCD Data Plus Decimal Point Status Information.

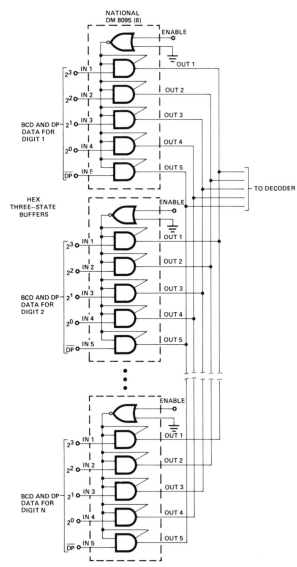

Figure 5.0.4-2 A Circuit for Character Parallel to Character Serial Data Conversion.

One possible approach is to have the display PC board contain all the necessary components that comprise the display subsystem. However, a more efficient and in many cases less costly approach is to use a display device that has an integrated circuit containing the subsystem logic functions, data latch/decoder/drivers, packaged on the same substrate as the display matrix. Such a device is termed an on-board integrated circuit (OBIC) display.

The cost per unit of an OBIC display device is more than that of a device which does not contain an integrated circuit. Even so, there are some specific advantages to an OBIC device that off-set this initial cost.

1. The pin connections between the data source and the display assembly are reduced due to the increased level of integration of the display subsystem.

2. The design of the display system is simplified and requires a minimum amount of engineering time.

3. The manufacturing costs and the time required to assemble each display system on a production basis are reduced to a minimum.

4. The reliability of the display system is significantly increased due to the reduction in circuit complexity and the reduced component count.

5. Less space is required to mount the display system assembly.

6. A pleasing font may be used, such as a modified 4x7 dot matrix, and need not conform to the font codes available with discrete commercial decoders.

The disadvantage of an OBIC display module is that as more functions are added, the module becomes unique in its configuration and a second source is usually not available.

A numeric OBIC display should provide a designer with a device that operates directly from a 5.0 volt supply, is compatible with commonly used data source logic such as TTL, has the electrical characteristic guaranteed over a wide temperature range, latches and decodes BCD data and directly drives a large pleasing character that is easily recognized at a distance.

The construction of an LED OBIC display typically begins with a ceramic substrate that has printed circuit metallization on the face and external leads brazed to the back surface. The integrated circuit and LEDs are die attached and wire bonded to the metallization. A hermetic device has a glass window covering the face of the display with a hermetic seal at the substrate rim wall-to-glass interface, as illustrated in Figure 5.1-1. A plastic device uses a coating of silicone gell to protect the wire bonds and reduce thermal stress to a minimum, then the substrate assembly is encapsulated in tinted undiffused epoxy that acts as an integral contrast filter and forms the display package.

As mentioned earlier, OBIC displays are usually unique in their design, thus the assembly technique described above will necessarily be different depending upon the package, character size and font, available functions incorporated within the integrated circuit, pin out arrangement and lead spacing. For this reason, Section 5.1.1 discusses a particular OBIC LED display family that incorporates the features described above.

5.1.1 The HP 5082-7300 OBIC Display Family

The HP 5082-7300 OBIC family offers a series of plastic or hermetic devices that display either decimal or hexidecimal numeric information. The plastic devices are epoxy encapsulated and are designed for use in consumer and commercial equipment, such as: control units for household appliances, electronic office equipment, computers and electronic measuring instruments. The hermetic devices, available in either industrial or military grade packages, are designed for use in high reliability applications involving the possible exposure to an adverse environment, such as: the instrument panels of transportation vehicles, controllers in numerical control milling machines, chemical processing equipment, military and aerospace systems. The construction of the hermetic OBIC display is pictured in Figure 5.1-1.

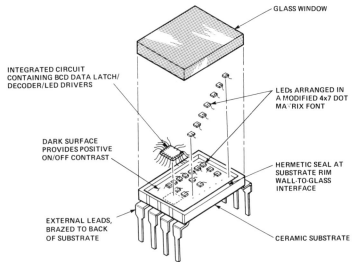

Figure 5.1-1 Construction Features of a Hermetic OBIC LED Display.

5.1.1.1 Character Font

The decimal devices are available with either right hand or left hand decimal point and display the numeric characters 0-9, and a minus sign (−). The hexidecimal devices display the characters 0-9 and A-F, and incorporate a blanking feature that allows the display to be blanked (turned off) without affecting the information stored in the data latch. A companion plus/minus one overrange display with right hand decimal point (±1.) is available in either of the hermtic or plastic packages. The character font is a 7.4mm (.29 inch) high modified 4x7 dot matrix as shown in Figure 5.1.1.1-1. The advantage of this font is that shaped characters are formed, so that a "3" does not resemble a reversed "E", a "B" looks different than an "8" and a "D" can be distinguished from a "0". The result is a font that forms characters which are easily recognized by an observer standing at a distance of 4.6 meters (15 feet).

TRUTH TABLE						
BCD INPUT				DECIMAL NUMERIC	HEXI-DECIMAL	OVER-RANGE
X_8	X_4	X_2	X_1			±1.
L	L	L	L	0	0	
L	L	L	H	1	1	
L	L	H	L	2	2	
L	L	H	H	3	3	
L	H	L	L	4	4	
L	H	L	H	5	5	
L	H	H	L	6	6	
L	H	H	H	7	7	
H	L	L	L	8	8	
H	L	L	H	9	9	
H	L	H	L	8 **2.**	A	
H	L	H	H	(Blank)	B	
H	H	L	L	(Blank)	C	
H	H	L	H	...	D	
H	H	H	L	(Blank)	E	
H	H	H	H	(Blank)	F	

NOTES:

1. H = LOGIC HIGH; L = LOGIC LOW. WITH THE ENABLE INPUT AT LOGIC HIGH CHANGES IN BCD INPUT LOGIC LEVELS HAVE NO EFFECT UPON DISPLAY MEMORY OR DISPLAYED CHARACTER.

2. TEST PATTERN FOR DECIMAL NUMERIC DEVICES.

Figure 5.1.1.1-1 Truth Table and Character Font for an OBIC LED Display.

5.1.1.2 The On-Board Integrated Circuit

The integrated circuit is composed of a 5-bit data latch for the decimal display (4-bit latch for the hexidecimal display) a decoder and LED matrix drivers. The block diagram for the integrated circuit is shown in Figure 5.1.1.2-1. On the hexidecimal devices, the decimal point is replaced by the blanking function. The data latch accepts logic high true BCD information and logic low true decimal point status. Information is loaded into the data latch when the enable input is at logic low. This information is then latched when the enable input is returned to a logic high. Changes at the data inputs will then have no effect upon the display memory or the displayed character. The blanking input controls only the LED drivers and has no effect upon the display memory. When the blanking input is at logic low, the display is on (character is illuminated) and at logic high, the display is blanked (character is not illuminated). A summary of the logic input functions is given in the truth table of Figure 5.1.1.1-1.

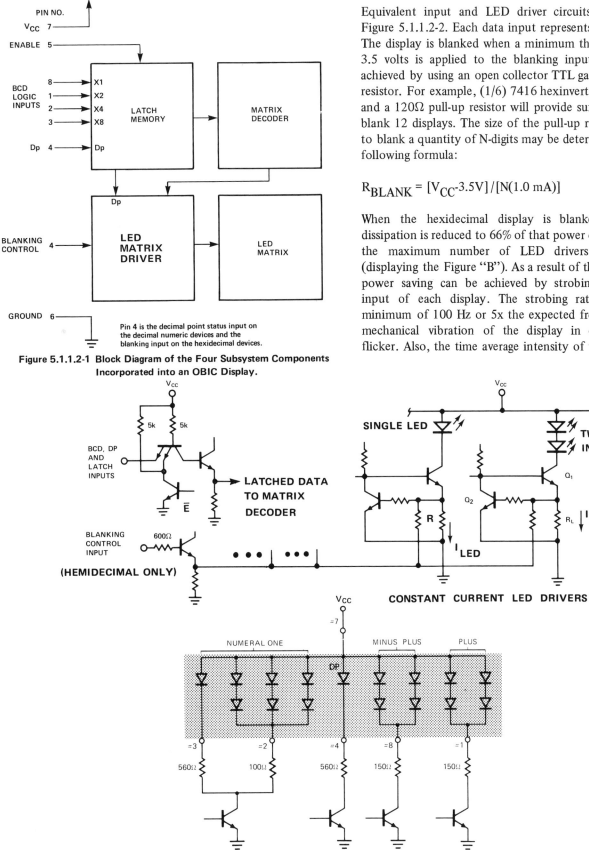

Equivalent input and LED driver circuits are shown in Figure 5.1.1.2-2. Each data input represents one TTL load. The display is blanked when a minimum threshold level of 3.5 volts is applied to the blanking input. This may be achieved by using an open collector TTL gate and a pull-up resistor. For example, (1/6) 7416 hexinverter/buffer/driver and a 120Ω pull-up resistor will provide sufficient drive to blank 12 displays. The size of the pull-up resistor required to blank a quantity of N-digits may be determined from the following formula:

$$R_{BLANK} = [V_{CC}\text{-}3.5V]/[N(1.0\text{ mA})] \tag{5.1.1.2}$$

When the hexidecimal display is blanked, the power dissipation is reduced to 66% of that power dissipated when the maximum number of LED drivers are operating (displaying the Figure "B"). As a result of this, a significant power saving can be achieved by strobing the blanking input of each display. The strobing rate should be a minimum of 100 Hz or 5x the expected frequency of any mechanical vibration of the display in order to avoid flicker. Also, the time average intensity of the hexidecimal

Figure 5.1.1.2-1 Block Diagram of the Four Subsystem Components Incorporated into an OBIC Display.

Pin 4 is the decimal point status input on the decimal numeric devices and the blanking input on the hexidecimal devices.

Figure 5.1.1.2-2 Equivalent Input and LED Driver Circuits for the Decimal, Hexidecimal and Overrange Displays.

5.7

display can be adjusted by blanking to comply with changing ambient lighting conditions. This is done by using pulse width modulation of the blanking control to vary the display on-time.

Constant current sources are used to drive the LED matrix. Each constant current driver is connected to either a single LED or two LEDs in series; the pattern is shown in Figure 5.1.1.2-3. A minimum V_{CC} of 4.5 volts is required to maintain sufficient compliance within the constant current drivers to produce an acceptably illuminated character. As V_{CC} falls below 4.5 volts, those drivers with two LEDs in series no longer have sufficient compliance to source current through two LED forward voltage drops, while those drivers servicing a single LED will still have sufficient compliance to source current through one LED forward voltage drop. The consequence of this may be insufficient illumination of the series diodes.

A real benefit to a designer using an OBIC display is the fact that the electrical parameters are guaranteed on the data sheet over a wide temperature range. This allows for worst case designing using known minimum and maximum parameter values at the temperature extremes. In operating a plastic OBIC device, it is the package temperature that is of concern. Therefore, the electrical parameters for the plastic OBIC devices are guaranteed over a case temperature range from -20°C to +85°C. The hermetic OBIC devices are characterized with respect to the operating ambient temperature, with the electrical parameters for the industrial devices guaranteed over the ambient temperature range from 0°C to +70°C, the same as for standard 7400 series TTL. The military grade devices are characterized over the full operating temperature range, guaranteeing to the designer the electrical parameters from -55°C to +100°C.

As is the case with the design of any data transmission system, a designer should take into account the timing requirements of both the data source and the OBIC display in order to insure that the correct information is latched and displayed. The timing requirements for this series of OBIC displays is given in Figure 5.1.1.2-4. These timing requirements are valid over the full operating temperature range for either a plastic or hermetic device. Enable rise times greater than 200 nsec may result in the latching of erroneous information. Data may be clocked into these OBIC displays at data rates up to 10 MHz.

5.1.1.3 Temperature Considerations

As indicated above, it is necessary to control the package temperature of a plastic OBIC device, as measured at the top of display pin number 3. For either a plastic or hermetic device, the primary thermal path for power

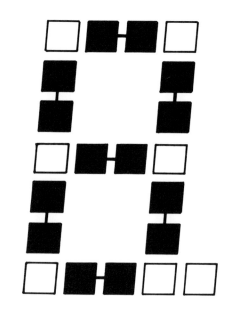

☐ = Seven constant current sources driving one LED.

◼H◼ = Seven constant current sources driving two LEDs in series. Maximum power dissipation for the decimal numeric device occurs with nine LED drivers operating, when numeral 5 and dp are displayed. Maximum power dissipation for the hexidecimal device occurs with ten LED drivers operating, when digit B is displayed.

Figure 5.1.1.2-3 Arrangement for LED Matrix Drivers.

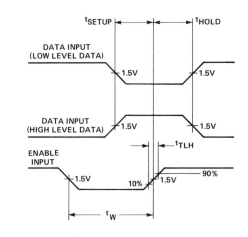

DESCRIPTION	SYMBOL	MIN.	MAX.	UNIT
Enable Pulse Width	t_W	100		nsec
Time data must be held before positive transition of enable line	t_{SETUP}	50		nsec
Time data must be held after positive transition of enable line	t_{HOLD}	50		nsec
Enable pulse rise time	t_{TLH}		200	nsec

Figure 5.1.1.2-4 Data Input and Enable Timing Requirements.

dissipation is through the device leads. The thermal resistance junction-to-lead is $15°C/W$. The maximum allowed junction temperature for a plastic OBIC device is $100°C$. Therefore, a designer should establish the thermal resistance to ambient of the display mounting structure in order to determine if the plastic OBIC devices may be operated in the maximum expected ambient temperature without heat sinking. As an example, a mounting structure consisting of DIP sockets that have been soldered onto a printed circuit board which has sufficient metallization to give a combined thermal resistance to ambient of $25°C/W$ per package, will permit the operation of plastic OBIC displays in an ambient temperature of $60°C$ without the use of external heat sinking. A mounting structure that has a thermal resistance to ambient of $35°C/W$ per package is sufficient to allow the operation of the hermetic OBIC displays in ambients up to $+100°C$, without the need of external heat sinking.

5.1.2 Intensity Control for Hexidecimal Displays Using Pulse Width Modulation

The optimal design operates a display at a light level that, when compared to the ambient or background, is bright enough to be seen without being so bright as to cause eye strain and fatigue. The fatigue becomes particularly noticeable when an operator must view an overbright display for long periods of time. An overbright display may be defined as one whose light level exceeds the optimal value by more than 10 to 1. If the ambient light conditions remain relatively constant, the selection of a good operating light level is straight forward. However, there are applications where the ambient can change by as much as 100:1, such as in the instrument panels of transportation vehicles, portable instrumentation, aircraft control tower or shipboard applications. For widely changing conditions such as these, the circuit designer may wish to add a dimming feature. Pulse width modulation can be used to control the percent time that a display is on (i.e., its "duty cycle"). By varying this duty cycle, the average light level can be varied over a wide range. Since the drive level is not changed (only the duty cycle); close matching among segments and digits is maintained even at very low light levels.

It is convenient to have the display intensity automatically track the ambient. In Figure 5.1.2-1 are two methods for pulse width modulating the displays blanking input. If a system clock is available, a photoresistor, R_x, can be used to control the duty cycle of a monostable multivibrator. The duty cycle varies from 80% in a bright ambient to 13% in a dim ambient.

If a system clock is not available, an alternative method employs a 555 timer as a free running oscillator. The photoresistor, R_x, controls the output pulse width resulting in a 90% duty factor in a bright ambient and a 2% duty factor in a dim ambient.

The frequency at which the blanking input is strobed to turn the display off and on should be fast enough, regardless of pulse width, to insure an observer sees a continuous, flicker free, display. This minimum blanking frequency should be at least 100 Hz or 5x the vibration frequency, if mechanical vibration of the display is expected.

5.1.3 Interfacing a Microprocessor to an OBIC Numeric Display

An OBIC decimal or hexidecimal display can interface very easily to a microprocessor system. Since the OBIC numeric display contains a latch, the only extra circuitry that is required to interface to a microprocessor is some external gating. An eight bit data word can be configured as two BCD or hexidecimal characters. This allows the desired display information to be stored compactly in a RAM and easily manipulated by the microprocessor. Since the OBIC display is dc driven, the microprocessor needs to update the display only when the information needs to be changed. Thus, the OBIC display requires only a minimum amount of microprocessor time. If hexidecimal information is to be displayed, an HP 5082-7340 display or equivalent should be specified. This display has a blanking input which does not affect the latch. When decimal information is to be displayed, an HP 5082-7300 display or equivalent should be specified. This display has a latched decimal point and has the capability to display a minus sign, blank, or lamp test.

If a moveable decimal point on an OBIC decimal display or selective digit blanking on an OBIC hexidecimal display is required, some additional circuitry may be necessary. Since the OBIC hexidecimal display does not have a latched blanking control, an external latch is needed. The eight bit data word can be configured as one character (four bits) plus one additional bit for a decimal point or blanking control. An example of this technique is shown in Figure 5.1.3-1. Each digit would be addressed by a different eight bit address code. With the Intel 8080A microprocessor, an output instruction is used to load the contents of the accumulator into the specified OBIC numeric display. In this example, the lowest order four bits (D_3, D_2, D_1, D_0) are used for the numeric information and D_4 is used for blanking control or decimal point control. When it is desired to pack two hexidecimal or BCD characters in a single eight bit word, a couple of bits from the address bus can be decoded as decimal point or blanking information. An example of this technique is shown in Figure 5.1.3-2. Each pair of digits is addressed by four possible address codes: n, n+1, n+2, or n+3. The two lowest order address bus lines, A_1 and A_0 are decoded as decimal point or blanking information. The second byte of the output instruction will determine which pair of characters is to be updated and whether either digit is to be blanked or have its decimal point turned on. When the decimal point inputs or the blanking inputs do not have to be controlled by the microprocessor, only a single output address code would uniquely specify each pair of digits. This can simplfy programming.

An OBIC hexidecimal display can also be used to simplify microprocessor program debugging. Hexidecimal information is much easier to decode than binary information by the software designer. Figure 5.1.3-3 shows a complete debugging system for an Intel 8080A

microprocessor. At the beginning of each machine cycle, the 8080A microprocessor strobes eight bits of information into a status latch. These eight bits uniquely specify which one of ten machine cycles will be executed by the microprocessor. The sixteen address lines specify where the information is to be read or written. The eight data lines hold the contents of that information. If the microprocessor is operated in a single step mode, then after the end of each machine cycle, one of the ten LED lamps will indicate which machine cycle was last executed and the hexidecimal displays will indicate what was on the address and data buses during that machine cycle. In this example, a "Fetch" machine cycle was just executed and a JMP instruction was read from $(00AF)_{16}$. Other microprocessor systems can be debugged in a similar fashion, by using the appropriate control signals from the microprocessor to load the contents of the address and data buses into OBIC hexidecimal displays and decoding the other control signals to indicate what type of machine cycle was executed.

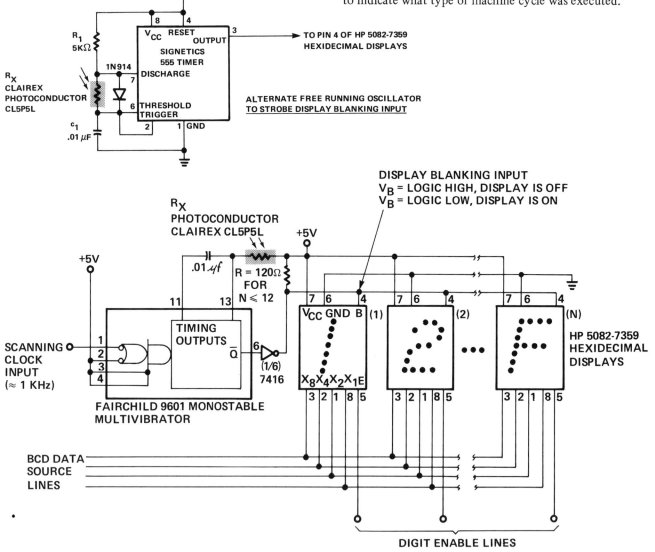

Figure 5.1.2-1 An Intensity Control Circuit for Hexidecimal Displays.

HP 5082-7340 USED AS HEXIDECIMAL OUTPUT FOR INTEL 8080A MICROPROCESSOR

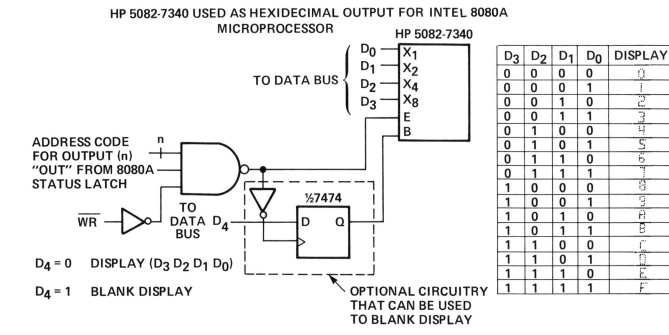

D_3	D_2	D_1	D_0	DISPLAY
0	0	0	0	0
0	0	0	1	1
0	0	1	0	2
0	0	1	1	3
0	1	0	0	4
0	1	0	1	5
0	1	1	0	6
0	1	1	1	7
1	0	0	0	8
1	0	0	1	9
1	0	1	0	A
1	0	1	1	B
1	1	0	0	C
1	1	0	1	D
1	1	1	0	E
1	1	1	1	F

$D_4 = 0$ DISPLAY (D_3 D_2 D_1 D_0)

$D_4 = 1$ BLANK DISPLAY

HP 5082-7300 USED AS DECIMAL OUTPUT FOR INTEL 8080A MICROPROCESSOR

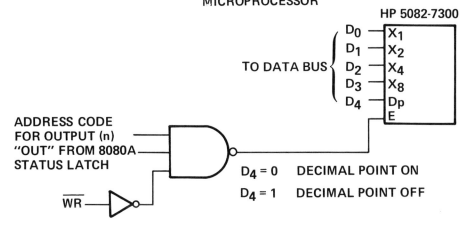

$D_4 = 0$ DECIMAL POINT ON

$D_4 = 1$ DECIMAL POINT OFF

D_3	D_2	D_1	D_0	DISPLAY
0	0	0	0	0
0	0	0	1	1
0	0	1	0	2
0	0	1	1	3
0	1	0	0	4
0	1	0	1	5
0	1	1	0	6
0	1	1	1	7
1	0	0	0	8
1	0	0	1	9
1	0	1	0	B
1	0	1	1	(BLANK)
1	1	0	0	(BLANK)
1	1	0	1	.....
1	1	1	0	(BLANK)
1	1	1	1	(BLANK)

Figure 5.1.3-1 Interfacing Obic Numeric Displays to a Microprocessor.
Five Bit Word Specifies BCD Data and Blanking/DP.

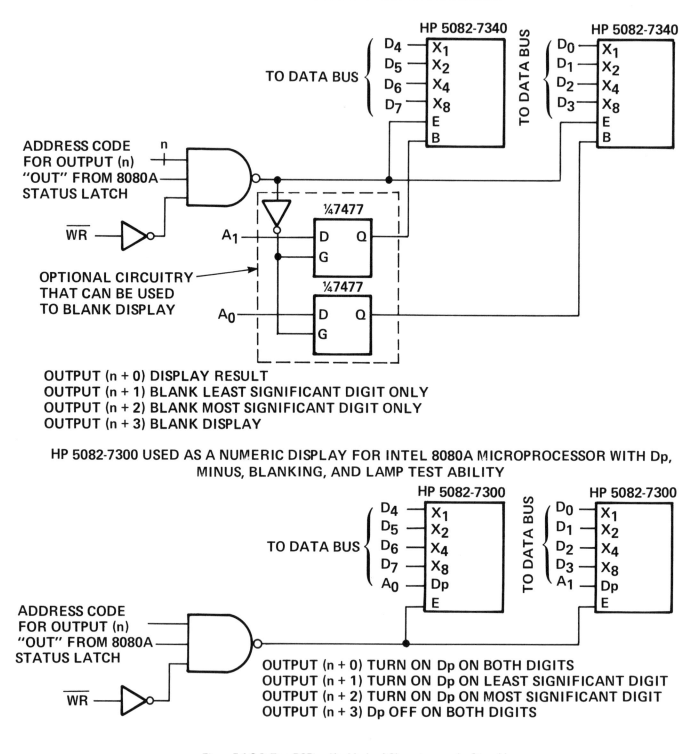

HP 5082-7340 USED AS NUMBERIC/HEXIDECIMAL OUTPUT FOR INTEL 8080A MICROPROCESSOR

OUTPUT (n + 0) DISPLAY RESULT
OUTPUT (n + 1) BLANK LEAST SIGNIFICANT DIGIT ONLY
OUTPUT (n + 2) BLANK MOST SIGNIFICANT DIGIT ONLY
OUTPUT (n + 3) BLANK DISPLAY

HP 5082-7300 USED AS A NUMERIC DISPLAY FOR INTEL 8080A MICROPROCESSOR WITH Dp, MINUS, BLANKING, AND LAMP TEST ABILITY

OUTPUT (n + 0) TURN ON Dp ON BOTH DIGITS
OUTPUT (n + 1) TURN ON Dp ON LEAST SIGNIFICANT DIGIT
OUTPUT (n + 2) TURN ON Dp ON MOST SIGNIFICANT DIGIT
OUTPUT (n + 3) Dp OFF ON BOTH DIGITS

Figure 5.1.3-2 Two BCD or Hexidecimal Characters can be Stored in
an Eight Bit Word. Address Bus is Decoded as
Blanking/DP.

5.12

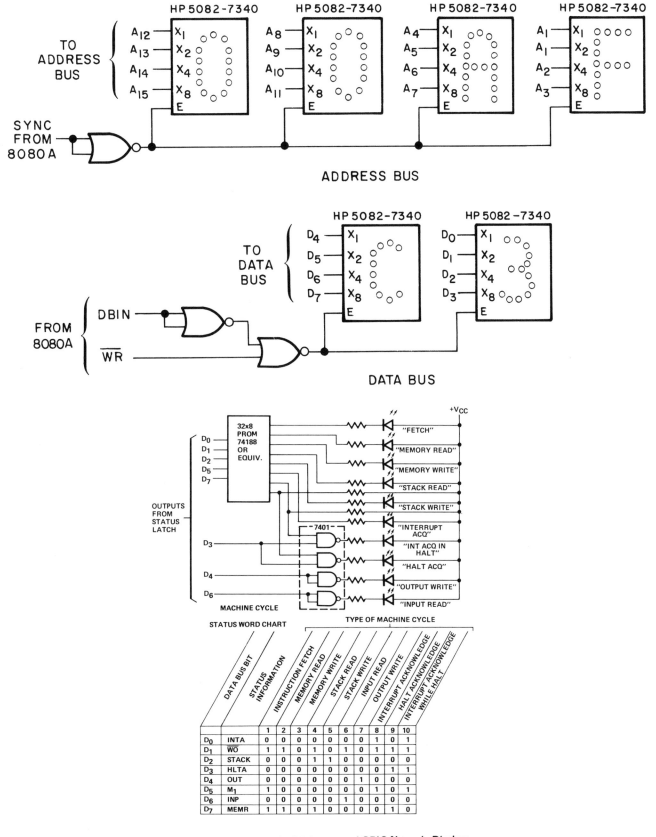

Figure 5.1.3-3   Use of LED Lamps and OBIC Numeric Displays
To Aid In Debugging A Microprocessor System.

5.2 Strobable-Seven-Stretched Segment Displays

The first seven segment LED displays to enter the marketplace were generally single digit devices with a small character height constructed on a thick-film ceramic substrate with round pins, either swaged or epoxyed into holes in the ceramic. Large rectangular slivers of GaAsP were either epoxy or eutectic die attached to the substrate in the seven segment format. Standard IC bonding techniques were then used to connect the anodes of the diodes to other traces on the substrate. The entire assembly was then encapsulated in a clear epoxy to provide protection for the bonds and chips. This construction technique requires large amounts of relatively expensive GaAsP, making the cost of the final product substantially higher than other display technologies. In addition, poor alignment of the diodes, visible metallization patterns both on the diode and on the substrate, and other problems, made most of these larger digits less than desirable in appearance by comparison to other designs.

Recently, a new stretched segment assembly technique has been developed which combines advantages of substantially lower manufacturing costs with a markedly improved display appearance. In this stretched segment design technique, a lead frame is utilized as the mechanical support for and makes electrical contact with the LED chips. Instead of encapsulating the assembly as before, a cone shaped reflecting cavity is cast inside a rectangular package above each LED, using glass filled epoxy. The top of this cavity forms the stretched segment. Light from an LED is uniformly emitted from the stretched segment resulting in a pleasing uniformly lighted rectangular area. Utilizing this technique, characters of larger size in the new high-efficiency red, yellow and green colors, as well as standard red, are assembled using LED chips an order of magnitude smaller than was used in the earlier design. Thus, the stretched segment display offers a variety of colors, sizes, improved appearance, good on/off contrast, and improved reliability at lower cost.

5.2.1 Description

The large digit LED display devices are manufactured using the concept of stretching the light from an LED by diffusion and reflection. This is accomplished by encapsulating the LED into a special cavity. This cavity is a rectangular cone with the small end down at the LED, and the large top end forming a segment of the display as shown in Figure 5.2.1-1. The area of the top end surface of the cavity may be 30 times or more the surface area of the LED. The emitted light from the LED is diffused as it passes through to the top surface of the cavity producing an evenly lighted segment. Optically, this segment will appear as an area source with a near lambertian radiation pattern.

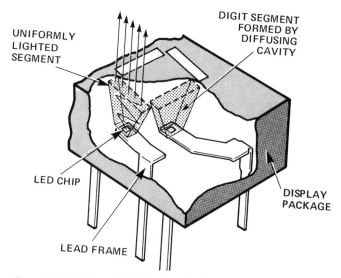

Figure 5.2.1-1 Assembly Technique of a Stretched Segment Display.

5.2.1.1 Construction

Figure 5.2.1.1-1 is a cross section of a strobable-seven-stretched-segment display. The Lexan® housing, called a "scrambler", forms the display package and contains the segment cavities. The sides of the cavities are made to have as near perfect reflection as possible to reduce light loss. The external top and side surfaces of the scrambler are coated with an epoxy paint to match the color of the LED. The colored scrambler helps provide good segment on/off contrast.

The lead frame base metal is covered with a 50 microinch copper flash and a 200 microinch silver plating. The pin size and spacing match that of a standard 14 pin DIP.

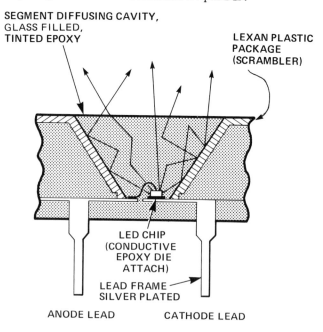

Figure 5.2.1.1-1 Cross-Sectional Diagram of One Segment in a Stretched Segment Display.

The construction of a stretched segment display is as follows: the LED chip is die attached to the cathode pin of the lead frame with an electrically conductive epoxy. The top contact of the LED is wire bonded to the anode pin; a ball bond is formed on the LED top contact and a wedge bond is formed on the lead frame.

The lead frame is inserted into the scrambler and the scrambler is then filled with a glass filled tinted epoxy. This glass filled epoxy forms the rectangular cone segments, the glass acting as the light diffusing agent. The tinting in the epoxy works in conjunction with the colored scrambler to enhance segment on/off contrast.

5.2.1.2 Data Sheet Parameters

The data sheet for a stretched segment display is divided into the following basic sections:

- Package dimensions and circuit diagrams
- Absolute maximum ratings
- Electrical/optical characteristics
- Operational curves
- Operational considerations

The data sheet is configured to give a display designer as much helpful information as possible. The operational considerations gives application information covering such topics as determining V_F, selecting a set of strobed operating conditions, calculation of time average luminous intensity, contrast enhancement suggestions and post solder cleaning suggestions.

The absolute maximum ratings give those limits beyond which the device should not be operated. The device is capable of being operating at the maximum ratings, however, for best reliability it is suggested that the device be operated below the maximum ratings.

The maximum dc forward current and dc power dissipation are the maximum dc drive conditions, not temperature derated, and are used in determining maximum strobed operating conditions. These limits have been set based on a projected operating life up to 100K hours. The maximum temperature limits are based on the display package capability.

The electrical/optical characteristics list typical values for various device parameters at an ambient temperature of $T_A = 25°C$.

The operational curves are used to determine display drive conditions.

5.2.2 Determining Display Drive Conditions

The process of establishing display drive conditions involves the examination of LED junction temperature, based on derated operating currents and device thermal resistance, and the calculation of the LED forward voltage. If the display is strobed, the process also includes the determination of refresh rate, duty factor, pulse width and peak current.

5.2.2.1 Maximum DC Current (I_{DC}), Junction Temperature (T_J) and Package Thermal Resistance (θ_{JA})

The end of operating life of an LED is defined to be when the light output has degraded to 50% of its initial value. The maximum dc current is established from reliability testing results which predict an operating life of 100K hours when the LED is continuously operated at that dc current in an ambient temperature of 25°C.

To keep within the package temperature limitations, the maximum junction temperature for stretched segment displays is T_J max = 100°C. This value of T_J is based on a thermal resistance junction-to-ambient of θ_{JA} = 100°C/W per package for a device soldered into a typical PC board. One exception to this is the .3 inch common cathode display. The thermal resistance per package is 110°C/W and the T_J max = 105°C.

The junction temperature is calculated assuming all eight segment are illuminated.

$$T_J = T_A + \Delta T_J \qquad (5.2.2.1)$$
$$\Delta T_J = \theta_{JA} \text{ (8 SEGMENTS) } (P_{SEGMENT})$$
$$\Delta T_J = \theta_{JA} \text{ (8) } [I_{AVG} (V_{TURN-ON} + I_{PEAK} R_S)]$$

Example: The maximum average LED junction temperature of a high-efficiency red device, that has a maximum dynamic resistance of 33Ω, being strobed at a refresh rate of 1 kHz with a peak current of 60 mA and an average current of 12 mA in an ambient of 50°C is 84°C.

$$T_J = 50°C$$
$$+ (100°C/W) \text{ (8) } [.012A (1.55V + .060A (33\Omega))]$$
$$T_J = 84°C$$

5.2.2.2 Forward Voltage (V_F) and LED Dynamic Series Resistance (R_s)

When calculating LED power dissipation and junction temperature, the variation in forward voltage with forward current must be taken into account. Each data sheet contains a graph of the typical variation of V_F with respect to I_F (the familiar diode curve). A linear approximation can be derived from this curve to form an equivalent circuit for an LED, see Figure 5.2.2.2-1.

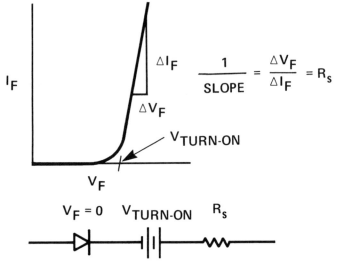

Figure 5.2.2.2-1 Equivalent Circuit for an LED.

The typical forward voltage for an LED may be scaled off of I_F vs. V_F curve or calculated. For example, the forward voltage of a standard red display may be calculated from the following formula, $V_{turn-on} = 1.55$ volts:

$$V_F = 1.55V + I_{PEAK} R_S \qquad (5.2.2.2-1)$$
$$\text{where:} \quad R_{S\,TYP} = 3\Omega$$
$$R_{S\,MAX} = 7\Omega$$

The forward voltage characteristics of a GaP transparent substrate device varies considerably from lot to lot. This is due to the wide range of resistivity throughout the GaP crystal and the variation in conductance of the reflective back contact. Figure 5.2.2.2-2 shows the measured forward voltage variation for the high-efficiency red display. From this kind of examination, the following table is derived:

Device	$R_{S\,MIN}$	$R_{S\,TYP}$	$R_{S\,MAX}$	$V_{TURN-ON}$
High-Efficiency Red	17Ω	21Ω	33Ω	1.55V
Yellow	15Ω	25Ω	37Ω	1.60V
Green	12Ω	19Ω	29Ω	1.75V

The maximum forward voltage for a given peak forward current may be calculated as follows:

$$V_{F\,MAX} = V_{TURN-ON} + I_{PEAK} R_{S\,MAX} \qquad (5.2.2.2-2)$$

The typical forward voltage is more accurately calculated by measuring the typical R_s value from the V_F point at $I_F = 5$ mA.

$$V_{F\,TYP} = V_{5mA} + R_{S\,TYP} (I_{PEAK} - 5\,mA) \qquad (5.2.2.2-3)$$

Where $V_{5\,mA}$ and R_s are obtained from the following table:

Device	V_{5mA}	$R_{S\,TYP}$
High-Efficiency Red	1.65V	21Ω
Yellow	1.75V	25Ω
Green	1.85V	19Ω

Example: For a high-efficiency red device operating at 60 mA peak, $V_F = 2.81$ volts:

$$V_{F\,TYP} = 1.65V + 21\Omega\,(.060-.005A) = 2.81V$$

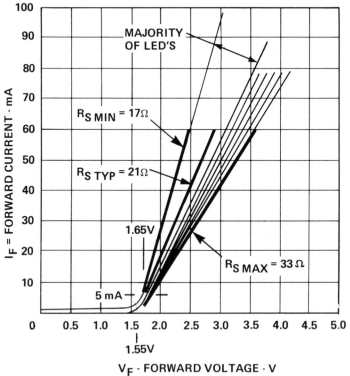

Figure 5.2.2.2-2 Measured Forward Voltage Variation for a High-Efficiency Red Display (8 Segments x 25 Devices = 200 LEDS)

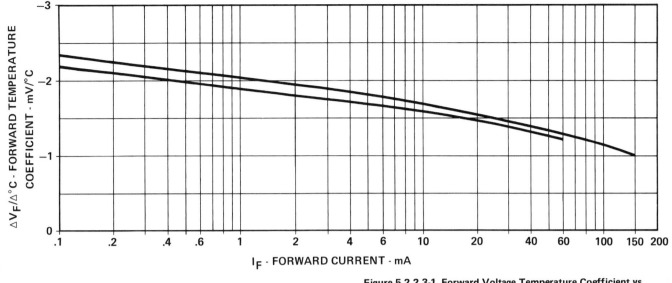

Figure 5.2.2.3-1 Forward Voltage Temperature Coefficient vs Forward Current.

5.2.2.3 Variation of Forward Voltage with Change in Temperature

The forward voltage temperature coefficient ($\Delta V_F/°C$ = $-mV/°C$) is given in Figure 5.2.2.3. The magnitude of this coefficient decreases in value with increasing forward current. The change in forward voltage with temperature should be taken into account when doing a worst case design.

5.2.2.4 Operational Curves for Strobing an LED Device

A power limiting criterion needs to be established in order to define the tolerable limitations with which an LED device may be operated in the strobed mode. The criterion which has been selected is the LED junction temperature. Specifically, the maximum tolerable strobe mode operating conditions are limited to a peak current (I_{PEAK}), pulse duration (t_p) and refresh rate (f) which produce a peak junction temperature ($T_{J\ PEAK}$) equal to the junction temperature obtained when the device is driven by the maximum temperature derated DC current ($I_{DC\ MAX}$). A graph of the curves which define the maximum tolerable strobe mode operating conditions for a high-efficiency red display is reproduced in Figure 5.2.2.4-1.

The thermal time constant of an LED mounted on a lead frame is about one millisecond. For this reason, refresh rates less than 1 KHz produce significant peaks and valleys in the absolute junction temperature with the average junction temperature less than the dc junction temperature. As the refresh rate increases, the average junction temperature approaches the dc junction temperature. This is illustrated by the constant duty factor line drawn across the face of Figure 5.2.2.4-1. As the refresh rate increases, the allowable operating peak current per segment also increases. It is, therefore, worthwhile to use a refresh rate of 1 KHz or faster.

5.2.2.5 Maximum DC Current ($I_{DC\ MAX}$) and Temperature Derating

The operational curves in Figure 5.2.2.4-1 are used in conjunction with the temperature derated maximum dc current to determine the maximum tolerable peak operating current per segment ($I_{PEAK\ MAX}$). The dc current derating curve for the high-efficiency red device is reproduced in Figure 5.2.2.5-1. Figure 5.2.2.5-2 shows the maximum tolerable dc junction temperature and the maximum tolerable dc power dissipation for the high-efficiency red device.

5.2.3 Sample Calculation of a Typical and Worst Case Design, Strobed Operation

Given:

1. High-efficiency red devices in a 4-digit display.

2. Maximum ambient temperature as measured between two digits just above PC board is T_A = 60°C (140°F).

3. Duty factor to be 25%.

Calculate:

1. Maximum tolerable operating conditions per segment.

2. Peak and average power dissipation.

3. Average junction temperature.

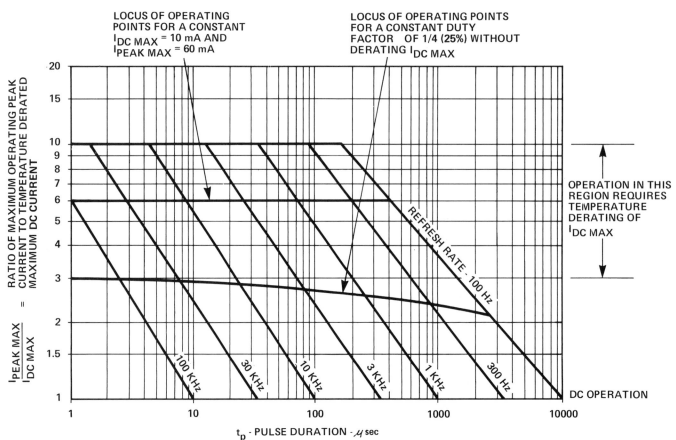

LOCUS OF OPERATING POINTS FOR A CONSTANT $I_{DC\ MAX}$ = 10 mA AND $I_{PEAK\ MAX}$ = 60 mA

LOCUS OF OPERATING POINTS FOR A CONSTANT DUTY FACTOR OF 1/4 (25%) WITHOUT DERATING $I_{DC\ MAX}$

OPERATION IN THIS REGION REQUIRES TEMPERATURE DERATING OF $I_{DC\ MAX}$

REFRESH RATE - 100 Hz

RATIO OF MAXIMUM OPERATING PEAK CURRENT TO TEMPERATURE DERATED MAXIMUM DC CURRENT = $\frac{I_{PEAK\ MAX}}{I_{DC\ MAX}}$

100 KHz   30 KHz   10 KHz   3 KHz   1 KHz   300 Hz   DC OPERATION

t_p - PULSE DURATION - μ sec

Figure 5.2.2.4-1 Maximum Tolerable Peak Current vs Pulse Duration for a High-Efficiency Red Display.

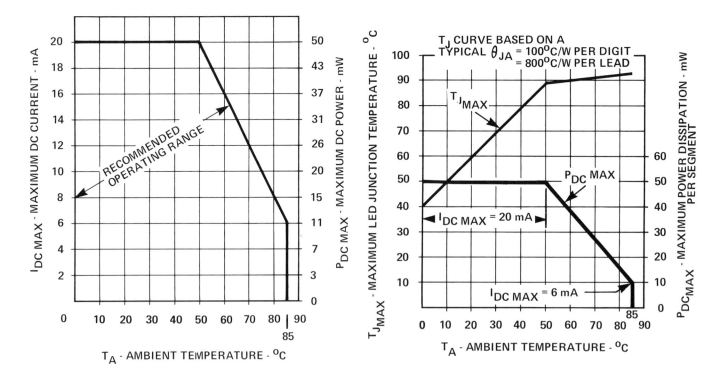

$I_{DC\ MAX}$ - MAXIMUM DC CURRENT - mA

$P_{DC\ MAX}$ - MAXIMUM DC POWER - mW

RECOMMENDED OPERATING RANGE

T_A - AMBIENT TEMPERATURE - °C

Figure 5.2.2.5-1 Maximum Allowable DC Current and DC Power Dissipation per Segment as a Function of Ambient Temperature for a High-Efficiency Red Display.

T_J CURVE BASED ON A TYPICAL θ_{JA} = 100°C/W PER DIGIT = 800°C/W PER LEAD

$T_{J\ MAX}$

P_{DC} MAX

$I_{DC\ MAX}$ = 20 mA

$I_{DC\ MAX}$ = 6 mA

$T_{J\ MAX}$ - MAXIMUM LED JUNCTION TEMPERATURE - °C

$P_{DC\ MAX}$ - MAXIMUM POWER DISSIPATION PER SEGMENT - mW

T_A - AMBIENT TEMPERATURE - °C

Figure 5.2.2.5-2 Maximum Tolerable LED Junction Temperature and Maximum Tolerable Power Dissipation for a High-Efficiency Red Display.

Step 1:

The duty factor is set at 1/4. It is necessary to select a refresh rate which will allow the maximum ratio of $I_{PEAK\ MAX}/I_{DC\ MAX}$ and stay within the timing constraints of the system. Refering to Figure 5.2.2.4-1, a minimum possible refresh rate of 100 Hz gives a ratio of 2.5 for a pulse duration of $t_p = 2.5$ msec, a refresh rate of 1 KHz gives a ratio of 2.6 for $t_p = 250$ μsec. and 10 KHz gives a ratio of 2.8 for $t_p = 25$ μsec. A good refresh rate is 3125 Hz which gives a ratio of 2.7 for $t_p = 80$ μsec.

In selecting a minimum refresh rate in order to avoid observable flicker, a designer should not use less than 100 Hz or less than 5X the expected vibration frequency.

Step 2:

Next determine $I_{DC\ MAX}$ from Figure 5.2.2.5-1, $I_{DC\ MAX}$ at $T_A = 60°C$ is 16 mA.

Calculate $I_{PEAK\ MAX}$:

$I_{PEAK\ MAX} = (2.7)\ (16\ mA) = 43.2\ mA$

Determine I_{AVG} = (Duty Factor) $(I_{PEAK\ MAX})$:

$I_{AVG} = (1/4)\ (43.2\ mA) = 10.8mA$

Step 3:

Now determine the power dissipation (P) and expected T_J:

$$P_{AVG\ SEG} = I_{AVG}\ V_F$$

The typical segment power may be calculated using the V_F from equation 5.2.2.2-3:

$V_{F\ TYP} = 1.65V + 21\Omega\ (.0432 - .005A)$
$V_{F\ TYP} = 2.452V$

The value $V_F = 2.452$ volts assumes $T_J = 25°C$. The junction temperature is above 60°C so the change in forward voltage due to junction temperature is now taken into account. This is an iterative calculation, however, one iteration is usually sufficient.

From Figure 5.2.2.3-1, $\Delta V_F/°C = 1.4$ mV/°C for $I_F = 43$ mA. An approximate T_J is obtained and used to calculate an adjusted V_F:

$T_{J\ APPROX} = (100°C/W)(8\ SEG)(P_{SEG\ APPROX}) + T_A$
$\qquad\qquad = (100°C/W)(8\ SEG)(.0108A)(2.452V) + 60°C$
$T_{J\ APPROX} = 21.2°C + 60°C = 81.2°C$

The adjusted forward voltage is 2.373 volts:

$\Delta V_F \approx (-.0014V/°C)(81.2 - 25°C) = -.079V$
$V_F = 2.452 - .079 = 2.373V$

The typical power dissipation and typical junction temperature can now be calculated with good accuracy:

$P_{SEG} = (.0108A)(2.373V) = .026\ W/SEG$
$P_{DIGIT} = (8\ SEG)(.026\ W/SEG) = .208W$
$T_{J\ TYP} = (100°C/W)(.208W) + 60°C \approx 81°C$

Now the worst case is calculated using the maximum R_s value. Equation 5.2.2.2-2 is used to calculate $V_{F\ MAX}$. The same iterative process, as above, is used to take into account the -1.4mV/°C change in $V_{F\ MAX}$.

$V_{F\ MAX} = 1.55V + (.0432A)(33\Omega) = 2.976V$
$T_{J\ MAX} = (100°C/W)(8\ SEG)(.0108A)(2.976V) + 60°C$
$T_{J\ MAX} = 25.7 + 60 = 85.7°C$
$V_{F\ MAX} = 2.976V - (.0014V/°C)(85.7 - 25°C) = 2.891V$
$P_{SEG\ MAX} = (.0108A)(2.891V) = .0312W$
$P_{DIGIT\ MAX} = (8\ SEG)(.0312\ W/SEG) = .250W$
$T_{J\ MAX} = (100°C/W)(.250W) + 60°C$
$T_{J\ MAX} = 25 + 60 = 85°C$

Comparing $T_{J\ MAX}$ and $T_{J\ TYPICAL}$ with the curve in Figure 5.2.2.5-2, the worst case T_J is 5°C below the limit and $T_{J\ TYP}$ is 9°C below the allowed limit for $T_A = 60°C$.

5.2.4 Relative Efficiency and Light Output

The efficiency of an LED (photons/electrons) increases with forward current until the junction reaches saturation. If the efficiency at a given current is taken as a reference, then the efficiency at other currents may be compared to the reference and a plot of relative efficiency can be generated as shown in Figure 5.2.4-1. A relative efficiency curve is actually generated by measuring the time average luminous intensity at various peak currents while maintaining a constant average current. The curve is normalized to a relative efficiency of 1.00 at that data sheet current where the minimum luminous intensity is specified.

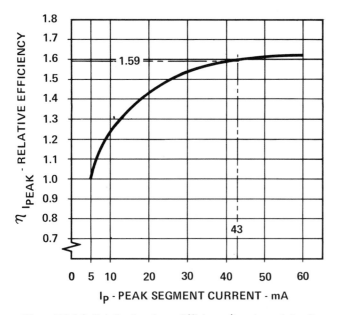

Figure 5.2.4-1 Relative Luminous Efficiency (Luminous Intensity Per Unit Current) vs Peak Current Per Segment for a High-Efficiency Red Display.

The time average luminous intensity can be calculated from the following equation:

(5.2.4-1)

$$I_v \text{ TIME AVG} = \left[\frac{I_{AVG}}{I_{AVG\ SPEC}}\right]\left[\frac{\eta_{I_{PEAK}}}{\eta_{I_{PEAK\ SPEC}}}\right]\left[I_{v_{SPEC}}\right]$$

where:

I_v = Operating point average current

$I_{AVG\ SPEC}$ = Average current for data sheet luminous intensity value, $I_{v_{SPEC}}$

$\eta_{I\ PEAK}$ = Relative efficiency at operating peak current.

$\eta_{I\ PEAK\ SPEC}$ = Relative efficiency at data sheet peak current where luminous intensity $I_{v_{SPEC}}$ is specified.

$I_{v_{SPEC}}$ = Data sheet luminous intensity, specified at $I_{AVG\ SPEC}$ and $I_{PEAK\ SPEC}$.

The light output of an LED also varies with temperature. As the temperature of an LED increases, the light output decreases according to the following exponential relationship:

$$I_v\ (T_A) = I_v\ (25°C)\ e^{K(T_A - 25°C)} \qquad (5.2.4-2)$$

where:

$I_v\ (T_A)$ = Luminous intensity at operating ambient temperature.

$I_v\ (25°C)$ = Luminous intensity at $T_A = 25°C$

T_A = Operating ambient temperature, °C

K = Exponent constant 1/°C, LED material dependent

e = 2.7183

LED	K
Standard Red	−.0188
High-Efficiency Red	−.0131
Yellow	−.0112
Green	−.0104

For an increase in temperature of 1°C, the light output change of the LED is as follows:

LED	+1°C Factor	% Change
Standard Red	$.9814 I_v$	−1.86
High-Efficiency Red	$.9870 I_v$	−1.30
Yellow	$.9889 I_v$	−1.11
Green	$.9897 I_v$	−1.03

Equation 5.2.4-1 applies to luminous intensity as perceived by the eye and is not applicable to radiant power output.

5.2.4.1 Sample Calculation of Time Average Luminous Intensity

The time average luminous intensity for the example in Section 5.2.3 can be calculated using the value of $I_v = 300$ mcd at $I_F = 5$ mA. The relative efficiency calculation does take into account junction temperature rise above ambient within reasonable error. The resultant I_vTIME AVG is then corrected for the increased ambient above $T_A = 25°C$.

The relative efficiency for $I_P = 43$ mA is 1.59, as obtained from Figure 5.2.4-1, and the time average luminous intensity at $T_A = 25°C$ is calculated to be 1030 microcandelas/segment.

$$I_v \text{ TIME AVG} = \left[\frac{10.8\ mA}{5\ mA}\right]\left[\frac{1.59}{1}\right][300\ \mu cd]$$

$$= 1030\ \mu cd/\text{SEGMENT}$$

At $T_A = 60°C$, the light output decreases to 651 microcandelas/segment.

$$I_v\ (60°C) = 1030\ (.0131)\ (60 - 25°C)$$

$$I_v\ (60°C) = 1030\ (.6322) = 651\ \mu cd/\text{SEGMENT}$$

5.2.4.2 Digit-to-Digit and Segment-to-Segment Luminous Intensity Ratio

The luminous intensity value listed on a 7-segment display data sheet is a digit average luminous intensity for the seven segments. The ratio of the maximum I_v segment to the minimum I_v segment within a digit needs to be held to a reasonable value. Also, the digit to digit luminous intensity ratio needs to be held to a reasonable value so that digits may be stacked side by side with a pleasing appearance.

5.20

Tests have shown that when a display is lighted with a string of 8's, an observer may scrutinize the digits with a critical eye, finding all sorts of faults. However, this is not a normal mode of operation. The exact same display lighted with a string of random numbers may well be acceptable to the same observer. In fact, the observer reading the same display to obtain information during normal usage will NOT notice the minor faults he detected when scrutinizing the string of 8's. This is true when the segment-to-segment and digit-to-digit ratios are kept within certain limits.

An observer scrutinizing an LED display is able to detect a difference in light output at a threshold level of about 1.6:1. A change in light output becomes quite noticeable at the ratio of 2.3:1.

5.2.5   Driving a Seven Segment Display

A typical logic system will produce output information in the form of BCD data. This BCD data must then be converted to the 7-segment matrix code format of the display. Drivers are then required to drive each segment of the display. The two methods of driving a 7-segment display are 1) direct dc drive of a single digit and 2) strobing a string of digits.

5.2.5.1 Direct dc Driving a Display

The simplest method of driving a 7-segment display is to have one decoder/driver for each digit. This is usually cost effective up to four or five digits depending upon the trade-off between component cost and circuit complexity. Figure 5.2.5.1-1 illustrates the direct dc drive concept. Each digit has its own decoder/driver and is continuously illuminated. Though LED displays are more efficient when strobed, the advantage with direct drive is that the drivers need not handle high current levels. Also, the display circuit complexity is simple as extra timing is not required.

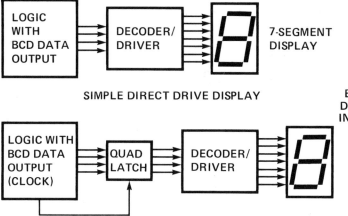

SIMPLE DIRECT DRIVE DISPLAY

Figure 5.2.5.1-1 Block Diagram for Direct DC Drive Scheme for 7-Segment Displays.

The forward current through an LED must be limited in some fashion to obtain the desired illumination and to prevent high current damage. The two most popular methods of accomplishing these two objectives is to (1) drive the display from a decoder/driver with switching outputs that control the flow of LED current through external limiting resistors and (2) drive the display from a decoder/driver with constant current outputs. Figure 5.2.5.1-2 illustrates both methods.

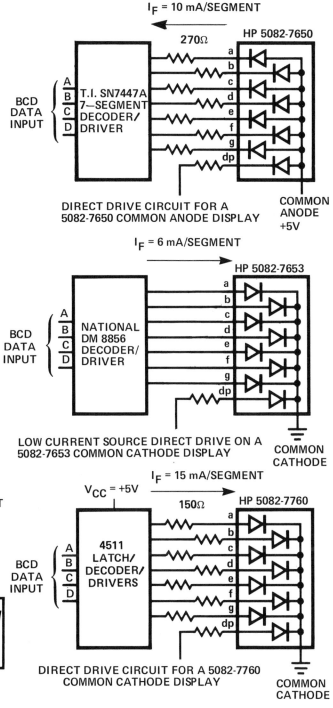

Figure 5.2.5.1-2 Example Methods for Direct DC Driving 7-Segment Displays.

When using resistor current limiting, the current limiting resistor may be determined from the following equation:

$$R = \frac{V_{CC} - V_F - V_{CE}}{I_F} \qquad (5.2.5.1\text{-}1)$$

where:

R = Series current limiter
V_{CC} = Supply voltage
V_F = LED forward voltage at I_F
V_{CE} = Saturation voltage drop across driver transistor
I_F = LED forward dc current

If a decoder/driver with constant current outputs is being used and the ambient temperature around the display requires dc current derating, shunt resistors at the driver outputs may be used to bleed off the excess current.

$$R_{SHUNT} = \frac{V_{OH}}{I_{OH} - I_F} \qquad (5.2.5.1\text{-}2)$$

As an example, the Fairchild 9368 7-segment latch/decoder/driver will source a maximum of 22 mA over a wide temperature range. A display is to be composed of standard red common cathode displays operating in an ambient temperature of 70°C. The data sheet derating is 0.43 mA dc/°C above $T_A = 50°$C. Therefore, at $T_A = 70°$C, $I_{DC\,MAX} = 16.4$ mA. A worst case design requires shunting 5.6 mA from each output of the 9368. The LED forward voltage is 1.60 volts. The shunt resistor value is 286Ω:

$$R_{SHUNT} = \frac{1.60V}{(.022 - .0164)A} = 286\Omega$$

5.2.5.2 High Speed Counter Using DC Drive

Multiple function devices that will direct drive LED displays are becoming more available. An example is the Texas Instrument SN74143 BCD counter/4 bit latch/BCD to 7-segment decoder/driver. The outputs are 15 mA constant current sink. The SN74144 is the same device with a 25 mA sink capability with external current limiting resistors. A 4-digit high speed counter can be made using the SN74143 and the yellow .43 inch common anode displays as pictured in Figure 5.2.5.2-1. The 15 mA LED drive current is sufficient for the display to be read in most bright ambients.

The data strobe line may be held low to obtain a continuous display of the counter state, or strobed to latch specific counts. A decimal point input is provided in each SN74143 so that any desired decimal point position may be obtained by providing a logic high true input on the appropriate dp select line.

5.2.5.3 Concept of Strobed (Multiplexed) Operation

The most efficient method of driving an LED display is to strobe it at a high peak current and a low duty factor. This takes advantage of the increased efficiency of an LED at high peak currents and the reduction in average power dissipation when compared to dc operation. A multiplexed display design becomes cost effective when the decoder/driver and display PC board size and complexity required for dc drive become significant when compared to the timing circuitry and digit driver cost and size of PC board needed for strobed operation. The power saving that is attainable with strobed operation may show up as a need for a smaller sized power supply, contributing to cost reduction.

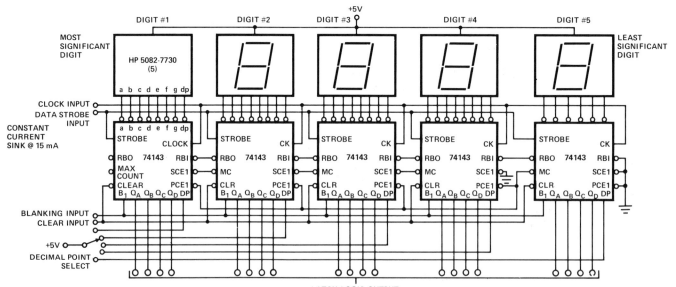

Figure 5.2.5.2-1 High Speed 5-Digit Counter Using Direct DC Driven LED Displays.

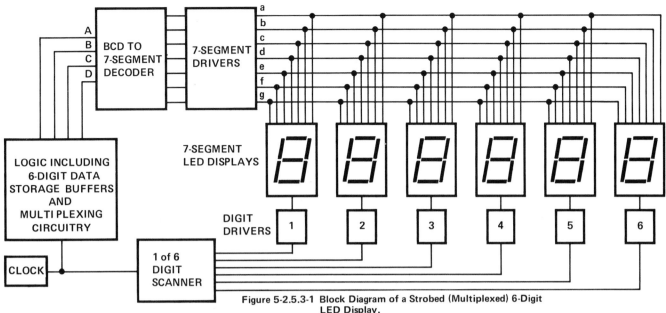

Figure 5-2.5.3-1 Block Diagram of a Strobed (Multiplexed) 6-Digit LED Display.

A strobed six digit display is illustrated in block diagram form in Figure 5.2.5.3-1. The like segments of each digit in the display are tied to a common segment bus line which is connected to one of seven segment drivers. For example, all of the "a" segments are tied common to the segment "a" driver and all of the "b" segments are tied common to the segment "b" driver. This allows one BCD to 7-segment/decoder driver to be time shared between all digits, with BCD data for each digit stored in one of six digit data storage buffers. Each digit is enabled by its own digit driver.

The operation of this multiplexing scheme is as follows: First Clock Pulse: The multiplexing logic selects digit data buffer number 1. This first digit BCD data is presented to the decoder/driver and the required segment lines are activated. The digit scanner enables only digit driver number 1 and the first number is displayed. Second Clock Pulse: The multiplexing logic selects digit data buffer number 2. This second digit BCD data is presented to the decoder/driver and again the required segment lines are activated. The digit scanner now enables only digit driver number 2 and the second number is displayed. This process is continued through digits 3, 4, 5 and 6, and then the whole process repeats.

The strobing rate to produce a flicker free display should be a minimum of 100 Hz or 5X the expect mechanical vibration. However, to optimize the overall performance of a 7-segment display, it is best to use a refresh rate of 1 KHz or faster.

When designing a multiplexed display system, timing considerations become important. The timing parameters to take into account when designing a display of a string of N-digits are:

N = quantity of digits

f = refresh rate;

t_p = on-pulse duration

t_b = blanking time between digits

The first consideration is to select a desired refresh rate. The reciprocal of the refresh rate is the refresh time period in which all N—digits are enabled, one at a time:

$$\textbf{Refresh Period} = \tau = \frac{1}{f} \qquad \textbf{(5.2.5.3-1)}$$

The maximum allotted digit on-pulse duration is the refresh period divided by the quantity of digits to be strobed:

$$\textbf{Maximum On-Pulse} = t_{p\ max} = \frac{\tau}{N} \qquad \textbf{(5.2.5.3-2)}$$

Figure 5.2.5.3-2 illustrates a basic timing diagram for strobing N=6 digits. Each digit is enabled in sequence without any overlapping of the digit enable time, $t_{p\ max}$. A segment line, however, may remain active if it is required in two or more successive digits. A variation of this basic timing scheme is to use Pulse Width Modulation (PWM) to vary the operating on-pulse duration, t_p, to values less than the allotted pulse duration, $t_{p\ max}$. PWM may effectively be used to vary the apparent brightness of a display to match changing ambient lighting conditions. As an example, for low level ambient conditions, t_p is reduced to dim the display. The segment peak current remains constant to take advantage of the increased LED efficiency at high peak currents, yet the average current is decreased which reduces the time average luminous intensity of the display, see equation 5.2.4-1.

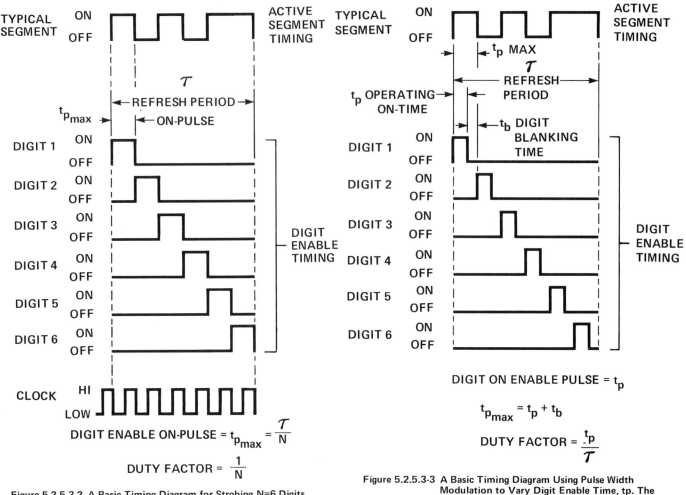

DIGIT ENABLE ON-PULSE = $t_{p_{max}}$ = $\dfrac{\tau}{N}$

DUTY FACTOR = $\dfrac{1}{N}$

Figure 5.2.5.3-2 A Basic Timing Diagram for Strobing N=6 Digits. (The Typical Segment is Shown Active When Digits 1, 2, 5 and 6 are Enabled.)

DIGIT ON ENABLE PULSE = t_p

$t_{p_{max}}$ = t_p + t_b

DUTY FACTOR = $\dfrac{t_p}{\tau}$

Figure 5.2.5.3-3 A Basic Timing Diagram Using Pulse Width Modulation to Vary Digit Enable Time, tp. The Active Segment Pulse Duration is not Modulated and Remains at tpMAX.

Figure 5.2.5.3-3 illustrates a basic timing diagram employing PWM. The digit enable on-pulse, t_p, is varied to obtain the desired display time average luminous intensity. The maximum allotted pulse duration is now composed of the digit operating on-pulse, t_p, and a blanking time, t_b:

Allotted Pulse Width = $t_{p\ max}$ = t_p + t_b   (5.2.5.3-3)

The duty factor is now defined as the ratio of the operating on-pulse to the refresh period:

Duty Factor (PWM) = tp/τ   (5.2.5.3-4)

The active segment on-pulse duration need not be modulated and can remain at $t_{p\ max}$. It is usually sufficient and easier to modulate only the digit enable operating time, though a designer may choose to modulate either the segment on-time, digit on-time or both.

An additional use of PWM is applied when a fast strobing rate in the neighborhood of 10 KHz is used. At this fast refresh rate, the turn-off time of driving transistors may not be fast enough to prevent overlapping of digit and segment on-pulse durations. This timing overlapping is called ghosting, since a digit that should be off is still partially turned-on. A timing overlap of approximately 3% is usually sufficient to have perceptible ghosting in moderate ambient lighting conditions. To prevent this situation from occuring, PWM is employed to effect inter-digit blanking. In this mode of operation, a fixed digit (and possibly segment) blanking time, t_b, is built into the display timing to insure sufficient time for the driving transistors for one digit to turn completely off prior to turning on the driver transistors for the next digit.

The value of the segment current limiting resistors may be calculated from the generalized schematic shown in Figure 5.2.5.3-4.

5.24

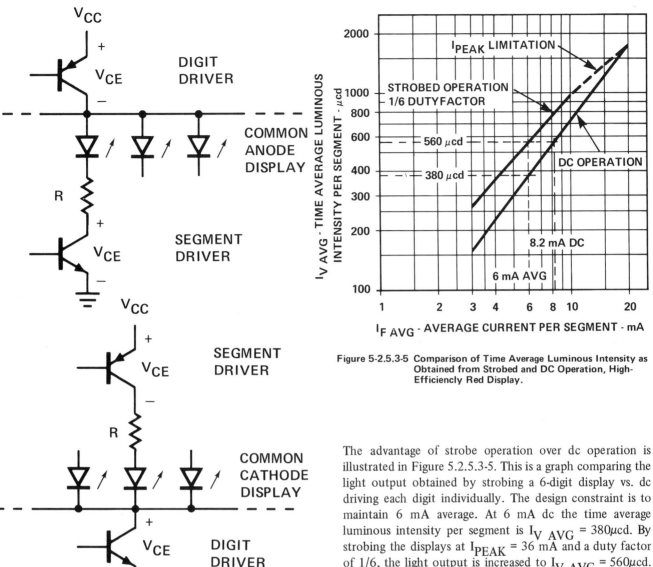

Figure 5-2.5.3-5 Comparison of Time Average Luminous Intensity as Obtained from Strobed and DC Operation, High-Efficiencly Red Display.

Figure 5.2.5.3-4 Generalized Drive Circuits for Strobed Operation.

From Figure 5.2.5.3-4, the following equation may be used to calculate the vaue of the current limiting resistor, R:

$$R = \frac{V_{CC} - (V_F + V_{CE\ SEG} + V_{CE\ DIGIT})}{I_{PEAK}} \qquad (5.2.5.3\text{-}5)$$

where:

V_{CC} = Supply voltage

V_F = LED forward voltage at I_{PEAK}

$V_{CE\ SEG}$ = Saturation voltage drop across segment driver at I_{PEAK}

$V_{CE\ DIGIT}$ = Saturation voltage drop across digit driver at (I_{PEAK} × the number of segments multipled)

I_{PEAK} = Peak LED segment current

The advantage of strobe operation over dc operation is illustrated in Figure 5.2.5.3-5. This is a graph comparing the light output obtained by strobing a 6-digit display vs. dc driving each digit individually. The design constraint is to maintain 6 mA average. At 6 mA dc the time average luminous intensity per segment is $I_{V\ AVG}$ = 380μcd. By strobing the displays at I_{PEAK} = 36 mA and a duty factor of 1/6, the light output is increased to $I_{V\ AVG}$ = 560μcd. This is an increase in light output of 1.47X. over dc operation for the same average current.

At 6 mA average, the strobed operation produces an average power dissipation within the display of 10.3 mW per segment vs. 10.0 mW per segment for dc operation. To obtain the light output of 560 μcd per segment by using dc operation would require a forward current of 8.2 mA dc and a power dissipation of 14.1 mW per segment. The total power saving obtained by strobing is considerable. This power saving is realized by the reduction of the required IC components needed to drive the display. For example, in this illustration, only one decoder/driver is required for strobed operation vs. six needed for dc operation.

The human eye is a time average detector and, therefore, will easily discern the increased light output obtained by strobing.

5.2.6 Interfacing Microprocessors to Seven Segment Displays

Seven segment displays can be interfaced to a microprocessor with only a few external components. The microprocessor can be used to control the refreshing of a multiplexed display or simply to periodically up-date a dc driven display. In either case, one or more eight bit latches are required to hold the the seven segment and digit information. If the seven segment display is multiplexed, then some timing circuitry is also required to periodically request new information from the microprocessor. The timing circuitry could consist of either a monostable multivibrator or an oscillator. This circuitry would request an interrupt and use the already available hardware to handle the proper interrupt decoding. Several commercially available seven segment decoder/drivers are available that can directly drive the seven segment LED displays.

When the seven segment display is driven on a dc basis, a seven segment decoder/driver is required for each display. Figure 5.2.6-1 shows an example of a seven segment-microprocessor interface. Each display is driven by its own seven segment driver. The Fairchild 9374 has current source outputs that drive each LED segment at 15 mA dc. The decimal points are driven by a National DS8859, a hex latch with current source outputs. The Intel 8080A microprocessor updates each display, with an OUTPUT instruction. The second byte of the OUT instruction specifies an eight bit address which determines the display that is to be updated. Upon execution of the OUT instruction, the lowest four bits of the accumulator are loaded into the appropriate latched-decoder/driver. A separate OUT instruction also updates the decimal points and overflow digit. Since the upper six states of the Fairchild 9374 are decoded as (-, E, H, L, P, and blank), it is possible to blank each digit selectly under microprocessor control, indicate an overflow by a row of minus signs, and use the word "HELP" to catch someone's attention.

A microprocessor can also be used to multiplex a seven segment display. Figures 5.2.6-2 and 5.2.6-3 show two ways by which this can be accomplished. In Figure 5.2.6-2, the microprocessor outputs two bytes per digit. The first byte contains segment information for the display. Since the seven segment information is decoded by the microprocessor, the programmer can program the character font to include any special symbols that are desired. The second byte turns on the proper digit driver. At the same time, the RCA 4047 monostable multivibrator is triggered and requests another interrupt in 2 milliseconds. Upon requesting an interrupt, the interrupt circuitry (not shown) must force an RST(7) instrument to be executed by the Intel 8080A microprocessor. Then the refresh program shown in Figure 5.2.6-2b is executed. The program reads a pointer from RAM and then outputs two bytes of information to the display. The pointer is incremented by one and compared to the final address of the data file. If they are equal, the pointer is reset to the address of the first byte in the file, otherwise the pointer is set to the address of the next byte in the file. As shown, the circuit refreshes five seven segment displays on a 20% duty factor at 60 mA peak/segment. This technique can be expanded to eight displays without additional hardware. The time required to refresh the display can be determined as shown below:

$$\text{REFRESH TIME} = \frac{(165N + 7)R}{\text{MICROPROCESSOR CLOCK RATE}} \quad \text{(5.2.6-1)}$$

where N is the number of digits to be refreshed, and R is the strobing rate.

For example, suppose five digits are refreshed on a 100 Hz rate and the microprocessor uses a 2MHz clock, then 4.2% of microprocessor time is required to refresh the display. For the remaining 95.8% of the time, the microprocessor can be performing countless other tasks.

Figure 5.2.6-3 shows another technique that can be used to multiplex a seven segment display. With this technique, the microprocessor outputs only a single byte of information to the display. Four bits contain the BCD character to be displayed, one bit contains the decimal point information, and the remaining three bits determine which digit should be turned on. The Motorola MC 14511 seven segment latched decoder driver can source up to 25 mA per segment to a common cathode LED display. The high efficiency red displays that are shown can be satisfactorily operated at 25 mA/segment on a 20% duty duty factor. This power requirement is considerably below those required for standard red seven segment displays. The interrupt request circuitry is similar to the previous example. After receiving an interrupt request, the interrupt hardware must force a RST(7) instruction to be executed by the Intel 8080A microprocessor. Following the RST(7) instruction, the program shown in Figure 5.2.6-3b will be executed. This program is similar to the program shown in Figure 5.2.6-2b except that only a single byte is outputed to the display. This technique can also be expanded to eight displays without additional hardware. The time required to refresh the display can be determined by equation 5.2.6-2:

$$\text{REFRESH TIME} = \frac{(143N + 7)R}{\text{MICROPROCESSOR CLOCK RATE}} \quad \text{(5.2.6-2)}$$

where N is the number of digits to be refreshed, and R is the strobing rate.

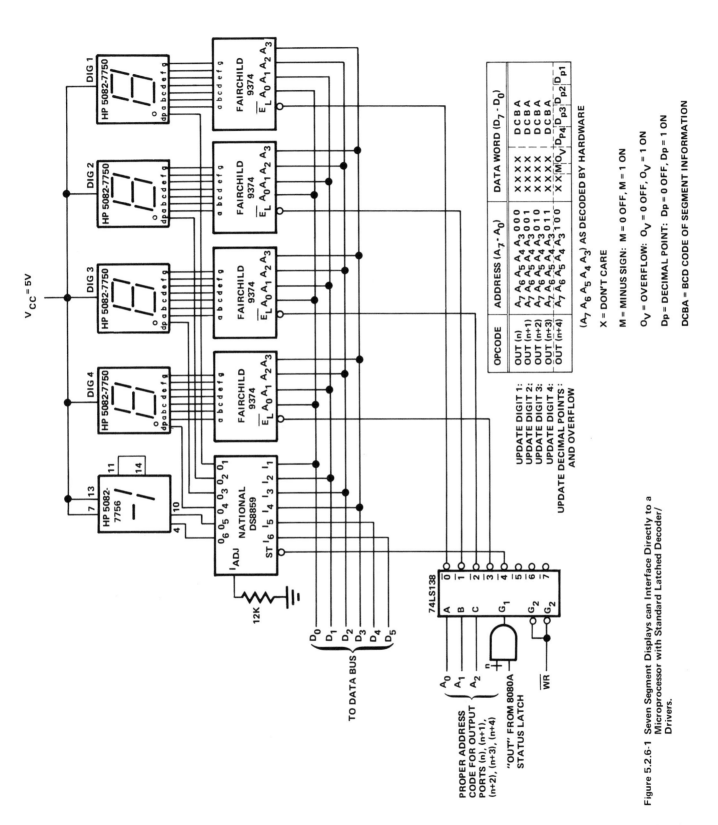

OPCODE	ADDRESS ($A_7 \cdot A_0$)	DATA WORD ($D_7 \cdot D_0$)
		D C B A
OUT (n)	$A_7\ A_6\ A_5\ A_4\ A_3\ 0\ 0\ 0$	X X X X $\quad$ D C B A
OUT (n+1)	$A_7\ A_6\ A_5\ A_4\ A_3\ 0\ 0\ 1$	X X X X $\quad$ D C B A
OUT (n+2)	$A_7\ A_6\ A_5\ A_4\ A_3\ 0\ 1\ 0$	X X X X $\quad$ D C B A
OUT (n+3)	$A_7\ A_6\ A_5\ A_4\ A_3\ 0\ 1\ 1$	X X X X $\quad$ D C B A
OUT (n+4)	$A_7\ \bar{A_6}\ A_5\ A_4\ A_3\ 1\ 0\ 0$	X X $|M|O_V|D_{p4}|D_{p3}\ p2|D\ p1$

($A_7\ A_6\ A_5\ A_4\ A_3$) AS DECODED BY HARDWARE

X = DON'T CARE

M = MINUS SIGN: M = 0 OFF, M = 1 ON

O_V = OVERFLOW: O_V = 0 OFF, O_V = 1 ON

Dp = DECIMAL POINT: Dp = 0 OFF, Dp = 1 ON

DCBA = BCD CODE OF SEGMENT INFORMATION

UPDATE DIGIT 1:
UPDATE DIGIT 2:
UPDATE DIGIT 3:
UPDATE DIGIT 4:
UPDATE DECIMAL POINTS:
AND OVERFLOW

TO DATA BUS

PROPER ADDRESS
CODE FOR OUTPUT
PORTS (n), (n+1),
(n+2), (n+3), (n+4)

"OUT" FROM 8080A
STATUS LATCH

Figure 5.2.6-1 Seven Segment Displays can Interface Directly to a
Microprocessor with Standard Latched Decoder/
Drivers.

5.27

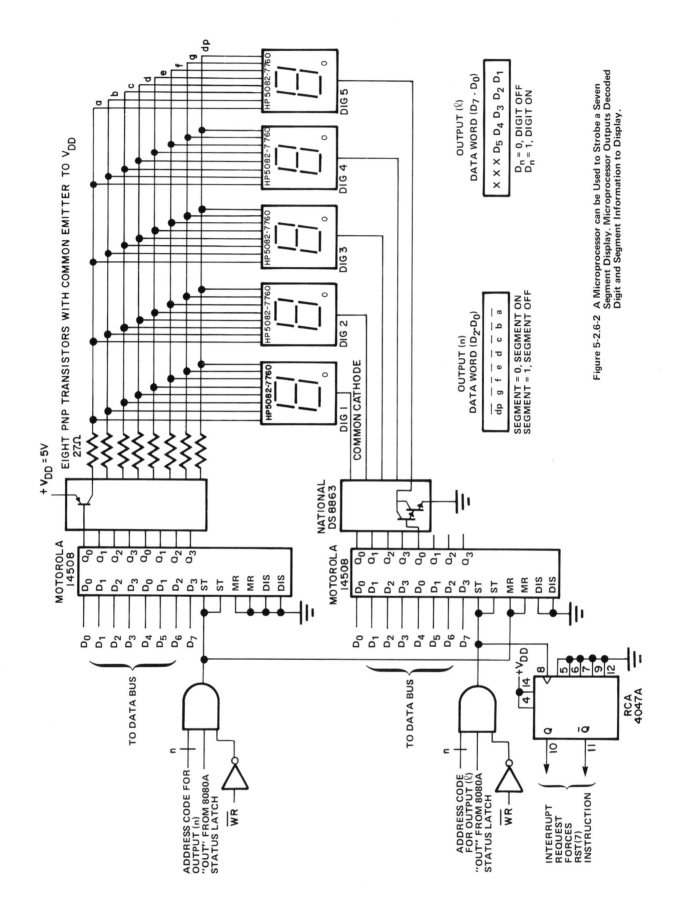

Figure 5-2.6-2 A Microprocessor can be Used to Strobe a Seven Segment Display. Microprocessor Outputs Decoded Digit and Segment Information to Display.

5.28

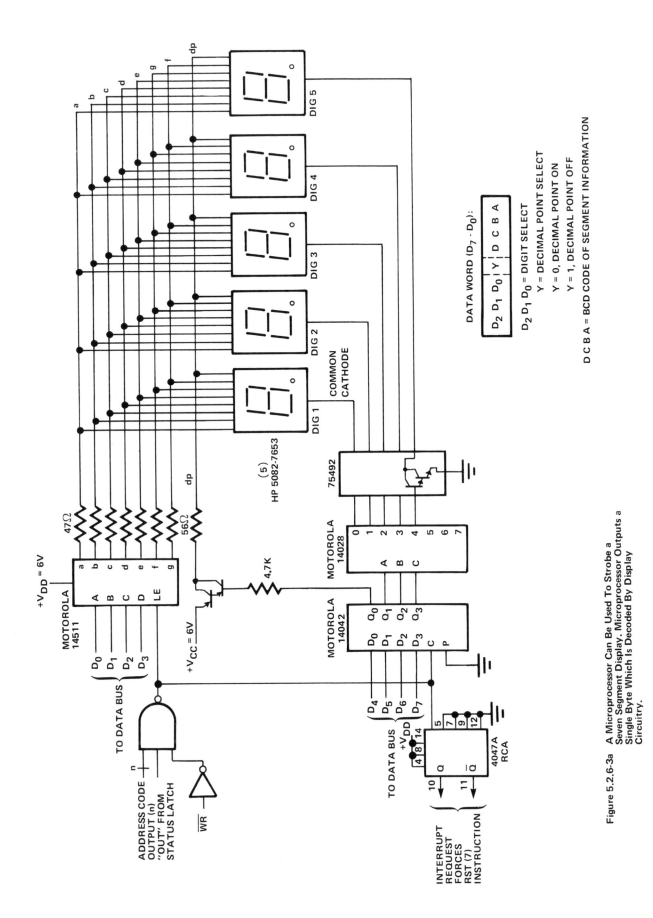

Figure 5.2.6-3a A Microprocessor Can Be Used To Strobe a Seven Segment Display. Microprocessor Outputs a Single Byte Which Is Decoded By Display Circuitry.

ADDRESS	OP CODE	CLOCK CYCLES	COMMENTS
(0038)$_{16}$	PUSH PSW.	(11)	
	PUSH HL	(11)	
	LHLD	(16)	HL = POINTER
	A$_L$	------	
	A$_H$	------	
	MOV A, M	(7)	A = (HL)
	OUT	(10)	STORES NEW SEGMENT INFORMATION
	n	------	
	INX HL	(5)	HL = HL + 1
	MOV A, M	(7)	A = (HL)
	OUT	(10)	TURNS ON DIGIT
	I	------	DRIVER
	MOV A, L	(5)	A = L
	CPI	(7)	COMPARE L TO ADDRESS OF LAST MEMORY
	(19)$_{16}$	------	LOCATION
	JNZ LOOP	(10)	JUMP IF A ≠ (19)$_{16}$
	(ADDRESS OF	------	
	LOOP)		
	MVI, L	(7)	L = (OF)$_{16}$
	(OF)$_{16}$	------	
LOOP	INX HL	(5)	HL = HL + 1
	SHLD	(16)	POINTER = HL
	A$_L$	------	
	A$_H$	------	
	POP HL	(10)	
	POP PSW	(10)	
	EI	(4)	
	RET	(10)	

MEMORY ADDRESS	CONTENTS (D$_7$ - D$_0$)	COMMENTS
A$_H$ A$_L$ A$_H$ A$_L$ + 1	POINTER$_L$ POINTER$_H$	POINTER = NEXT DIGIT TO BE DISPLAY
(X X 1 0)$_{16}$	dp g f e d c b a	DIGIT 1
(X X 1 1)$_{16}$	0 0 0 0 0 0 0 1	
(X X 1 2)$_{16}$	dp g f e d c b a	DIGIT 2
(X X 1 3)$_{16}$	0 0 0 0 0 0 1 0	
(X X 1 4)$_{16}$	dp g f e d c b a	DIGIT 3
(X X 1 5)$_{16}$	0 0 0 0 0 1 0 0	
(X X 1 6)$_{16}$	dp g f e d c b a	DIGIT 4
(X X 1 7)$_{16}$	0 0 0 0 1 0 0 0	
(X X 1 8)$_{16}$	dp g f e d c b a	DIGIT 5
(X X 1 9)$_{16}$	0 0 0 1 0 0 0 0	

Figure 5.2.6-2b Refresh Program Used to Strobe the Display Shown in Figure 5.2.6-2.

ADDRESS	OP CODE	CLOCK CYCLES	COMMENTS
(0038)$_{16}$	PUSH PSW	(11)	
	PUSH HL	(11)	
	LHLD	(16)	HL = POINTER
	A$_L$	------	
	A$_H$	------	
	MOV A, M	(7)	A = (HL)
	OUT	(10)	DISPLAY IS UPDATED
	n	------	
	MOV A, L	(5)	A = L
	CPI	(7)	COMPARE L TO ADDRESS OF LAST MEMORY
	(14)$_{16}$	------	LOCATION
	JNZ LOOP	(10)	JUMP IF A ≠ (14)$_{16}$
	(ADDRESS OF	------	
	LOOP)		
	MVI, L	(7)	L = (OF)$_{16}$
	(OF)$_{16}$	------	
LOOP	INX HL	(5)	HL = HL + 1
	SHLD	(16)	POINTER = HL
	A$_L$	------	
	A$_H$	------	
	POP HL	(10)	
	POP PSW	(10)	
	EI	(4)	
	RET	(10)	

MEMORY ADDRESS	CONTENTS (D$_7$ - D$_0$)		COMMENTS
A$_H$ A$_L$ A$_H$ A$_L$ + 1	POINTER$_L$ POINTER$_H$		POINTER = NEXT DIGIT TO BE DISPLAY
(X X 1 0)$_{16}$	0 0 0 Y		DIGIT 1
(X X 1 1)$_{16}$	0 0 1 Y	BCD	DIGIT 2
(X X 1 2)$_{16}$	0 1 0 Y	DATA	DIGIT 3
(X X 1 3)$_{16}$	0 1 1 Y		DIGIT 4
(X X 1 4)$_{16}$	1 0 0 Y		DIGIT 5

X = DON'T CARE
Y = DECIMAL POINT INFORMATION
    Y = Q DECIMAL POINT ON
    Y = 1, DECIMAL POINT OFF

Figure 5.2.6-3b Refresh Program Used to Strobe the Display Shown in Figure 5.2.6-3.

For example, 3.6% of the microprocessor's time would be required to refresh five displays at a 100 Hz refresh rate with a 2 MHz microprocessor clock. While these examples use an Intel 8080A microprocessor, these techniques can be used with other microprocessors by modifying the hardware and programs to reflect a different instruction set and different ways to output information to peripheral devices.

5.2.7 Detection and Indication of Segment Failures in Seven Segment LED Displays

The failure of a single element in a seven segment display device could result in the presentation of erroneous information which may not be detectable to the user. Such failures may merely be annoying in some equipment applications but may, on the other hand, be quite serious in the case of point of sales terminals or medical instrumentation.

Segment failures in seven segment LED displays, although rare, do have a measurable probability of occurrence. Characteristically, the failures are electrical opens or unusual degradation of light intensity. Shorted segments have such an extremely low probability of occurrence that they do not need to be considered. Fortunately, it is a rather simple matter to detect the occurrence of an open segment in a seven segment display matrix.

The circuits on the following pages depict two possible techniques for the detection and indication of such failures.

5.2.7.1 Seven-Segment Self-Test Circuit for Common Anode Displays

The circuit in Figure 5.2.7.1-1 will detect an open segment in any of the digits of a multi-digit 7 segment display. Basically, the test consists of turning on all segments for a short period not detected by the eye. Should the segment current corresponding to an "ON" segment not flow, a logic circuit blanks the entire display until reset. The portion of the circuit inside the labeled dashed line represents the failure detection system.

The scanning circuitry for the strobed display consists of a clock, a counter and a decoder. The -9601 one-shot multivibrator produces a 1 microsecond pulse out for each clock period input to the scan counter. The Q output of the one shot enables the lamp test function of the BCD to seven segment decoder driver, turning on all of the segment outputs. The status of each of the segment lines at nodes A is monitored by the 8 input -7430 NAND gate. If all segments are intact, the voltage of all the nodes A will be a logic high $(V_{CC} - V_{CE\ SAT} - V_F\ LED) = \sim 2$ volts. The output of the 8 input $\overline{NAND}$ will be a logic low and therefore no clock pulse will be gated through to the -7474 flip flop. If a

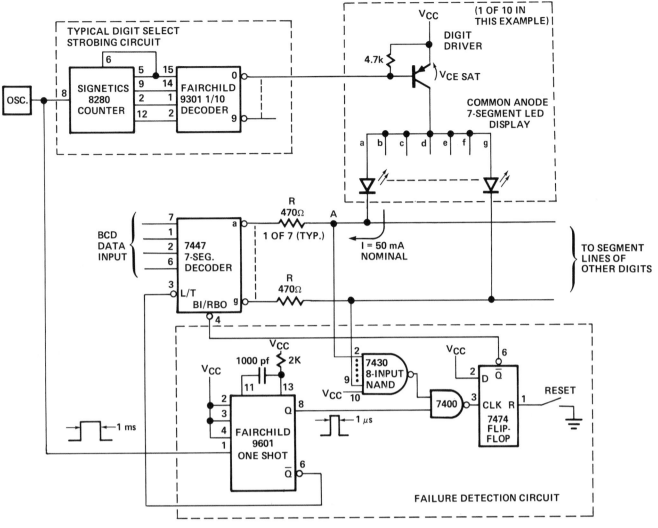

Figure 5.2.7.1-1 Self Test Circuit For 7-Segment Displays.

segment should be open, the low true output of the -7447 decoder will produce a logic 0 at the node A corresponding to the open segment. The resultant logical 1 output of the 8 input NAND will be gated through the 2 input NAND by the Q ouput of the one shot, thereby clocking the flip-flop. The 470 ohm segment current limiting resistor R is the largest value which can be used and still product a logic low at node A when using a standard TTL 8 input NAND gate.

In the circuit shown, the output of the flip-flop is used to blank the display on the occurrence of an open. The flip-flop is externally reset.

5.2.7.2 Self Test Circuit for Seven-Segment Displays and Associated Decoder Driver

The circuit in Figure 5.2.7.2-1 will detect an open segment in any of the digits of a multidigit seven segment display. Also, through the use of a redundant BCD to seven segment decoder, a malfunction of the decoder can be detected.

The circuit functions through the use of exclusive OR gating. The 8 gates labeled U_1 will only produce a logical 1 output when a segment is designated "ON" (-7447 decoder output low) and current is flowing through the segment so as to produce a logical 1 at node A and a logical 0 at node B. This is condition 1. The other two conditions are: (2) a segment turned off; nodes A and B are high, and (3) a segment is turned on by the decoder but no current flows due to an open; nodes A and B are low. Both of these conditions will result in a logical 0 at the output of gates U_1. Failure detection occurs at gates U_2. One input of each of these exclusive OR gates is the output of the gates U_1. The other input is the respective output of the -7448 BCD to 7 segment decoder. The -7448 has inverted logic outputs when compared to the -7447 (high true instead of low true). The three conditions defined above will produce the following results at the outputs of U_2.

Condition:

1.   C and D both logical 1; U_2 output logical 0.

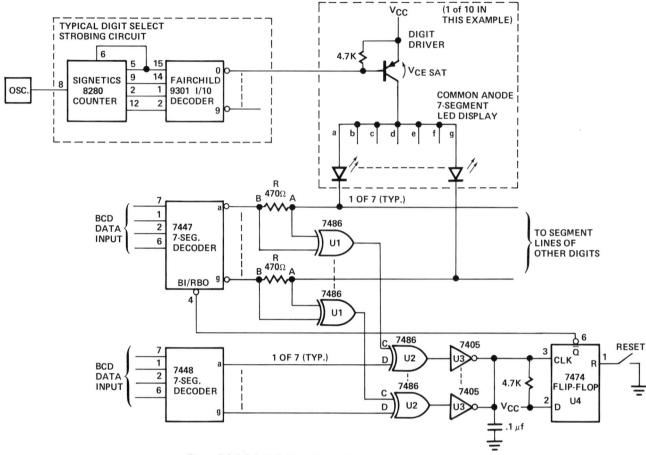

Figure 5.2.7.2-1 Self Test Circuit for 7-Segment Displays and
Associated Decoder/Driver.

2. C and D both logical 0; U_2 output logical 0.

3. C at logical 0; D at logical 1; U_2 output at a logical 1.

Condition 3 will produce a clock input pulse to U_4, thereby setting the flip-flop; Q to logical 1.

A fourth condition will occur if either one of the decoders fails to function properly. Unless the respective segment outputs of the two decoders are always at opposite logic states, a logical 1 will be produced at the output of one of the gates U_2, causing the flip-flop to set; Q to logical 1.

In the circuit shown U_2 and U_3 are combined to form a 2 input exclusive NOR with open collector limiting resistors R to be as great as 1.1K. A capacitor is included on the clock input of U_4 to filter logic transition spikes off of the clock line allowing only a true change in the state of U_2 to clock the flip-flop. In this circuit, the Q output of the flip-flop is used to blank the display. The flip-flop must be manually reset; bringing Q to logical 0.

5.2.8.1 Suggested Drive Currents for Stretched Seven Segment Displays Used in Various Ambient Light Levels

The level of ambient light and the display contrast filter must be considered together when establishing the proper LED drive current. The LED drive current must be of sufficient magnitude to illuminate a display segment so that it can be easily recognized behind the contrast filter in the expected user ambient. As the ambient light level increases, the LED drive current must also increase to offset any masking of the display emitted light due to elevated ambient.

Table 5.2.8.1-1 lists suggested drive currents for 7.62mm (.3 inch) and 10.92mm (.43 inch) stretched seven segment displays when used in a given ambient light level with a specific filter. The currents listed should be considered as starting points and should be adjusted as necessary for each specific application. The currents listed allow the 7.62mm devices to be read from a distance of 4 meters (6 meters for the 10.92mm devices) in the corresponding ambient.

AMBIENT LIGHTING CONDITION	LED DRIVE MODE	.3 STD RED CATEGORY[2] C	.43 STD RED CATEGORY[2] B	.43 HER CATEGORY[2] AA	.43 YELLOW CATEGORY[2] B	.43 GREEN CATEGORY[2] B
DIM (HOME OR LOW LIGHT LEVEL CONTROL ROOM) (10–100 lux)	DC DRIVE	4 mA	5 mA	3 mA	5 mA	7 mA
	STROBED DRIVE[1]	3.4 mA	4.4 mA	2.1 mA	3.8 mA	4.9 mA
	FILTERS	RH 2423 P60 H1605 3MV 3M655	RH 2423 P60 H1605 3MV 3M655	P65 H1670 3M625	P27 H1720 3M590	P48 H1440 3M565
MODERATE (TYPICAL OFFICE) (100–1000 lux)	DC DRIVE	6 mA	9 mA	6 mA	11 mA	11 mA
	STROBED DRIVE[1]	5.25 mA	8.35 mA	4.5 mA	8.0 mA	7.6 mA
	FILTERS	RH2423 P60 H1605 3MV 3M655	RH2423 P60 H1605 3MV 3M655	P65 H1670 3M625	P27 H1720 3M590	P48 H1440 3M565
BRIGHT (OUTDOORS, VERY BRIGHT OFFICE) (1000–10,000 lux)	DC DRIVE	15 mA	21 mA	9 mA	16 mA	20 mA
	STROBED DRIVE[1]	14.25 mA	20.5 mA	6.25 mA (7.9 mA)	12 mA	13.5[3] mA
	FILTERS	P63 3M655L 3MNDL	P63 3M655L 3MNDL	P60 P65 3M625L 3MNDL	P27 H1720 3M590L 3MNDL	H1425 3M565L 3MNDL

NOTES: [1] FIGURE IS AVERAGE CURRENT. ASSUME $I_{pk} = 5 \times I_{AVG}$

[2] DRIVE CURRENT FOR OTHER CATEGORIES SHOULD BE VARIED BY A RATIO OF 1.5:1 PER CATEGORY WITH ADDITIONAL CORRECTION FOR RELATIVE EFFICIENCY.

[3] $I_{pk} = 4 \times I_{AVG}$

PANELGRAPHIC	SGL HOMALITE	3M COMPANY (L = WITH LOUVERS)
P63 – DARK RED 63	H1605 – RED H100-1605	3MV – VIOLET
P60 – RUBY RED 60	H1670 – RED H100-1670	3M655 – RED 655
P65 – SCARLET RED 65	H1720 – AMBER H100-1720	3M625 – RED 625
P27 – YELLOW 27	H1440 – GREEN H100-1440	3M590 – AMBER 590
P48 – GREEN 48	H1425 – GREEN H100-1425	3M565 – GREEN 565
		3MND – NEUTRAL DENSITY

RH2423 = ROHM & HASS PLEXIGLAS 2423 OR 2444

Figure 5.2.8.1-1

5.3   MONOLITHIC DISPLAYS

5.3.1   Introduction

The monolithic seven segment display represents one of the largest segments of the optoelectronics industry today. Monolithic LED displays are found in many handheld calculators and digital watches. Monolithic LED displays are used because of their low cost, small size, and low power requirements. Monolithic displays have a very high sterance which allow them to be used outdoors and in other areas of high ambient lighting. While monolithic displays have generally been used for low cost applications, improvements in character size and viewing angle have also made monolithic displays suitable for portable instruments and other high quality applications. Table 5.3.1-1 compares the relative merits of a premium monolithic display to other large seven segment displays that could be used in a portable instrument application.

Monolithic displays differ from other types of LED displays in that the individual light emitting segments are formed by diffusing separate LED junctions on a single chip of GaAsP. Usually the GaAsP substrate is n doped material and each LED junction is formed by a p+ diffusion, so monolithic displays are normally of common cathode configuration. Although most monolithic displays are constructed with

	MONOLITHIC	LARGE SEVEN SEGMENT	
1) CHARACTER SIZE	4.45 mm (.175 inches)	7.62 mm (.300 inches)	10.92 mm (.430 inches)
2) DIGIT SPACING	5.84 mm (.230 inches)	10.16 mm (.400 inches)	12.70 mm (.500 inches)
3) MAXIMUM VIEWING DISTANCE	2 m	4 m	6 m
4) VIEWING ANGLE	$\pm 38°$	$\pm 75°$	$\pm 75°$
5) TYPICAL STERANCE	300 cd/m^2 (90 fL) @2mA avg/seg 1/5 DF	120 cd/m^2 (35 fL) @15 mA avg/seg 1/5 DF	130 cd/m^2 (38 fL) @6 mA avg/seg 1/5 DF
6) POWER REQUIREMENTS	1.5–2.0 mA avg/seg	10–15 mA avg/seg (Standard Red)	4–6 mA avg/seg (High Efficiency Red)
7) DECIMAL LOCATION	RHD$_p$, CD$_p$ Colon Available	RHD$_p$, LHD$_p$	RHD$_p$, LHD$_p$
8) EASE OF USE	Single Multidigit Package Inherent Digit Alignment Inherent Luminous Intensity/Matching Must be Multiplexed Common Cathode	Single Digit Package DC or Multiplexed Operation Common Anode or Common Cathode	Single Digit Package DC or Multiplexed Operation Common Anode or Common Cathode

TABLE 5.3.1–1 COMPARISON OF MONOLITHIC AND LARGE SEVEN SEGMENT DISPLAYS

the standard seven segment format, almost any number of segments or character shapes can be diffused into the GaAsP substrate. Figure 5.3.1-1 shows the construction of a monolithic GaAsP chip. Since several segments are diffused into a single GaAsP chip, monolithic displays cannot be constructed on a GaP substrate. GaP is relatively transparent to red light, so the contrast between on and off segments would be inadequate. Because GaAsP is relatively expensive, most monolithic displays are magnified to keep chip sizes small, while attaining a viewable character size. In most cases, the monolithic display is magnified by an external lens. The resulting apparent character height is equal to the actual character height times the magnification of the lens.

5.3.2 Effect of External Lens

The external lens has a very important impact in the appearance of the final display. Besides increasing the apparent character height, the lens influences the radiation pattern, the axial luminous intensity, and the viewing angle of the display. A lens increases the axial luminous intensity by the square of the magnification:

$$I_{V\ (MAG)} = m^2\ I_{V\ (UNMAG)} \qquad (5.3.2\text{-}1)$$

The addition of a lens may also increase the fresnel loss of the optical system. Fresnel loss, which is explained in more detail in Section 2.1.2, is the loss of light due to reflection as light passes from a medium with one index of refraction to a medium with a different index of refraction. The efficiency of light transmission between two mediums can be expressed as:

$$T = \frac{4}{2 + n_2/n_1 + n_1/n_2} \qquad (5.3.2\text{-}2)$$

where T is the transmission coefficient, n_1 is the index of refraction of the first medium, and n_2 is the index of refraction of the second medium.

Lenses for monolithic displays fall into two categories -- immersion lenses and non immersion lenses. Immersion lenses are formed by molding a lens directly over the LED chip, non immersion lenses have at least one layer of air between the LED chip and the lens assembly. The axial luminous intensity for either type of lens is equal to:

$$I_{V\ (MAG)} = (\prod_{i=1}^{n} T_i)\ m^2\ I_V \qquad (5.3.2\text{-}3)$$

where $I_{V\ (MAG)}$ is the magnified axial luminous intensity, T_i is the transmission coefficient from medium i-1 to medium i, n is the total number of medium interfaces, m is the magnification of the lens, and I_V is the axial luminous intensity just beneath the surface of the LED chip.

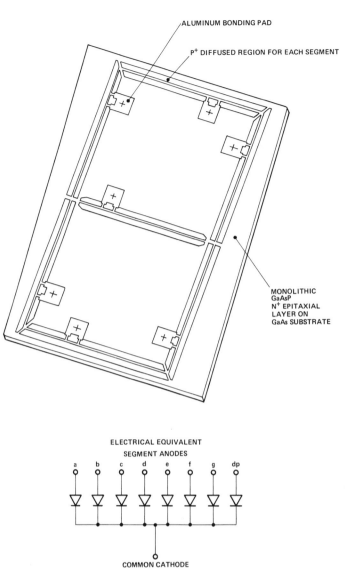

Figure 5.3.2-1 shows examples of immersion and non immersion lenses and compares bare chip axial luminous intensity to magnified axial luminous intensity for immersion and non immersion lenses. This increase in axial luminous intensity allows a reduction in display drive currents to achieve the same luminous intensity as an unmagnified display.

The maximum off axis viewing angle at which any display can be observed is limited either by the radiation pattern of the display or by the maximum amount of distortion before the display becomes unreadable. An unmagnified LED device generally has very close to a lambertian radiation pattern, that is the luminous intensity varies as the cosine of the off axis angle:

$$I_V\ (\theta) = I_V\ \cos\theta \qquad (5.3.2\text{-}4)$$

Figure 5.3.1-1 Construction of LED Monolithic Chip

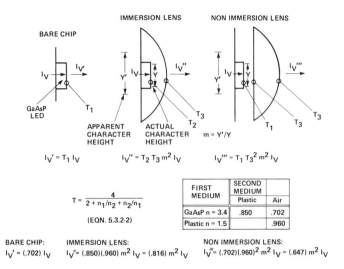

$$I_V' = T_1 I_V \qquad I_V'' = T_2 T_3 m^2 I_V \qquad I_V''' = T_1 T_3^2 m^2 I_V$$

$$T = \frac{4}{2 + n_1/n_2 + n_2/n_1}$$

(EQN. 5.3.2-2)

FIRST MEDIUM	SECOND MEDIUM	
	Plastic	Air
GaAsP n = 3.4	.850	.702
Plastic n = 1.5		.960

BARE CHIP:
$I_V' = (.702) I_V$

IMMERSION LENS:
$I_V'' = (.850)(.960) m^2 I_V = (.816) m^2 I_V$

NON IMMERSION LENS:
$I_V'' = (.702)(.960)^2 m^2 I_V = (.647) m^2 I_V$

Figure 5.3.2-1 Comparison of the Axial Luminous Intensity of Bare Chip, Immersion Lens, and Non-Immersion Lens Systems.

where $I_V (\theta)$ is the off axis luminous intensity, θ degrees off axis, and I_V is the axial luminous intensity.

Thus at an angle of $60°$ (total included angle of $120°$) the luminous intensity will be 50% of the axial luminous intensity. The maximum viewing angle for a display of this type is limited for a practical purposes to about $\pm75°$ because of the requirement that the display be recessed behind a filter. The radiation pattern of a magnified system may or may not be lambertian. However, in a magnified display, the viewing angle is generally limited by character distortion. The maximum viewing angle of a seven segment display can be defined as the off axis angle when the display is rotated that causes any of the segments to narrow or disappear. The maximum viewing angle can further be defined as the vertical viewing angle (segments a, d), the horizontal viewing angle (segments b, c, e, f), and the decimal point viewing angle. The maximum viewing angle is determined by magnification, type of lens, shape of lens, distance between the lens and the monolithic GaAsP chips and the spacing between characters. The calculation of maximum viewing angle is beyond the scope of this text. Figure 5.3.2-2 summarizes the relationships between magnification, luminous intensity, and viewing angle for magnified displays.

The earliest magnified monolithic displays used cast epoxy immersion lenses. The immersion lens served to increase the character height of the display, increase the axial luminous intensity and still provide a very acceptable viewing angle. Because of the relatively high cost of manufacturing this type of display, manufacturers began to experiment with other lower cost methods to manufacture a magnified monolithic display. The most commonly used technique that was developed was to die attach the monolithic GaAsP chips to some kind of substrate and then fasten an injection molded lens to the substrate. This technique was able to

reduce the manufacturing cost, improve the monolithic chip alignment, and improve the optical consistency between individual lenses in the display. The lens shape has evolved from spherical immersion lenses to plano-convex non immersion lenses, and then to positive meniscus non immersion lenses. The first non immersion lens consisted of several spherically shaped plano-convex "bubble" lenses that were molded into a linear array such that each monolithic digit was magnified by a separate "bubble" lens. Lens designers discovered that a positive meniscus lens with two spherically shaped surfaces allowed a much wider viewing angle than the simple plano-convex lens. The present state of the art is a positive meniscus lens with a spherical inner surface and an aspheric outer surface. This lens design can be optimized to allow almost as large a viewing angle as an immersion lens with the same magnification or to reduce off axis distortion of the monolithic character. An additional cylindrical magnifier can also be attached over the primary magnifying lens to further increase character height with the reduction of some viewing angle.

5.3.3 Construction of Monolithic Displays

Monolithic displays can be classified into two basic types according to whether the lens is an immersion or non immersion lens. Monolithic displays constructed with immersion lenses are manufactured by die attaching the monolithic GaAsP chips to either a ceramic substrate or to a lead frame. The die attach pad also forms the common cathode electrical connection to the monolithic chip. The aluminum contact for each segment on the monolithic chip is then wirebonded to the appropriate anode contact. A ball bond is formed on the LED top contact and a wedge bond is formed on the lead frame. The completed device is then encapsulated in epoxy. The magnifying lens is formed during encapsulation. Figure 5.3.3-1 shows the typical mechanical construction of a monolithic display constructed on a lead frame with a cast epoxy immersion lens.

Monolithic displays with non immersion lenses are usually constructed by epoxy die attaching the monolithic GaAsP chips to a special high temperature printed circuit board. The length of the printed circuit board depends on the desired number of digits and the desired digit spacing, since this type of display is generally not end stackable. Then the aluminum contact for each segment on the monolithic chip is wire bonded to the appropriate anode trace on the printed circuit board. A precision injection molded lens is then aligned and attached to the printed circuit board using holes that were drilled during fabrication of the printed circuit board. This insures accurate alignment of the lens over the LED chips. A secondary cylindrical lens can also be attached for added character height. Figure 5.3.3-2 shows the typical mechanical construction of a monolithic

EFFECTS OF MAGNIFICATION ON INTENSITY, VIEWING ANGLE, AND APPARENT SIZE

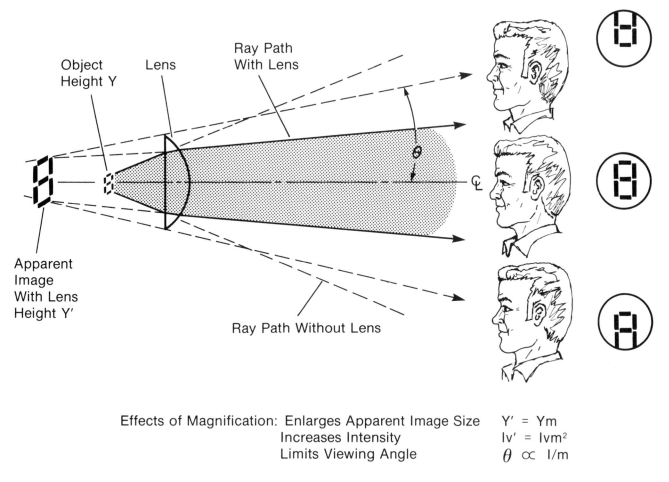

Effects of Magnification: Enlarges Apparent Image Size $Y' = Ym$
Increases Intensity $Iv' = Ivm^2$
Limits Viewing Angle $\theta \propto 1/m$

Figure 5.3.2-2 Effects of Magnification on Luminous Intensity
Viewing Angle, and Apparent Character Size.

display manufactured with these techniques. Displays constructed with these techniques are commonly available in digit strings from five to fifteen digits. Table 5.3.3-1 compares the differences between these two packaging techniques with respect to manufacturing costs, ease of use by the customer, optics, and package reliability.

5.3.4 Electrical-Optical Characteristics

Because of their construction, seven segment monolithic displays are generally connected as an 8xN x-y addressable array, where N refers to the numer of digits in the display. Each of the eight anode lines addresses a different segment (a to dp) and each of the N cathode lines addresses a particular digit. This allows a five digit cluster to be constructed on a 14 pin lead frame and a fifteen digit monolithic display to be constructed with only 23 active pins. While substantially reducing the number of wires used to interconnect to the display, it also requires that the display be strobed on a 1 of N or lower duty factor.

The voltage and current characteristics of each monolithic LED are similar to the voltage and current characteristics of a single diode LED lamp. Each monolithic LED is effectively isolated from adjacent LED segments below the V_{BR} of the display. Because monolithic displays are constructed on a GaAsP substrate, they have a very low dynamic resistance in the forward region. This allows the monolithic display to be multiplexed at relatively high peak currents for higher luminous efficiency with forward voltages typically less than 1.8 volts.

Every monolithic display has upper and lower limits on the peak and average currents at which the device can be operated. The maximum peak forward current per segment has been chosen by the manufacturer to limit the maximum current density through the semiconductor junction. The average current per segment is limited by the ambient temperature and the thermal resistance of the display. In general, the junction temperature should be prevented from exceeding the maximum allowable storage temperature for

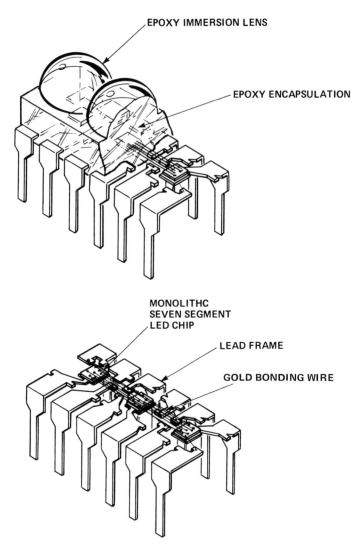

Figure 5-3.3-1 Mechanical Construction of a Monolithic Display with an Immersion Lens.

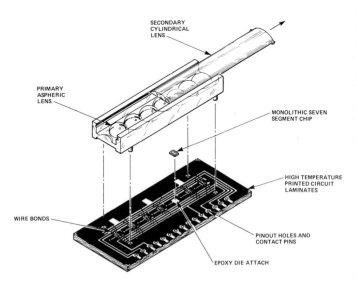

Figure 5-3.3-2 Mechanical Construction of a Monolithic Display Constructed on a PC Board With a Non-Immersion Lens.

the particular display. Typically monolithic displays with cast epoxy immersion lenses are limited to 100°C storage temperature and monolithic displays with non immersion lenses are limited to 85°C storage temperature. Above these temperatures, the reliability of the display may be impaired, or the lens may permanently be deformed. In most applications, monolithic displays can be driven at considerably lower average currents than this maximum average current. One exception is in high ambient applications where it is desirable to have a very high sterance display and a very dense filter to achieve an acceptable contrast ratio. The minimum peak current at which the display can be operated is determined by the junction area of each segment. Below a minimum current density, the luminous intensity matching between segments and digits may be unacceptable.

On most monolithic displays, the decimal point is located either in the center of the digit or in the lower right hand corner of the digit. For low cost applications, the right hand decimal point is recommended. The center decimal point requires a full digit location and time frame to be allocated to display the decimal. The center decimal point is an asset in quality instruments because it reduces the chances of an error caused by misreading the decimal point location. The center decimal point also allows a slightly wider viewing angle for the display since the monolithic chip can then be centered under the magnifying lens.

Most monolithic displays emit light with a peak wavelength of 655 nm. This wavelength represents the best compromise between the human eye response curve and the quantum efficiency of the GaAsP material.

Monolithic displays are usually characterized for luminous intensity under drive conditions that simulate typical usage conditions. A graph showing the relative luminous efficiency vs. peak forward current allows the designer to determine the time averaged luminous intensity for other peak and average forward currents. An example of a "relative luminous efficiency" curve is shown in Figure 5.3.4-1. The time averaged luminous intensity can be determined for any other drive condition by equation 5.3.4-1:

$$I_V \text{ TIME AVG} = \frac{[I_{PEAK}] [DUTY\ CYCLE] [\eta(I_{PEAK})] [I_{V\ SPEC}]}{[I_{SPEC}] [DUTY\ CYCLE\ SPEC] [\eta(I_{SPEC})]} \qquad (5.3.4\text{-}1)$$

where I_{PEAK} is the desired peak current, DUTY CYCLE is the ratio of time the LED is "on" to total time, η is the relative luminous efficiency of the LED at I_{PEAK} or at I_{SPEC} and I_{SPEC} and DUTY CYCLE SPEC are the test conditions under which $I_{V\ SPEC}$ was originally characterized.

	Lead Frame Construction Cast Immersion Lens	Printed Circuit Construction Injection Molded Non-Immersion Lens
1) MANUFACTURING COST	Medium	Low
2) EASE OF USE		
a) Digit Number	2, 3, 4, 5 — end stackable (multiple clusters)	5, 8, 9, 12, 14, 15 — single module
b) Socketing	.3" DIP	Special sockets available
c) Solderability	Wave, or Hand	Hand Solder
d) Cleaning	Vapor Degreasing	Hand Clean Only
e) Display Alignment	Clusters should be aligned	Pre-alignment included
f) Luminous Intensity	Clusters should be matched	Intensity Matching included
g) Repairability	Replace Cluster	Replace Display
3) OPTICS		
a) Viewing Angle	Good	Depends on Lens design — can be as good as Immersion Lens
b) Character Height	Varies slightly due to lens casting 2.79 mm (.11")	Precision (may be increased by Optional Magnifier) 2.54 mm (.10") to 4.45 mm (.175")
4) RELIABILITY		
a) Temp Storage	-40°C to $+100^\circ$C	-20°C to $+85^\circ$C
b) Temp Cycling	Good	Very Good
c) Moisture Resistance	Good	Condensing atmosphere should be avoided
d) Mechanical Vibration	Very Good	Reasonably Good

Table 5.3.3-1 Comparison Between Monolithic Displays Constructed
with Cast Immersion Lenses and Monolithic Displays
with Non-Immersion Lenses.

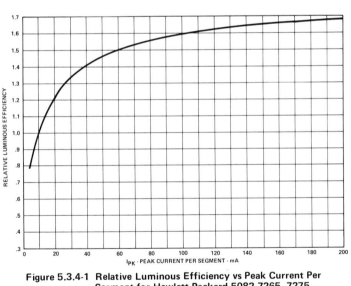

Figure 5.3.4-1 Relative Luminous Efficiency vs Peak Current Per
Segment for Hewlett-Packard 5082-7265, 7275,
7285, 7295 Monolithic Displays.

For example, suppose a monolithic device is tested for luminous intensity at 10 mA peak per segment, 20% duty cycle and the typical time averaged luminous intensity is 70 μcd/segment. If the same display is strobed at 80 mA peak per segment on a 2.5% duty cycle, the typical time averaged luminous intensity will be:

$$I_V = \frac{(80)(.025)(1.55)(70)}{(10)(.20)(1.00)} \qquad (5.3.4\text{-}2)$$

$$= 108 \mu\text{cd/segment}$$

Another representation of the "relative luminous efficiency" curve is shown in Figure 5.3.4-2. In this figure, equation 5.3.4-1 and Figure 5.3.4-1 have been used to calculate typical time averaged luminous intensities for a wide variety of drive conditions. This figure graphically shows the advantage of strobing at higher peak currents to obtain a higher time averaged luminous intensity for the same average current. For example suppose a five digit display is strobed on a 20% duty cycle at 1.5 mA average current per segment (7.5 mA peak/segment), then the typical time averaged luminous intensity would be 50 μcd/segment. This represents a very acceptable luminous intensity for this type of display under ambient conditions similar to an office. If the display is strobed at higher peak currents, the average current per segment can be reduced to get the same time averaged luminous intensity or the time averaged luminous intensity can be increased. At 200 mA peak current per segment, the average current can be reduced to .82 mA per segment and still obtain the same 50 μcd/segment; or by maintaining the same average current, the time averaged luminous intensity can be increased to 90 μcd/segment.

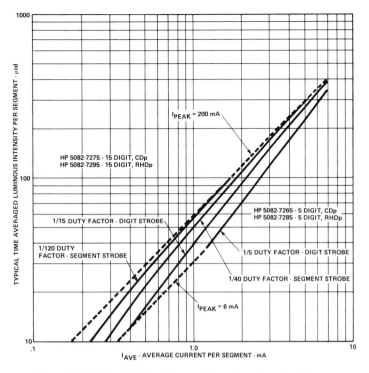

Figure 5-3.4-2 Typical Time Averaged Luminous Intensity Per Segment vs Average Current Per Segment for Hewlett-Packard 5082-7265, 7275, 7285, 7295 Monolithic Displays.

5.3.5 Driving Monolithic Displays

Monolithic displays are driven in the same way as any other common cathode strobed display. The advantage of monolithic seven segment displays is that they can be driven at considerably lower average currents than large seven segment displays for a satisfactory luminous intensity. For example, a monolithic display designed for a low cost calculator can be satisfactorily driven at average currents as low as .25 mA per segment and a somewhat larger monolithic display such as one designed for a portable instrument can be driven at average currents as low as 1 mA per segment. By comparison, a high efficiency red large seven segment display probably requires 4 to 6 mA average current per segment for a comparable luminous intensity. Since the monolithic LED display can be driven at such low average currents, it is particularly suitable for battery powered applications. Furthermore, a monolithic LED display can often be driven directly from many readily available integrated circuit devices.

Monolithic displays are commonly driven either with digit or segment strobe techniques. When a display is digit strobed, each digit is sequentially enabled and at the same time the appropriate combination of segments is turned on. For an N digit display, the maximum duty factor for a digit strobed display is 1/N. If a display is segment strobed, then each digit is sequentially enabled and while that digit is enabled, each segment is sequentially turned on. The maximum duty factor for a segment strobed system is

$1/(8N)$. Thus for the same time averaged luminous intensity, the segment strobed display must drive each LED segment at a higher peak current where the LED has a higher quantum efficiency. This allows the display to be driven at a lower average current for the same time averaged luminous intensity. While more logic is required for a segment strobed system, segment strobing may also be preferable in custom MOS designs because the peak digit driver currents can be reduced. For example, suppose a five digit monolithic display can be described by the "relative efficiency" curve shown in Figure 5.3.4-1. The typical time averaged luminous intensity for this display is shown in Figure 5.3.4-2. When this display is digit strobed at 7.5 mA peak current per segment on a 1/5 duty factor, the typical time averaged luminous intensity is 50 μcd per segment. Each segment driver must be able to source 7.5 mA and each digit driver must be able to sink 60 mA from the display. When the display is segment strobed on a 1/40 duty factor, the same time averaged luminous intensity can be achieved at 39 mA peak current per segment. Thus each segment driver would be required to source 39 mA and each digit driver would be required to sink 39 mA from the display, a 35% reduction in average current per segment.

The primary difference between driving a monolithic display with a center decimal point and driving a monolithic display with a right hand decimal point is that the segments surrounding the center decimal point should be blanked when the decimal point is turned on. With a right hand decimal point, each digit is displayed regardless of the decimal location. Figure 5.3.5-1 shows a typical CMOS logic interface to a five digit monolithic display. The outputs of the National 74C90 counter sequentially enable each of the five digit cathodes and at the same time enable circuitry inside the digital subsystem to output the desired segment and decimal information to the display. If a display with a center decimal point is used, the proper digit can be blanked either by outputting a BCD code that is decoded as a blank state by the seven segment decoder (state 15 for the National 74C48) or using the blanking input of the seven segment decoder. A fixed decimal point can be implemented simply by connecting the Dp input to the desired cathode driver. Then whenever the selected cathode driver is on, the decimal point will be turned on. Several LSI integrated circuits are available that combine all the circuitry shown in Figure 5.3.5-1 into one or a few integrated circuit packages.

5.3.6 Interfacing Microprocessors to Monolithic Seven Segment Displays

Microprocessors can be interfaced to monolithic displays with only a few external components. In general, these external components will include one or more latches to hold the information to be displayed and additional circuitry to periodically request new information from the display. For displays of up to eight digits, either of the

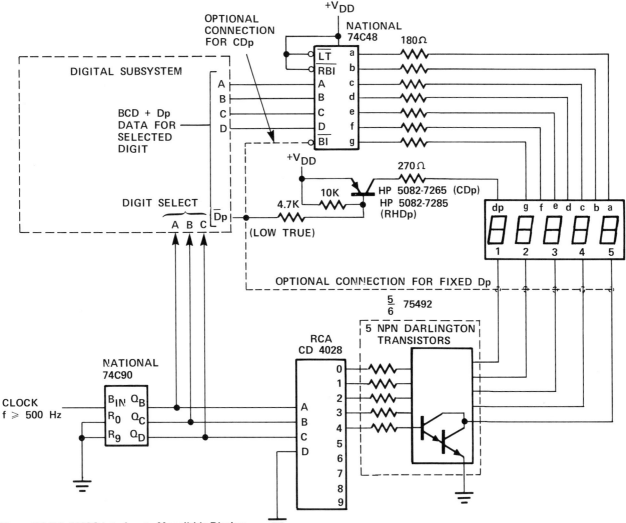

Figure 5.3.5-1 CMOS Interface to Monolithic Display.

circuits described in Figures 5.2.6-2 or 5.2.6-3 can be used. Monolithic displays of up to 16 digits can be driven from the circuit shown in Figure 5.3.6-1.

This circuit differs from any of the circuits shown in Section 5.2.6 in the way the decimal point and interrupts are handled. The microprocessor outputs only a single byte of information to the display. Four bits contain the segment and decimal point information and four bits select the proper digit enable. The upper six states (10 to 15) of the RCA CD4511 seven segment decoder are decoded as blank states. Four of these states are decoded externally to select the decimal point or minus sign. In this example, states 10 and 11 are decoded as a minus sign, states 12 and 13 are decoded as the decimal point select, and states 14 and 15 are decoded as blank states. Every 667µs, the refresh clock requests an interrupt from the Motorola 6800 microprocessor. Upon requesting an interrupt, the interrupt circuitry (not shown) must force a vectored interrupt to address "REFRESH". The interrupt request can also be decoded by a software polling program but will increase the microprocessor time that is required to service the display.

Then the refresh program shown in Figure 5.3.6-1b is executed. The program reads a pointer from RAM and then outputs the contents of the memory location specified by the pointer to the display. Then the pointer is compared to the address of the final byte of information in the data file. If they are equal, the pointer is reset to the address of the first byte of information in the data file, otherwise, the pointer is set to the address of the next byte of information in the file. Since the decimal point is decoded separately from the other seven segments, a separate time frame is required to display the decimal point. Thus, a 15 digit monolithic display with a right hand decimal point will require 16 bytes of information in the data file. The microprocessor time required to refresh any display of up to 16 digits can be determined as shown below:

$$(5.3.6-1)$$

$$\text{REFRESH TIME} = \frac{(52N+1)R}{\text{MICROPROCESSOR OR CLOCK RATE}}$$

5.41

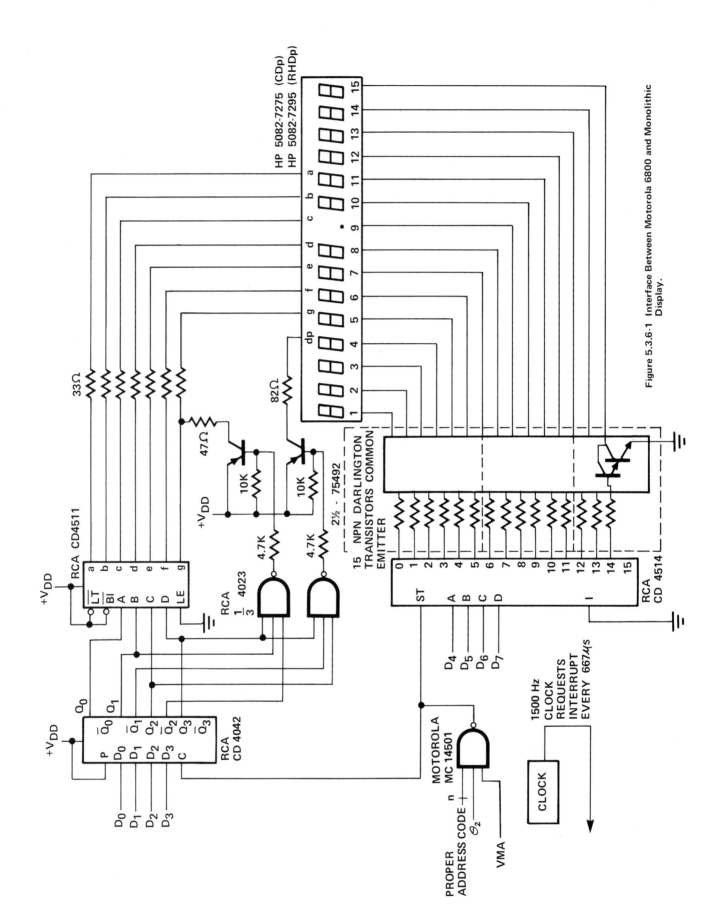

Figure 5.3.6-1 Interface Between Motorola 6800 and Monolithic Display.

5.42

ADDRESS	OP CODE	CLOCK CYCLES	COMMENTS
REFRESH	LDX	5	IX = POINTER
	A_H	----	
	A_L	----	
	LDAA, X	5	AA = (IX)
	O	----	
	STA A	5	STORES NEW SEGMENT AND DIGIT
	$DISP_H$	----	INFORMATION
	$DISP_L$		
	CPX #	3	IX − $(XXIE)_{16}$
	$(XX)_{16}$	----	
	$(1E)_{16}$	----	
	BNE	4	JUMP IF IX ≠ $(XXIE)_{16}$
	$(07)_{16}$	----	
	LDAA #	2	AA = $(10)_{16}$
	$(10)_{16}$	----	
	STA A	5	POINTER = $(XX10)_{16}$
	A_H	----	
	$A_L + 1$	----	
	CLI	2	CLEAR INT FLAG
	RTI	10	RETURN
	INC	6	POINTER = POINTER + 1
	A_H	----	
	$A_L + 1$	----	
	CLI	2	CLEAR INT FLAG
	RTI	10	RETURN

MEMORY ADDRESS	CONTENTS		BCD DATA TABLE	
$A_H A_L$	$POINTER_H$		BCD	OUTPUT
$A_H A_L + 1$	$POINTER_L$			
$(XX10)_{16}$	0000		0000	0
$(XX11)_{16}$	0001		0001	1
$(XX12)_{16}$	0010		0010	2
$(XX13)_{16}$	0011		0011	3
$(XX14)_{16}$	0100		0100	4
$(XX15)_{16}$	0101	BCD	0101	5
$(XX16)_{16}$	0110	DATA	0110	6
$(XX17)_{16}$	0111		0111	7
$(XX18)_{16}$	1000		1000	8
$(XX19)_{16}$	1001		1001	9
$(XX1A)_{16}$	1010		1010	−
$(XX1B)_{16}$	1011		1011	−
$(XX1C)_{16}$	1100		1100	.
$(XX1D)_{16}$	1101		1101	.
$(XX1E)_{16}$	1110		1110	.
			1111	

POINTER (NEXT DIGIT TO BE DISPLAYED)

X= DON'T CARE

Figure 5.3.6-1b Refresh Program Used to Strobe the Monolithic Display Shown in Figure 5.3.6-1.

where N is the number of digits on a CDp display or one plus the number of digits on a RHDp display, and R is the strobing rate.

For example, suppose a fifteen digit RHDp display is refreshed at a 100 Hz refresh rate and the microprocessor uses a 1 MHz clock, then 8.3% of the microprocessor's time is required to service the display. For the remaining 91.7% of the time, the microprocessor can be performing other functions. While this example used a Motorola 6800 microprocessor, this technique can be used with any other microprocessor by modifying the hardware and software to reflect a different instruction set and different ways to output information to peripheral devices.

5.4 Alphanumeric Displays

Seven segment display devices are capable of transmitting only a limited number of useful different states of information. Theoretically, up to 49 different information states could be depicted utilizing the 7 segment font, however, less than 1/3 of these have any meaning in normally utilized communications context. By increasing the number of elements in a display character from 7 to 9, it becomes possible to depict in rough form most of the Roman alphabet and arabic numerals. Utilization of a 14 or 16 segment character allows better symbol differentiation and use of some special symbols. However, the lower case alphabet cannot be represented, and asthetic quality and readability are still poor. By increasing the number of elements still further, it is possible to add further refinement to the display quality. By utilizing an infinite number of elements, virtually any 2 dimensional character and character shape could be achieved. In practice, it is found that 35 elements in a 5x7 array or 63 elements in a 7x9 array offer satisfactory resolution for most alphabetic language communications. The trade-off is one of the cost of supplying and addressing the individual elements vs. the improvement gained from higher resolution. For LED alphanumeric dot matrix displays, the industry standard product is the 35 element 5x7 array.

5.4.1 The 5x7 LED Array

Figure 5.4.1-1 depicts schematically one digit of a 5x7 array. The row lines are tied to the anodes of all of the diodes of a particular row and are designated by the Roman numerals I-VII. The column lines tie together the cathodes of each diode of a particular column and are designated A-E. In multidigit arrays, the columns of digit one are designated 1A-1E, digit two 2A-2E, etc. The row lines are extended in common to all digits in an array. The effect of this type of organization is to form a 7x5N crosspoint matrix (N = number of characters) of diodes where an individual diode is energized by selecting the appropriate row and column lines. Using this technique, a 4 digit array of 140 diodes may be addressed using only 27 lines (7 row lines + 4x5 digit lines). If decoding or data storage circuits are included inside the display package, the external pin count can be even further reduced.

5.4.2 Character Generation in 5x7 Arrays

The generation of character information in a 5x7 array is somewhat more complex than the techniques utilized for 7 segment displays. In the 7 segment display, all elements of a digit can be simultaneously addressed so that the entire character is formed during one address cycle. In a 5x7 array, each character must be made up of 5 or 7 subsets of data which are presented during sequential address cycles.

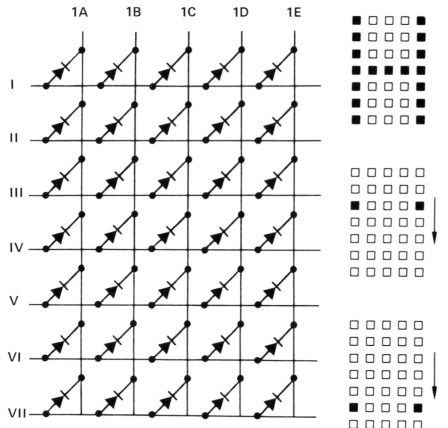

1A 1B 1C 1D 1E

I

II

III

IV

V

VI

VII

Figure 5.4.1-1 5 x 7 LED Matrix

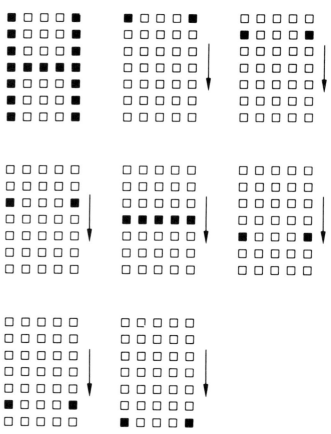

Fig. A.   Row Strobing

By the nature of its electronic organization, then, a 5x7 display character must be operated on a strobed basis. Operating the display on this basis, though it increases the system complexity somewhat, does offer advantages which clearly outweigh the negative aspects, even for a single character display.

- A minimum number of interconnects are used to interface between the data source and the display.

- The cost of the character generation and associated drive circuitry is shared over many digits.

- Clock, timing and data storage elements required for the strobed display can be shared with other portions of the system.

There are two common methods for addressing the diodes in the 5x7 array in order to generate character information. Both of these methods depend on the eye's response to a strobed display. Figures 5.4.2-1a,b depict the sequence of steps necessary to form the character "H" using either row or column strobing techniques.

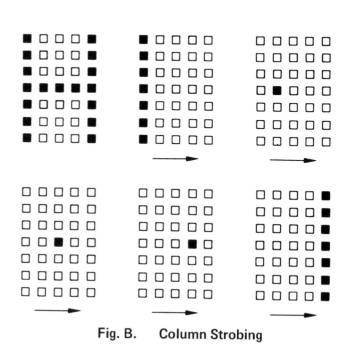

Fig. B.   Column Strobing

Figure 5.4.2-1a,b Character Generation Using Row (a) or Column (b) Strobe Methods.

5.44

In row strobing, data for one row of the desired character is applied to the column lines and the row is then energized. This process is repeated for each of seven rows. For row strobing of a multicharacter string, row data for each character in the string is decoded and stored in a data latch associated with each character. The row line is then enabled, displaying simultaneously one row of data for all of the characters in the string.

In column strobing, data for one column of the desired character is supplied to the row lines and the column is then energized. The process is repeated, varying the data, until all of the columns of the character have been selected. Multicharacter displays are treated as simply an extension of the single character case such that the number of subsets of data generated will be equal to 5N, where N is equal to the number of digits in the display.

If either of these sequences is repeated at a rate which insures that each of the appropriate matrix locations is re-energized a minimum of 100 times per second the eye will perceive a continuous image of the entire character. The apparent intensity of each of the display elements will be equal to the intensity of that element during the "ON" period multiplied by the ratio of "ON" time to REFRESH PERIOD. This ratio is referred to as the display DUTY FACTOR.

From a conceptual standpoint, column scanning is the easiest method to understand; however, it is more limited as to the number of digits which may be displayed using only one decoder. The block diagram in Figure 5.4.2-2 depicts the elements of a column strobing system. Circuit operation is as follows:

Coded character information from the keyboard is first stored in the input storage buffers, generally six bits per character. The timing circuitry then selects the first character word, from storage buffer number one, presenting this data to the seven line output ROM. The timing circuitry also inputs the column one select code to the ROM causing the ROM output states in turn to present column 1, character 1 data at its output terminals. These states turn on the row drivers for the appropriate diodes in the first column of the display. The first column is then enabled by its column driver for an appropriate "ON" period. The second column data is then selected and displayed, etc., until all five columns of the first character have been displayed.

The timing circuit then selects the data for the second character from buffer number two, and displays each of the columns. This procedure is repeated until all columns of all characters have been displayed. If each column is refreshed at 100 times per second, the eye will not perceive flicker in

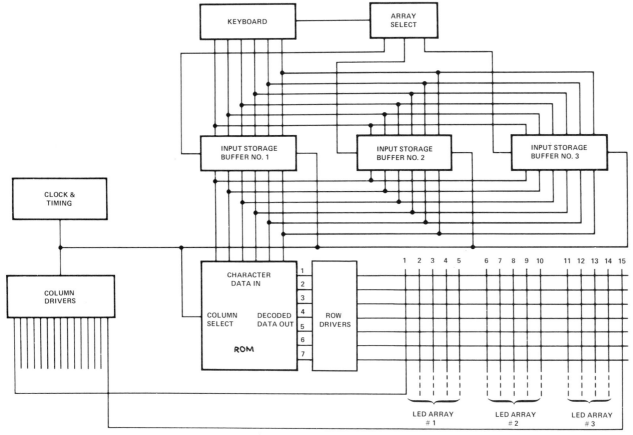

Figure 5.4.2-2 Block Diagram of a Column Strobed Alphanumeric Display.

5.45

the display. Herein lies the problem with column strobing. Since the maximum peak current per diode is limited to about 100 mA and the desired minimum average current per diode is around 1 mA, the maximum number of columns which could be displayed is 100 columns or 20 characters. Given the practicalities of timing and switching, this number is more realistically 16 characters.

The row strobing scheme eliminates the problem of high peak currents and consequent severe limitations on the number of digits which can be strobed in one display string utilizing only one decoder. However, there is a substantial increase in circuit complexity. The block diagram for a row strobing scheme is shown in Figure 5.4.2-3.

Coded alphanumeric information is sequentially entered and stored in the input storage buffers. Next, the timing circuitry enables the ROM and selects data word No. 1. An additional input code selects row I output data from the ROM. The first row of character information is stored in the five bit output storage buffer No. 1. Each of the remaining words are selected, decoded into first row character information and loaded into the five bit output storage registers corresponding to each of the characters of the display. At this point in the sequence, each of the five output storage buffers has been loaded with the appropriate display information for the first row of each character. The timing circuitry now turns on the top row driver so current flows through and lights all of the appropriate LEDs in the top row of all of the characters. The complete cycle is repeated to decode and display row II, then row III, and so on through the seventh row. Repeating this scan at a rate above 100 Hz in most applications gives a flicker free alphanumeric character at each of the digit locations. The only practical limitation to the number of characters which can be operated from a single character generator in this mode is imposed by the maximum clock rate of the logic. In general, character string lengths well in excess of 100 digits should not pose a problem.

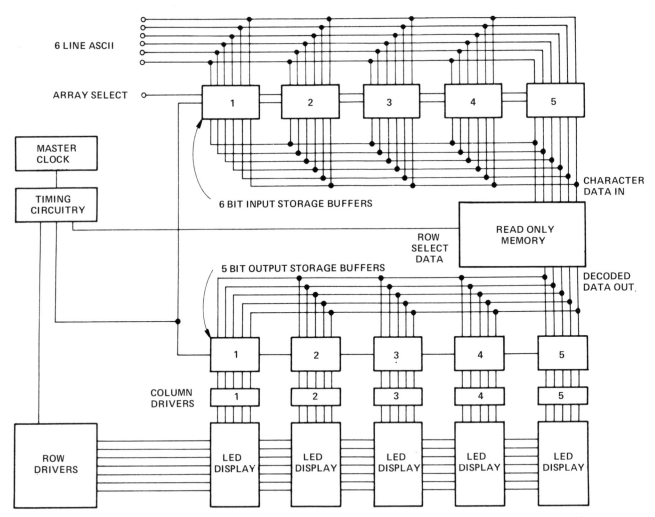

Figure 5.4.2-3 Block Diagram of a Row Strobed Alphanumeric Display.

5.4.3 Implementation of a 16 Character Row Scan Display

The circuit Figure 5.4.3-1 depicts a 16 character alphanumeric display utilizing the row strobing technique. One 7496 five bit shift register is used as the output data storage buffer for each character of the display. Data from the ROM is loaded into a parallel to serial shift register and then shifted into the data storage buffers on each positive transition of the clock. After five clock pulses a new input data word is selected from the ASCII data storage RAM (1½ Fairchild 93403). After all sixteen words (one word per character) have been decoded and shifted into the output data storage buffers, the shift register clock input is gated off and the row line is enabled to light the display. At the end of the display period, the row is turned off, the row counter is incremented by one count, and the second row of data is loaded into the output storage buffer. The row scanner is then enabled while the shift register clock is again gated off. This process is repeated for all rows of the display.

The high true output from the parallel to serial shift register is inverted before loading into the output data buffers. The shift register outputs will then be low for a dot "ON" condition. The output low state of the 7496 is rated to sink 16 mA and can, therefore, be used to directly drive the display columns at 16 mA peak. A 500 KHz input clock rate will give a refresh rate of about 93 Hz. The DUTY FACTOR expressed in Equation 5.4.3-1 is set by the ratio of the loading time to the display time multiplied by $1/m$ where m is equal to the number of rows to be addressed; seven in this case:

$$\text{D.F.} = (1 - \frac{2^4}{2^7}) \quad \frac{1}{m} = 12.5\% \qquad (5.4.3\text{-}1)$$

Average diode current is 2.0 mA.

5.4.4 Alphanumeric Displays with On Board Data Storage

The circuit in Figure 5.4.3-1 utilizes a shift register to store decoded row data during the display cycle. If the shift register and current limiting elements are included along with the display matrix in a single hybrid package, significant improvements are realized in the total cost of display implementation, packing density, and reliability. The HP HDSP-2000 display is designed to provide on-board storage of decoded column data plus constant current sinking row drivers for each of the 28 rows in the 4 character display. This approach allows the user to address each display package through just 11 active interconnections vs. the 176 interconnections and 36 components required to effect a similar function using conventional LED matrices.

Figure 5.4.4-1 is a block diagram of the internal circuitry of the display. The device consists of four LED matrices and two 14-bit serial-in-parallel-out shift registers. The LED matrix for each character is a 5x7 diode array organized with the anodes of each column tied in common and the cathodes of each row tied in common. The row cathode commons of each character are tied to the constant current sinking outputs of 7 successive stages of the shift register. The like columns of each of the 4 characters are tied together and brought to a single address pin (i.e., column 1 of all 4 characters is tied to pin 1, etc.). In this way, any diode in the four 5x7 matrices may be addressed by shifting data to the appropriate shift register location and applying a voltage to the appropriate column.

The two on-board shift registers act as a single serial-in-parallel-out (SIPO) register. The SIPO shift register has a constant current sinking output associated with each shift register stage. The output stage is a current mirror design with a nominal current gain of 10. The magnitude of the current to the reference diode is established from the output voltage of the V_B input buffer applied across the current reference resistors, R. The reference current flow is switched by a transistor tied to the output of the associated shift register stage. A logical 1 loaded into the shift register will turn the current source "ON" thereby sinking current from the row line. A voltage applied to the appropriate column input will then turn "ON" the desired diode.

Data is loaded serially into the shift register on the high to low transition of the clock line. The data output terminal is a TTL buffer interface to the 28th bit of the shift register (i.e., the 7th row of character 4 in each package). The Data Output is arranged to directly interconnect to the Data Input on a succeeding 4 digit display package. The Data, Clock and V_B inputs are all buffered to allow direct interface to any TTL or DTL logic family. The display is organized so that column strobing is utilized in the same manner as in a row strobed system; (i.e., the data for all of the like columns in the display string is loaded into shift register and then the column is energized to generate character information).

For a four character display, 28 bits representing the first column of each of the four characters are loaded serially into the on-board SIPO shift register and the first column is then energized for a period of time, T. This process is then repeated for columns 2 through 5.

If the time required to load the 28 bits into the SIPO shift register is t, then the duty factor is:

$$\text{D.F.} = \frac{T}{5(t + T)} \qquad (5.4.4\text{-}1)$$

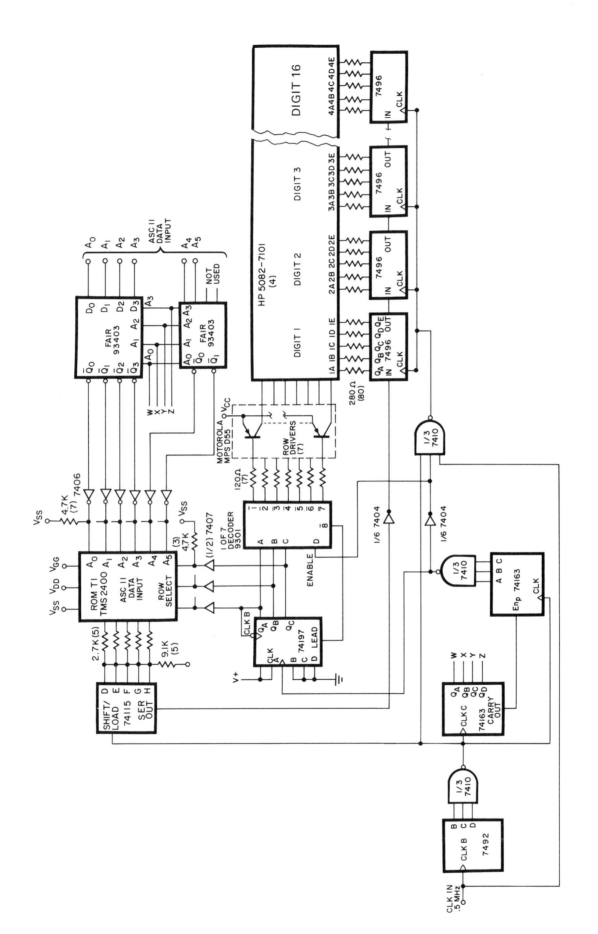

Figure 5.4.3-1   Practical Implementation of a Row Strobed Display.

5.48

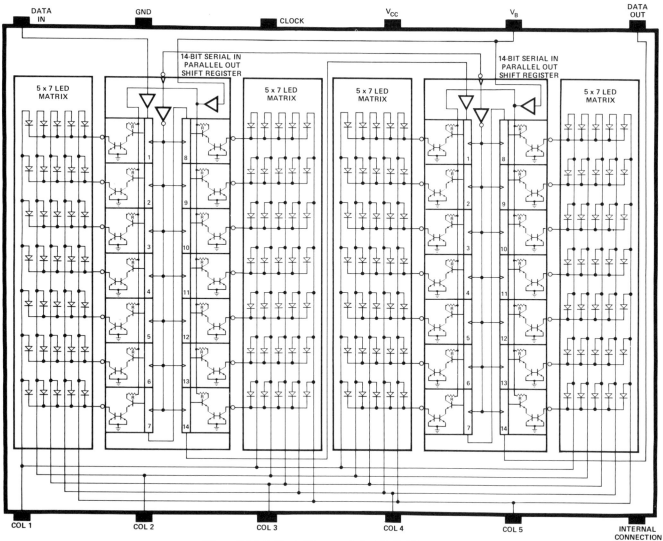

Figure 5.4.4-1 Block Diagram of the HDSP-2000.

the term 5(t+T) is then the refresh period. For a satisfactory display, the refresh period should be:

$$1/[5(t+T)] \geqslant 100 \text{ Hz} \qquad (5.4.4\text{-}2)$$

or conversely

$$5(t+T) \leqslant 10 \text{ ms} \qquad (5.4.4\text{-}3)$$

which gives

$$(t+T) \leqslant 2 \text{ ms} \qquad (5.4.4\text{-}4)$$

Two milliseconds then is the maximum time period which should be allowed for loading and display of each column location. For $t \ll T$, the duty factor will approach 20%. The number of digits which can be addressed in a single string is then dependent upon the minimum acceptable duty factor and the choice of clock rate. For instance, at 1 MHz clock rate, a 100 character string of 25 packages could be operated at a duty factor of

$$(5.4.4\text{-}5)$$

$$\text{D.F.} = \frac{(T+t) - (\# \text{ of bits to be loaded}) \cdot (1/1 \text{ MHz})}{5(t+T)}$$

$$= \frac{(2 \text{ ms}) - 700 \, (1 \, \mu\text{s})}{5 \times 2 \text{ ms}}$$

$$= 13\% \qquad (5.4.4\text{-}6)$$

For most applications, a duty factor of 10% or greater will provide more than satisfactory display intensity. In brightly illuminated ambient environments, a larger duty factor may be desirable whereas, in dim ambient situations, the duty factor may have to be reduced in order to provide a display with satisfactory contrast.

5.4.4.1 Drive Circuit Concept

A practical display system utilizing the HP HDSP-2000 display requires interfacing with a character generator and refresh memory. A block diagram of such a display system is depicted in Figure 5.4.4.1-1. In explanation, assume that this system is for a four character display. Therefore, the 1/N counter becomes a 1/4 counter where N is equal to the number of characters in the string.

The refresh memory is utilized to store the information to be displayed. Information can be coded in any one of several different standard data codes, such as ASCII or EBDIC, or the code and the display font can be customized through the use a custom coded ROM. The only requirement is the output data be generated as 5 subsets of 7 bits each. The character generator receives data from the refresh memory and outputs 7 display data bits corresponding to the character and the column select data input. This data is converted to serial format in the parallel to serial shift register for clocking into the display shift register. In the typical system, the right most character to be displayed is selected first and the data corresponding to the ON and OFF display elements in the first column is clocked into the first 7 shift register locations. In a similar manner, column 1 data for characters 3, 2, and 1 is selected by the 1/N counter, decoded and shifted into the display shift register. After 28 clock counts, data for each character is located in the shift register locations which are associated with the 7 rows of the appropriate LED matrix. The 1/N counter overflows, triggering the display time counter,

enabling the output of the 1/5 column select decoder and disabling the clock input to the display. The information now present in the shift registers will be displayed for a period, T, at the column 1 location. At the end of the display period, T, the divide by 5 counter which provides column select data for both the display and the character generator is incremented one count and column 2 data is then loaded and displayed in the same manner as column 1. This process is repeated for each of 5 columns which comprise the 5 subsets of data necessary to display the desired characters. After the fifth count, the 1/5 decoder automatically resets to "one" and the sequence is repeated. The only changes required to extend this interface to character strings of more than 4 digits are to increase the size of the refresh memory and to change the divide by four counter to a modulus equal to the number of digits in the desired string.

Since data is loaded for all of the like columns in the display string and these columns are then enabled simultaneously, only five column switch transistors are required regardless of the number of characters in the string. The column switch transistors should be selected to handle approximately 110mA per character in the display string. The collector emitter saturation voltage characteristics and column voltage supply should be chosen to provide a $3.0V \leq V_{COL} \leq V_{CC}$. To save on power supply costs and improve efficiency, this supply may be a fullwave rectified unregulated dc voltage as long as the PEAK value does not exceed the value of V_{CC} and the minimum value does not drop below 3.0 volts.

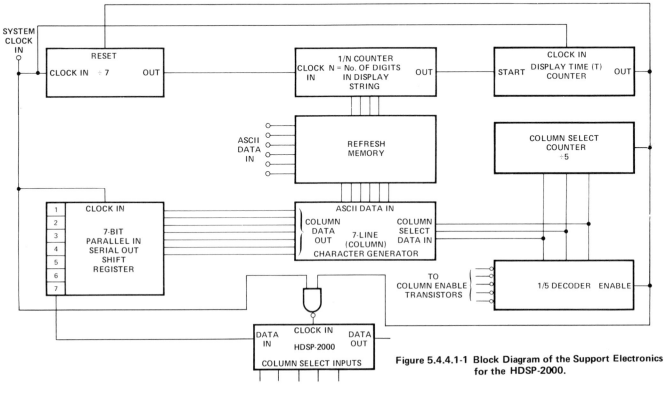

Figure 5.4.4.1-1 Block Diagram of the Support Electronics for the HDSP-2000.

5.50

Since large current transients can occur if a column line is enabled during data shifting operations, the most satisfactory operation will be achieved if the column current is switched off before clocking begins. I_{CC} will be reduced by about 10-15% if the clock is held in the logical 1 state during the display period, T.

5.4.4.2 Interface Circuits for HP HDSP-2000

There are many possible practical techniques for interfacing to the HP HDSP-2000 alphanumeric display. Two basic approaches will be treated here.

5.4.4.2.1 Instrumentation Interface Circuit

The circuit shown in Figure 5.4.4.2.1-1 is for a 16 character display and is designed to function primarily as a readout for general instrumentation systems. CMOS logic circuitry is utilized in this design; however, it should be a simple exercise to substitute TTL functions if CMOS is not desired. In this circuit, a CD 4022 and CD 4520 are combined to perform the functions of the divide by 7, divide by 16 (1/N) and display time counters as depicted in Figure 5.4.4.1-1. The timing diagram, Figure 5.4.4.2.1-2, demonstrates the relationship of the various critical outputs and inputs. The CD4022 actually acts here as a divide by 8 counter with the first count used to latch data into the parallel-in-serial-out (PISO) shift register and the other 7 counts shifting data out of the PISO and into the display shift register. The CD4520 is a dual 4 bit counter wired as an 8 bit binary ripple counter. The NAND gate, U_1, establishes the ratio of loading time to display time. In this case, loading will occur once in every 8×2^7 clock counts for a period of 8×2^4 clock counts. Duty factor is then from Equation 5.4.4-1

$$D.F. = \frac{(8 \times 2^7) - (8 \times 2^4)}{5 (8 \times 2^7)} = 17.5\% \qquad (5.4.4.2.1\text{-}1)$$

and the refresh period is

$$5 (8 \times 2^7) t, \qquad (5.4.4.2\text{-}1\text{-}2)$$

where t = clock period.

The four least significant bits of the CD4520 counter are used to continually address the CD4036 refresh memory. Data can be written into the desired memory address by strobing the WRITE ENABLE line when the appropriate memory address appears on the WRITE ADDRESS lines. This function can occur simultaneously with a read from memory.

Two counters, a CD4029 and a CD4022, are used for the column data generator and the column select decoder, respectively. Note that the Signetics 2516 character

generator requires column select inputs of binary coded 1 to 5 instead of 0 to 4. For this reason, the CD4029 is preset to a binary 1 by the same pulse which is used to reset the CD4022 column select decoder. To minimize I_{CC}, the V_B terminal is held low during data load operations, turning "OFF" the current mirror reference current. The column current switch is a PNP darlington transistor driven from a buffered NAND gate. The 1N4720 serves to reduce the column voltage by approximately 1 volt, thereby reducing on board power dissipation in the display devices. Due to maximum clock rate limitations of the CMOS logic, clock input should not exceed 1 MHz.

5.4.4.2.2 Microprocessor Interface for the HDSP-2000

The HDSP-2000 alphanumeric display is ideally suited for interface to a microprocessor. There are several different ways in which the hardware/software partitioning can be arranged in such a system. The choice of the technique will depend on how much of the microprocessor time the designer wishes to devote to supporting the display. The additional microprocessor software time required is traded off against additional hardware costs. Figures 5.4.4.2.2-1a,b,c illustrate three different partitioning techniques which can be utilized in interfacing the HP HDSP-2000 to a microprocessor. In 5.4.4.2.2-1a, the display and its interface hardware appear as an autonomous peripheral to the microprocessor. This system accepts data into a local RAM whenever the microprocessor updates the information to be displayed. A good example of this type of system is illustrated in Figure 5.4.4.2.1-1. This system, however, duplicates some of the functions available in the microprocessor system.

Figure 5.4.4.2.2-1b illustrates a technique in which the coded data RAM and character generator ROM have been removed from the display interface and included as a fraction of the microprocessor hardware. In this approach, the entire message is decoded and the data is sent to the display where it is stored and accessed by the display scanning logic. This approach is ideal where one microprocessor system may be utilized to service several displays.

The display interface may be even further simplified by developing a system in which the microprocessor supplies the display refresh information periodically in response to an interrupt request from the display. The HP HDSP-2000 due to the onboard storage of decoded data, is ideally suited to the implementation of this third type of interface. Depending on the number of digits in the display string, and the microprocessor clock rate, the display can be updated at a time burden to the microprocessor of less than 1% per digit. Figure 5.4.4.2.2-2 illustrates the practical implementation of the technique presented in Figure

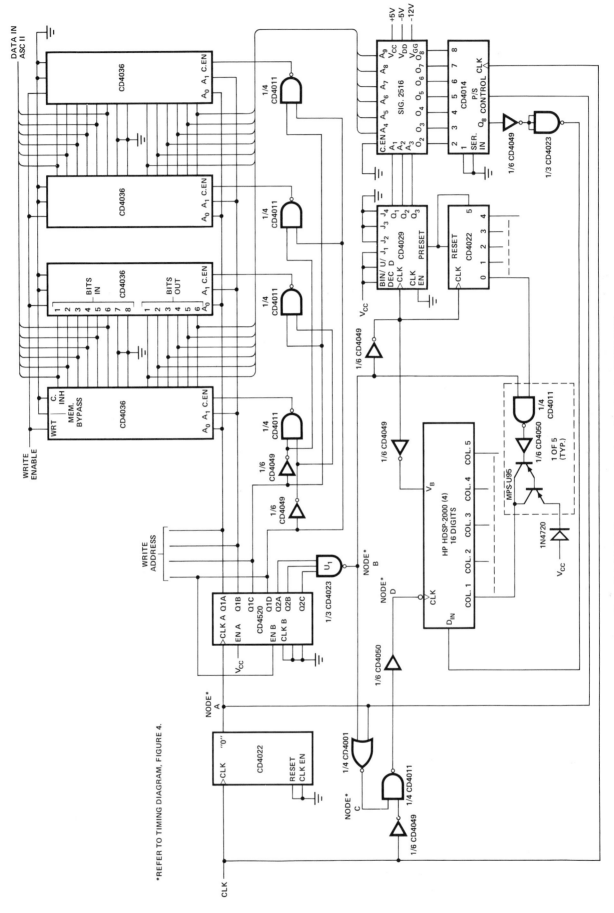

Figure 5.4.4.2.1-1 Instrumentation Interface for the HDSP-2000.

*REFER TO TIMING DIAGRAM, FIGURE 4.

5.4.4.2.2-1c, interfacing a 16 digit string of HP HDSP-2000 displays to an 8080A microprocessor. In this circuit, an astable timing element generates an interrupt request once every 2 msec. The interrupt subroutine outputs 16 bytes of data to the display via a parallel to serial converter. A local clock shifts this data into the display between output cycles of the microprocessor. Only 7 bits of each output word are shifted to the display. A column select word is sent to a separate address. Prior to data loading, this word is used to turn all column enable transistors off. After all of the data has been sent, this word gates the appropriate column data "ON". Decoded data for the display is stored in a memory stack in the microprocessor RAM. A memory pointer in the programmed subroutine is utilized to indicate the appropriate block of data to be sent to the display.

For a display refresh rate of 100 Hz (interrupt every 2 msec), the microprocessor time required to service the display as a percent of total time is expressed in the following equation:

Display strings of other lengths can be accomodated by changing the number of words which are written to the display from the RAM. This involves a simple one instruction software change.

In this situation, the microprocessor load time can be calculated from Equation 5.4.4.2.2-1 using

$$t_{LOAD} = (166 + 22N) \ \frac{1}{f} \qquad (5.4.4.2.2-2)$$

where:

f = Microprocessor clock frequency

N = Number of digits in the display string

$$\% \ t_{SERVICE} = \frac{t_{LOAD}}{t_{INTERRUPT}} \times 100\% \qquad (5.4.4.2.2-1)$$

$$= \frac{518 \times .5 \ \mu s}{2 \ ms} \times 100\%$$

$$= 12.55\%$$

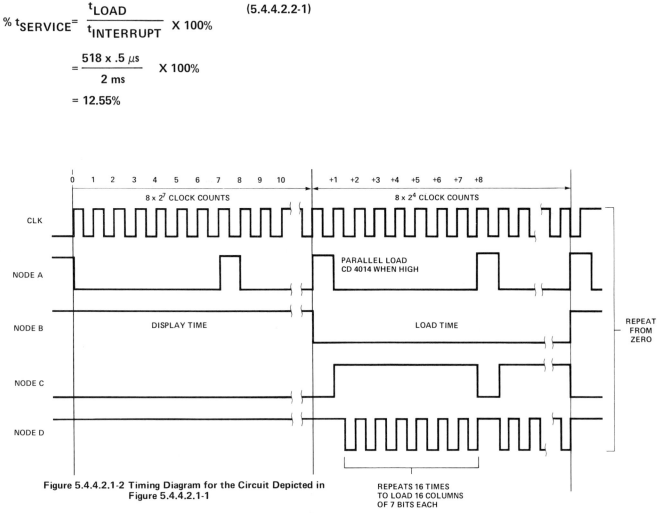

Figure 5.4.4.2.1-2 Timing Diagram for the Circuit Depicted in Figure 5.4.4.2.1-1

5.53

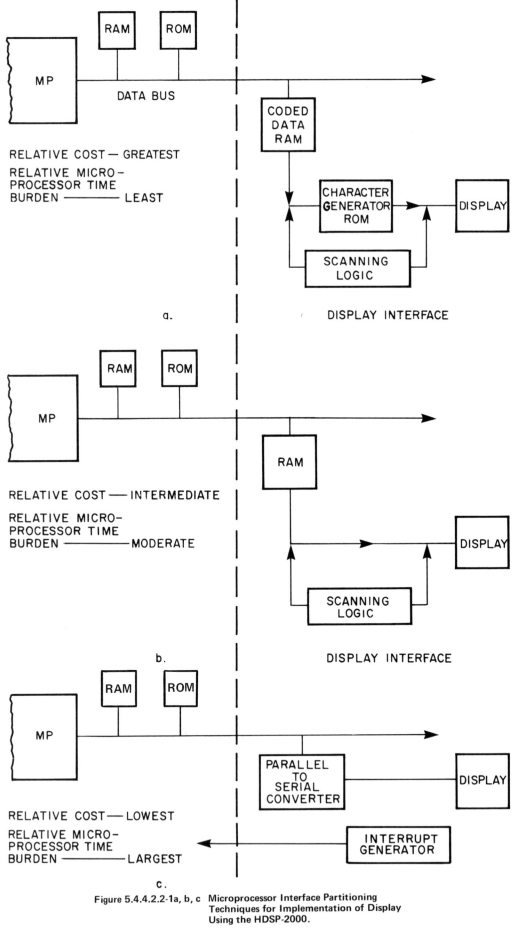

Figure 5.4.4.2.2-1a, b, c Microprocessor Interface Partitioning
Techniques for Implementation of Display
Using the HDSP-2000.

5.54

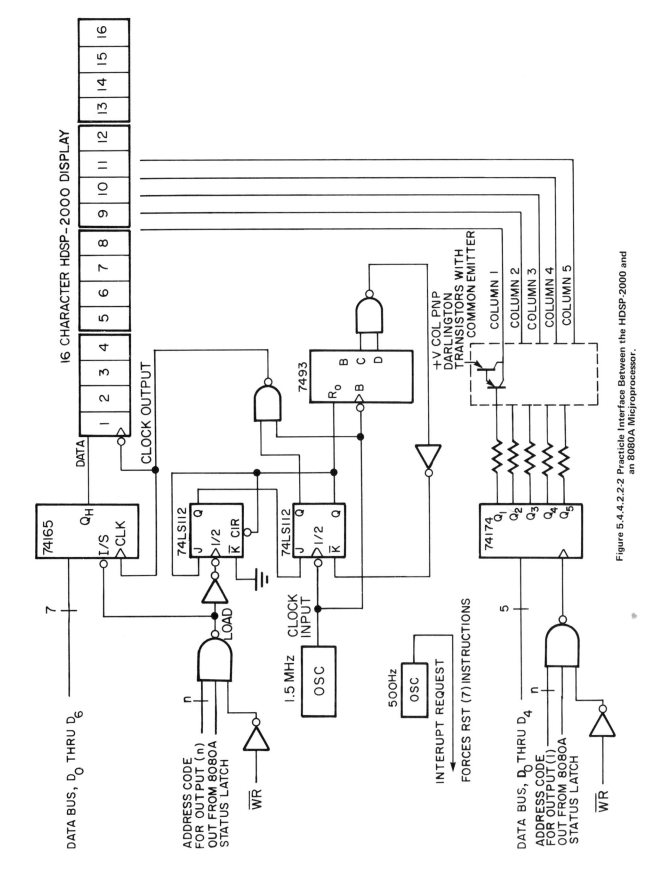

Figure 5.4.4.2.2.2 Practicle Interface Between the HDSP-2000 and an 8080A Microprocessor.

5.55

PROGRAM

ADDRESS	OP CODE	CLOCK CYCLES	COMMENTS
$(0038)_{16}$	PUSH PSW	(11)	
	PUSH HL	(11)	
	ORI	(7)	$A = (FF)_{16}$
	$(FF)_{16}$	------	
	OUT	(10)	TURNS OFF COLUMNS
	I	------	
	LHLD	(16)	HL = POINTER
	A_L	------	
	A_H	------	
	MOV A, M	(7)	A = (HL)
	OUT	(10)	DIGIT 16 = A
	n	------	
	INX HL	(5)	HL = HL + 1
	MOV A, M	(7)	A = (HL)
	OUT	(10)	DIGIT 15 = A
	n	------	
	INX HL	(5)	HL = HL + 1
	MOV A, M	(7)	A = (H2)
	OUT	(10)	DIGIT 14 = A
	n	------	
	⋮		
	INX HL	(5)	HL = HL + 1
	MOV A, M	(7)	A = (HL)
	OUT	(10)	DIGIT 1 = A
	n	------	
	INX HL	(5)	HL = HL + 1
	MOV A, M	(7)	A = (HL)
	OUT	(10)	TURNS A COLUMN ON
	I	------	
	MOV A, L	(5)	A = L
	CPI	(7)	COMPARE L TO ADDRESS
	$(64)_{16}$	------	OF LAST MEMORY LOCATION
	JNZ	(10)	JUMP IF A ≠ $(64)_{16}$
	(ADDRESS	------	
	OF LOOP)	------	
	MVI L	(7)	$L = (OF)_{16}$
	$(OF)_{16}$	------	
LOOP	INX HL	(5)	HL = HL + L
	SHLD	(16)	
	A_L	------	
	A_H	------	
	POP HL	(10)	
	POP PSW	(10)	
	EI	(4)	
	RET	(10)	

(b).

RAM

ADDRESS	CONTENTS (D_7-D_0)	COMMENTS
$A_H A_L$	POINTER L	
$A_H A_L + 1$	POINTER H	
POINTER → $(XX10)_{16}$	X R_1 R_2 R_3 R_4 R_5 R_6 R_7	DIGIT 16, COL 1
⋮	⋮	⋮
$(XX1F)_{16}$	X R_1 R_2 R_3 R_4 R_5 R_6 R_7	DIGIT 1, COL 1
$(XX20)_{16}$	X X X 1 1 1 1 0	COL 1 ENABLE
$(XX21)_{16}$	X R_1 R_2 R_3 R_4 R_5 R_6 R_7	DIGIT 16, COL 2
⋮	⋮	⋮
$(XX30)_{16}$	X R_1 R_2 R_3 R_4 R_5 R_6 R_7	DIGIT 1, COL 2
$(XX31)_{16}$	X X X 1 1 1 0 1	COL 2 ENABLE
$(XX32)_{16}$	X R_1 R_2 R_3 R_4 R_5 R_6 R_7	DIGIT 16, COL 3
⋮	⋮	⋮
$(XX41)_{16}$	X R_1 R_2 R_3 R_4 R_5 R_6 R_7	DIGIT 1, COL 3
$(XX42)_{16}$	X X X 1 1 0 1 1	COL 3 ENABLE
$(XX43)_{16}$	X R_1 R_2 R_3 R_4 R_5 R_6 R_7	DIGIT 16, COL 4
⋮	⋮	⋮
$(XX52)_{16}$	X R_1 R_2 R_3 R_4 R_5 R_6 R_7	DIGIT 1, COL 4
$(XX53)_{16}$	X X X 1 0 1 1 1	COL 4 ENABLE
$(XX54)_{16}$	X R_1 R_2 R_3 R_4 R_5 R_6 R_7	DIGIT 16, COL 5
⋮	⋮	⋮
$(XX63)_{16}$	X R_1 R_2 R_3 R_4 R_5 R_6 R_7	DIGIT 1, COL 5
$(XX64)_{16}$	X X X 0 1 1 1 1	COL 5 ENABLE

POINTER
(POINTS AT NEXT WORD TO BE SENT TO **HDSP-2000** DISPLAY)

X = DON'T CARE

(c.)

Figure 5.4.4.2.2-2 Continued. Microcade and RAM Contents Used to Support Microprocessor Display Interface.

5.56

DEVICE DESCRIPTION	BASIC DEVICE NUMBER	I_{MAX} mA	MANUFACTURER
Quad Segment Driver, MOS to LED Anode	75491	50 (sink/source)	FAIR, MOT, TI
	75493	25 (source)	NS, TI
	7895/8895	19 (source)	NS
	501	40 (source)	ITT
	503	34 (source)	
	507	18 (source)	
	517	16 (source)	
	518	14.5 (source)	
	522	12 (source)	
	523	10 (source)	
	491	50 (source)	
HEX Digit Driver, MOS to LED Cathode	75492	250 (sink)	FAIR, MOT, NS, TI
	75494	150 (sink)	NS, TI
	8870	350 (sink)	
	8877	50 (sink)	
	8892	200 (sink)	
	500	250 (sink)	ITT
	502	200 (sink)	
	506	200 (sink)	
	510	160 (sink)	
	492	250 (sink)	
7-Digit Driver, MOS to LED Cathode	75497	150 (sink)	TI
	8866	50 (sink)	NS
	546	50 (sink)	ITT
	552	500 (sink)	
	554	500 (sink)	
	556	500 (sink)	
8-Digit Driver, MOS to LED Cathode	8863/8963	500 (sink)	NS
	8865	50 (sink)	
	8871	40 (sink)	
	514/525	40 (sink)	ITT
9-Digit Driver, MOS to LED Cathode	526	40 (sink)	ITT
	548	60 (sink)	
	558	40 (sink)	
	75498	150 (sink)	TI
	8855	50 (sink)	NS
	8874/8876/ 8879	50 (sink)	
	8973/8974/ 8976	100 (sink)	
12-Digit Driver, MOS to LED Cathode	8868	110 (sink)	NS
	8973	40 (sink)	
Segment Driver, MOS to LED	8861 (5-Seg)	50 (source)	NS
	8877 (8-Seg)	14 (source)	

TABLE 5.5-1 List of LED Display to Logic Interface Devices

DEVICE DESCRIPTION	BASIC DEVICE NUMBER	I_{OUT} mA SOURCE/SINK	FONT: 6 AND 9 WITH OR WITHOUT TAILS	MANUFACTURER
BCD to 7-Segment Decoder/Driver OUTPUTS: Active High, Interal Resistive Pull-up	7448	6.4 (sink) 2 (source)	W/O	FAIR, ITT, MOT, NS, SIG, TI
	74248	6.5 (sink) 2 (source)	W	TI
	74LS48	6 (sink) 2 (source)	W/O	NS, TI
	8T05	15 (sink) 4 (source)	W/O	SIG
	7856	7.5 (source)	W/O	NS
BCD to 7-Segment Decoder/Driver OUTPUTS: Active High, Current Source	7857	60 (source)	W/O	NS
	7858	50 (source)	W/O	NS
BCD to 7-Segment Decoder/Driver OUTPUTS: Active High, Open Collector	7449	10 (sink)	W/O	FAIR, MOT
	74LS49	8 (sink)	W/O	NS, TI
	74249	10 (sink)	W	TI
	74LS249	8 (sink)	W	TI
BCD to 7-Segment Decoder/Driver OUTPUTS: Active Low, Open Collector	7447	40 (sink)	W/O	FAIR, ITT, MOT, NS, SIG, TI
	74L47	20 (sink)	W/O	TI
	74LS47	12 (sink)	W/O	NS, TI
	74247	40 (sink)	W	TI
	74LS247	24 (sink)	W	TI
	9317B	40 (sink)	W/O	FAIR
	9317C	20 (sink)	W/O	FAIR
	8T04	40 sink	W/O	SIG
BCD to 7-Segment Decoder, CMOS; OUTPUTS: N-channel sink and NPN Bipolar Sources	74C48	3.6 (sink) 50 (source)	W/O	HAR, NS
BCD to 7-Segment Latch/Decoder/ Driver; CMOS with Bipolar Outputs	4511 (14511)	25 (source)	W/O	FAIR, HAR, MOT, RCA, SSS
BCD to 7-Segment Decoder/Driver, CMOS	14558	0.28 (sink) 0.35 (source)	W/O	MOT
BCD to 7-Segment Latch/Decoder/ Driver OUTPUTS: Constant Current	9368	22 (source)	6 (W) 9 (W/O)	FAIR
	9374	18 (sink)	6 (W) 9 (W/O)	FAIR
	8673/ 8674	18 (sink)	6 (W) 9 (W/O)	NS
BCD to 7-Segment Latch/Decoder/ Driver OUTPUTS: Active Low, Open Collector	9370	40 (sink)	6 (W) 9 (W/O)	FAIR

LIST OF MANUFACTURERS

FAIR = FAIRCHILD SEMICONDUCTOR
HAR = HARRIS SEMICONDUCTOR
ITT = ITT SEMICONDUCTOR
MOT = MOTOROLA SEMICONDUCTOR
NS = NATIONAL SEMICONDUCTOR
RCA = RCA SOLID STATE DIVISION
SIG = SIGNETICS
SSS = SOLID STATE SCIENTIFIC
TI = TEXAS INSTRUMENTS

TABLE 5.5.9 List of LED Display to Logic Interface Devices

Section 6
Contrast Enhancement

6.0 CONTRAST ENHANCEMENT FOR LED DISPLAYS

The most important attribute of any equipment utilizing a digital readout is the ability to clearly display information to an observer. An observer must be able to quickly and accurately recognize the information being displayed by the instrument. The display, usually front panel mounted, must be visible without difficulty in the ambient light conditions where the instrument will be used.

Since most ambient light levels are sufficiently bright to impair the visibility of an LED display, it is necessary to employ certain techniques to develop a high viewing contrast between the display and its background. Since the quality of visibility is primarily subjective, it is not easily measured or treated by analytical means. Thus, human engineering plays a very important role in display applications. The best judge of the viewing esthetics of a display is the human eye, and the final display design must be pleasing to the eye when viewed in the end user ambient.

6.1 Contrast and Contrast Ratio

The objective of contrast enhancement is to maximize the contrast between display "ON" and display "OFF" conditions. This is accomplished by (1) reducing to a minimum the reflected ambient light from the face of the display and (2) allowing a maximum of the display's emitted light to reach the eye of an observer. The goal is to achieve a maximum contrast between "ON" segments and "OFF" segments as well as a minimum contrast between "OFF" segments and display package and background.

Both contrast and contrast ratio are used as the measure of the difference in the luminous sterance of a source with respect to the surrounding background. Contrast, C, for an LED display is defined as:

$$C = \frac{L_{vS} - L_{vB}}{L_{vS}} \; ; \; 0 \leqslant C \leqslant 1 \quad \frac{L_{vS} - L_{vB}}{L_{vS}} \qquad (6.1\text{-}1)$$

where: L_{vS} = Source luminous sterance (cd/m^2)

L_{vB} = Background luminous sterance (cd/m^2)

Contrast varies from zero, where $L_{vS} = L_{vB}$, to a value of one, where the source intensity is at a level such that $L_{vS} \gg L_{vB}$.

$$CR = \frac{L_{vS}}{L_{vB}} \; ; \; 1 \leqslant CR \leqslant \infty \qquad (6.1\text{-}2)$$

Contrast ratio, CR, is the ratio of the source luminous sterance to the background luminous sterance.

Contrast ratio ranges in value from one, where $L_{vS} = L_{vB}$, to a value tending towards infinity, when $L_{vS} \gg L_{vB}$. Contrast and contrast ratio are related by the following two expressions:

$$C = 1 - \frac{1}{CR} \qquad (6.1\text{-}3)$$

$$CR = \frac{1}{1-C} \qquad (6.1\text{-}4)$$

As shown in the curve of Figure 6.1-1, the range of attainable contrast for LED displays is between a useable minimum of 0.90 (CR=10) to 0.98 (CR=50) with a typical filter providing a contrast of 0.95 (CR=20).

A spot photometer is used to measure luminous sterance. The display should be located behind the contrast filter, installed in its mounting assembly and packaged as the final product in order to obtain realistic values. Ambient lighting, which represents the end use environment, should be incident to the filter during the measurement. Care should be taken to prevent shading of the spot being measured by the photometer.

With stretched segment displays, it is difficult to achieve a high value of segment on/off contrast while effectively concealing the display package from view. For example, a display with a black package is easily concealed from view, however, the "OFF" segments will be visible. This is due to the difference in reflectivity between the "OFF" segments and the black package.

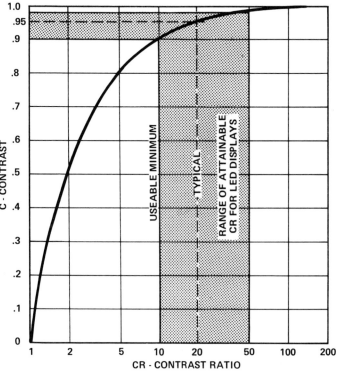

Figure 6.1-1 The Relationship Between Contrast and Contrast Ratio.

A reduction in the reflectivity difference between the "OFF" segments and the package of a stretched segment display may be obtained by adding a small amount of dye to color tint the segments, and the display package may be colored to match the "OFF" segment color. With the addition of an appropriate optical filter placed in front of the display, the "OFF" segments tend to be indistinguishable from the background. The trade-off is that a colored package is more visible than a black package. Because of this trade-off, a designer has to decide which is more important, concealing "OFF" segments or concealing the display package. Since the usual choice is to conceal "OFF", segments, Hewlett-Packard is using this colored package technique on its high-efficiency red, yellow and green stretched segment displays.

Contrast enhancement under artificial lighting conditions may be accomplished by use of selected wavelength optical filters. Under bright sunlight conditions contrast enhancement becomes more difficult and requires additional techniques such as the use of louvered filters combined with shading of the display. The effect of a wavelength optical filter is illustrated in Figure 6.1-2. The filtered portion of the display can be easily read while the "OFF" segments are not apparent. By comparison, reading the unfiltered portion of the display is difficult.

6.2 Eye Response, Peak Wavelength and Dominant Wavelength

The 1931 CIE (Commission Internationale De L'Eclairage) standard observer curve, also known as the photopic curve, is shown in Figure 6.2-1. This curve represents the eye response of a standard observer to various wavelengths of light. The vivid color ranges are also identified in Figure 6.2-1 to illustrate the sensitivity of the eye to the various colors. The photopic curve peaks at 555 nanometers (nm) in the yellowish-green region. This peak corresponds to 680 lumens of luminous flux (lm) per watt of radiated power (W).

Two wavelengths of the LED emission are important to a user of LED displays; Peak Wavelength and Dominant Wavelength. Peak wavelength (λ_p) is the wavelength at the peak of the radiated spectrum. The peak wavelength may be used to estimate the approximate amount of display emitted light that passes through an optical filter. For example, if an optical filter has a relative transmission of 40% at a given λ_p, then approximately 40% of the display emitted light at the peak wavelength will pass through the filter to the viewer while 60% will be absorbed. This gives a designer an initial estimate of the amount of loss of display emitted light he should expect.

Dominant wavelength (λ_d) is used to define the color of an

Figure 6.1-2 Effect of a Wavelength Optical Filter on an LED Display.

LED display. Specifically, the dominant wavelength is that wavelength of the color spectrum, which, when additively mixed with the light from the source CIE illuminant C, will be perceived by the eye as the same color as is produced by the radiated spectrum. CIE illuminant C is a 6500°K color temperature source that produces light which simulates the daylight produced by an overcast sky. A graphical definition of λ_d and color purity is given on the CIE chromaticity diagram in Figure 6.2-2. The dominant wavelength is derived by first obtaining the x,y color coordinates from the radiated spectrum. These color coordinates are then plotted on the CIE chromaticity diagram. A line is drawn from the illuminant C point through the x,y color point intersecting the perimeter of the diagram. The point where the line intersects the perimeter is the dominant wavelength.

The color purity, or saturation, is defined as the ratio of the distance from the x,y color point to the illuminant C point, divided by the sum of this distance and the distance from the x,y point to the perimeter. The x,y color coordinates for LEDs plot very close to the perimeter of the chromaticity diagram. Therefore, the color purity approaches a value of 1, typical of the color saturation obtained from a monochromatic light source.

The dominant wavelengths and corresponding colors for LEDs are shown on the CIE chromaticity diagram in Figure 6.2-3. As defined by λ_d, the color of a standard red LED is red, a high-efficiency red LED is reddish-orange, a yellow LED is yellowish-orange and a green LED is actually greenish-yellow.

It is of value to know the actual colors of each LED when selecting a contrast filter, as the optimum filter will have the same color as the device. Both λ_p and λ_d are listed on LED data sheets.

6.3 Filter Transmittance

The contrast filter must meet certain basic requirements. It should be mechanically stable with temperature, have

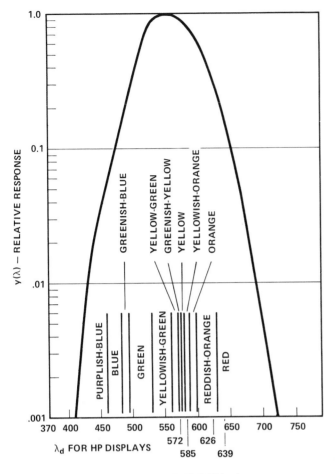

Figure 6.2-1 CIE Standard Observer Eye Response Curve (Photopic Curve), Including CIE Vivid Color Ranges.

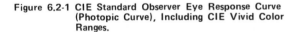
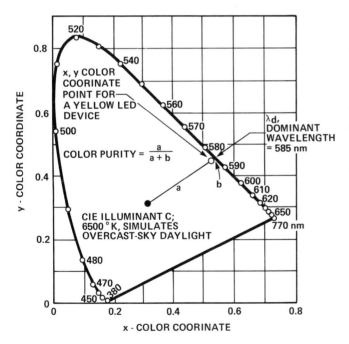

Figure 6.2-2 Definition of Dominant Wavelength and Color Purity, Shown on the CIE Chromaticity Diagram.

reasonable chemical resistance, be free from visual defects, have a homogeneous index of refraction and have sufficient transmission of the radiated spectrum of the LED. It is the last two requirements that are of primary optical importance.

The relative transmittance of an optical filter with respect to wavelength is defined to as: (6.3-1)

$$T(\lambda) = \frac{\text{Luminous Flux with Filter at Wavelength } \lambda}{\text{Luminous Flux without Filter at Wavelength } \lambda}$$

The relative transmittance is a function of the index of refraction of the filter material and the transmission through the material as determined by the coloring. Both are a function of wavelength. Specifically, the transmission through the filter is dependent upon the amount of incident light reflected at the filter/air interface and the amount of wavelength absorbtion within the filter material.

The index of refraction determines the amount of incident light that is reflected at the filter/air interface. The amount of reflected light is given by the following ratio:

$$R = \frac{(n_1 - n_2)^2}{(n_1 + n_2)^2} \qquad (6.3-2)$$

where: n_1 = Index of refraction of the filter material.

n_2 = Index of refraction for air = 1.0.

A plastic filter with an average index of refraction equal to 1.5, for the range of wavelengths encompassing the LEDs radiated spectrum, will reflect 4% of the normal incident light at each filter/air interface. Thus, 8% of the LED emitted light, passing through the filter to an observer, is lost due to reflection.

Inside the filter, light is lost due to absorbtion by the tinted material. The amount of absorbtion is a direct function of wavelength and is determined by the dye coloring and dye concentration. If the dye coloring is held at a constant density, the transmission through the filter material at any given wavelength, T_λ, is an exponetial function of the thickness of the material:

$$T_\lambda = e^{-ax} \qquad (6.3-3)$$

where: x = The quantity of unit thicknesses of filter material.

e = 2.71828

a = Absorbtion coefficient and is equal to $-\ln t$, where t is the transmission for a unit thickness.

As an example, the transmission through 1.0mm thickness of filter material is 0.875 at a wavelength of 655 nm. At x = 2.5, the thickness is 2.5mm and the transmission is 0.716:

$$-\ln (.875) = 0.1335$$

$$T_\lambda = e^{-(.1335)(2.5)} = .716$$

The relative transmission of a contrast filter at a particular wavelength, T(λ), may be calculated with reasonable accuracy by using the following relationship:

$$T_{(\lambda)} = \left[\frac{2n}{n+1}\right] T_\lambda \qquad (6.3\text{-}4)$$

where: n = The index of refraction of the filter material.
T_λ = Transmission through the filter material at wavelength λ.

For a 1.0mm thick plastic filter with n = 1.5 and T_λ = .875 at a wavelength of 655 nm, the relative transmission is 0.808:

$$T (\lambda = 655 \text{ nm}) = \left[\frac{2 (1.5)}{(1.5)^2 +1}\right] [.875] = .808$$

The same filter material at a thickness of 2.5 mm has a relative transmittance of 0.661, as shown in Figure 6.3-1.

6.3.1 Plastic Filters

Due to their low cost, ease in machining to size and resistance to breakage, plastic contrast filters are being used in a majority of display applications. The filter requirements for dim, moderate and bright ambients are different for each lighting condition. Therefore, it is advantageous to become familiar with the various relative transmittance characteristics that are available in plastic filters.

Most manufacturers of wavelength filters for use with LED displays provide relative transmittance curves for their products. Sample transmittance curves are presented in Figures 6.3.1-1, -2, -3 and -4. These curves represent approximate filter characteristics which may be used in various ambient light levels. The total transmittance curve shape and wavelength cut-off points have been chosen in direct relationship to the LED radiated spectrum. Each filter curve has been empirically determined and is similar to commercially available products. The higher the ambient light, the more optically dense the filter must be to absorb refected light from the face of the display. Because the display emitted light is also strongly absorbed, the display must be driven at a high average current to be readily visible. For dim ambient light, the filter may have a high value of transmittance as the ambient light will be at

levels much less than display emitted light. The display can now be driven at a low average current.

Dim ambients are in the range of 32 to 215 lx (3 to 20 footcandles), moderate ambients are in the range of 215 to 1076 lux (20 to 100 footcandles), and bright ambients are in the range of 1076 to 5382 lux (100 to 500 footcandles). Lux = (lm/m^2) and footcandle = (lm/ft^2).

Listed on each filter transmittance curve are empirically selected ranges of relative transmittance values at the peak wavelength which may give satisfactory filtering. For example, a filter to be used with a yellow display in moderate ambient lighting could have a transmittance value at the peak wavelength $[T(\lambda_p)]$ between 0.27 and 0.35. The filter wavelength cut-off should occur between 530 and 550 nm for best results.

When selecting a filter, the transmittance curve shape, attenuation at the peak wavelength and wavelength cut-off should be carefully considered in relationship to the LED radiated spectrum and ambient light level so as to obtain optimum contrast enhancement.

Three manufacturers of plastic wavelength filters are Panelgraphic Corporation (Chromafilter®), SGL Homalite and Rohm & Haas Company (Plexiglas®). The LED filters produced by these manufacturers are useable with all LED display and lamp devices. Table 6.9-2 lists some of the filter manufacturers and where to go for further information. Table 6.9-1 lists some specific wavelength filter products with recommended applications.

6.3.2 Optical Glass Fitlers

Optical glass filters are typically designed with constant density so it is the thickness of the glass that determines the optical density. This is just the opposite of plastic filters which are usually designed such that all material thicknesses have the same optical density.

The primary advantage of an optical glass contrast filter over a plastic filter is its superior performance. This is especially true for red LED filters. Figure 6.3.2-1 illustrates two thicknesses of a red optical glass filter for use with standard red LED displays. The relative transmittance is generally higher than that of a comparable plastic filter, and the slope of the relative transmittance curve is usually much steeper and follows more closely the shape of the radiated spectrum of the LED. This particular filter provides excellent contrast in a bright ambient. A reddish-orange optical glass filter which is suitable for use with a high-efficiency red display in a moderate ambient is shown in Figure 6.3.2-2. The relative transmittance of this filter follows almost exactly the leading edge of the shape of the LED's radiated spectrum.

A leading manufacturer of optical glass filters is the Schott Optical Glass, Inc. of Duryea, Pennsylvania and Munich, Germany.

6.4 Wavelength Filtering

The application of wavelength filters as described in the previous section is the most widely used method of contrast enhancement under artificial lighting conditions. However, they are not very effective in daylight due to the high level

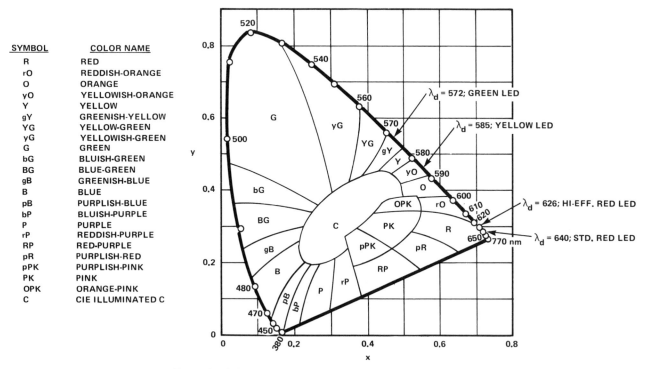

SYMBOL	COLOR NAME
R	RED
rO	REDDISH-ORANGE
O	ORANGE
yO	YELLOWISH-ORANGE
Y	YELLOW
gY	GREENISH-YELLOW
YG	YELLOW-GREEN
yG	YELLOWISH-GREEN
G	GREEN
bG	BLUISH-GREEN
BG	BLUE-GREEN
gB	GREENISH-BLUE
B	BLUE
pB	PURPLISH-BLUE
bP	BLUISH-PURPLE
P	PURPLE
rP	REDDISH-PURPLE
RP	RED-PURPLE
pR	PURPLISH-RED
pPK	PURPLISH-PINK
PK	PINK
OPK	ORANGE-PINK
C	CIE ILLUMINATED C

Figure 6.2-3 Dominant Wavelengths and Corresponding Colors for LEDs, Shown on the CIE Chromaticity Diagram.

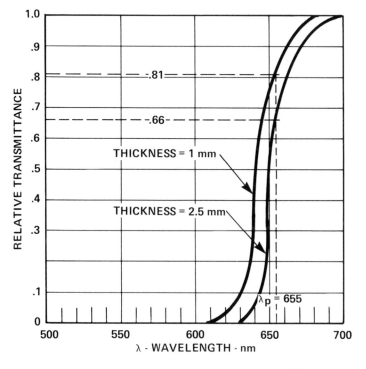

Figure 6.3-1 Variation in Relative Transmittance vs. Thickness for a Constant Density Filter Material.

ambient light. Filtering in daylight conditions is best achieved by using louvered filters (discussed in a later section).

The figures in sections 6.4.1 through 6.4.4 show the relationship between artificial lighting and the spectra of LED displays, both unfiltered and filtered. Figures 6.4.1-1 through 6.4.4-1 show the relationship between the various LED spectra and the spectra of daylight fluorescent and incandescent light. The photometric spectrum (shaded curve) is obtained by multiplying the LED radiated spectrum [$f(\lambda)$] by the photopic curve [$y(\lambda)$]. Thus, photometric spectrum = $f(\lambda) \cdot y(\lambda)$. Figures 6.4.1-2 through 6.4.4-2 demonstrate the effect of a wavelength filter. The filtered photometric spectrum is what the eye perceives when viewing a display through a filter (shaded curve). Thus, filtered photometric spectrum = $f(\lambda) \cdot y(\lambda) \cdot T(\lambda)$. The ratio of the area under the filtered photometric spectrum to the area under the unfiltered photometric spectrum is the fraction of the visible light emitted by the display which is transmitted by the filter:

$$\frac{\text{Fraction of Available}}{\text{Light from Filtered Display}} = \frac{\int f(\lambda) \cdot (\lambda) \cdot T(\lambda) \cdot d\lambda}{\int f(\lambda) \cdot y(\lambda) \cdot d\lambda}$$

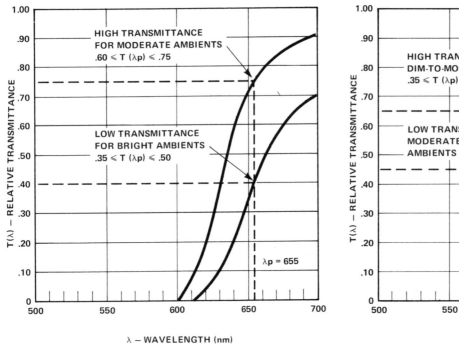

Figure 6.3.1-1   Typical Transmittance Curves for Filters to be Used with Standard Red Displays.

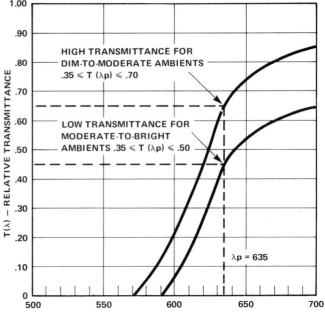

Figure 6.3.1-2   Typical Transmittance Curves for Filters to be Used with High-Efficiency Red Displays.

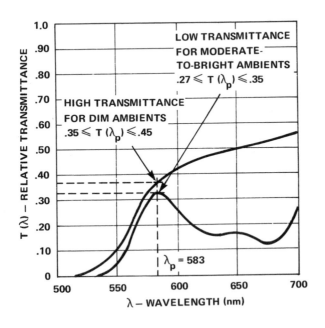

Figure 6.3.1-3   Typical Transmittance Curves for Filters to be Used with Yellow Displays.

Figure 6.3.1-4   Typical Transmittance Curves for Filters to be Used with Green Displays.

6.6

In addition to attenuating a portion of the light emitted by the display, a filter also shifts the dominant wavelength, thus causing a shift in the perceived color. For a given display spectrum, the color shift depends on the cut-off wavelength and shape of the filter transmittance characteristic. A choice among available filters must be made on the basis of which filter and LED combination is most pleasing to the eye. A designer must experiment with each filter as he cannot tell by transmittance curves alone. The filter spectra presented in Figures 6.3.1-1 through 6.3.1-4 are suggested starting points. Filters with similar characteristics are commercially available.

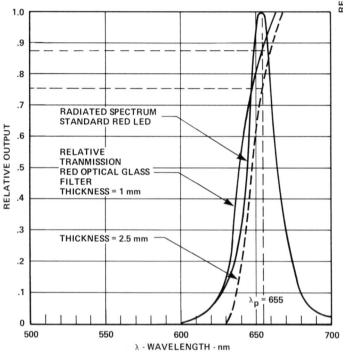

Figure 6.3.2-1   A Red Optical Glass Filter for use with Standard Red Displays.

6.4.1   Filtering Standard Red Displays (λ_p = 655 nm)

Filtering out reflected ambient light from red displays is easily accomplished with a long wavelength pass filter having a sharp cut-off in the 600 nm to 620 nm range (see Figures 6.3.1-1 and 6.4.1-2). Under bright fluorescent light, a red filter is very effective due to the low concentration of red in the fluorescent spectrum. The spectrum of incandescent light contains a large amount of red, and therefore, it is difficult to filter red displays effectively in bright incandescent light.

6.4.2   Filtering High-Efficiency Red Displays (λ_p = 635 nm)

The use of a long wavelength pass filter with a cut-off in the 570 nm to 590 nm range gives essentially the same results as is obtained when filtering red displays (see Figures

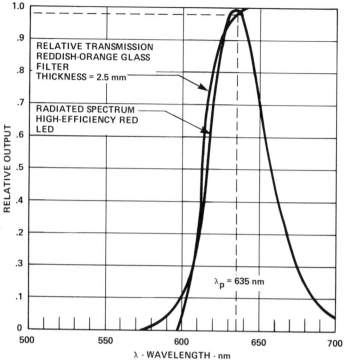

Figure 6.3.2-2   A Reddish-Orange Optical Glass Filter for Use with High-Efficiency Red Displays.

6.3.1-2 and 6.4.2-2). The resulting color is a rich reddish-orange.

6.4.3   Filtering Yellow Displays (λ_p = 583 nm)

The peak wavelength of a yellow LED display is in the region of the photopic curve where the eye is most sensitive (see Figure 6.4.3-1). Also, there is a high concentration of yellow in the spectrum of fluorescent light and a lesser amount of yellow in incandescent light. Therefore, filters that are more optically dense than red filters at the peak wavelength are required to filter yellow displays. The most effective filters are the dark yellowish-orange (or dark amber) filters as shown in Figure 6.3.1-3. The use of a low transmittance yellowish-orange filter, as shown in Figure 6.4.3-2 results in a similar color to that of a gas discharge display. Pure yellow filters provide very little contrast enhancement.

6.4.4   Filtering Green Displays (λ_p = 565 nm)

The peak wavelength of a green LED display is only 10 nm from the peak of the eye response curve (see Figure 6.4.4-1). Therefore, it is very difficult to effectively filter green displays. A long wavelength pass filter, such as is used for red and yellow displays, is no longer effective. An effective filter is obtained by combining the dye of a short wavelength pass filter with the dye of a long wavelength pass filter, thus forming a bandpass yellow-green filter which peaks at 565 nm as shown in Figure 6.3.1-4. Pure

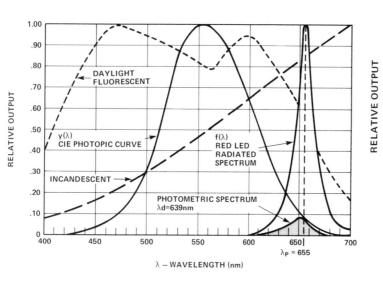

Figure 6.4.1-1   Relative Relationship Between the Standard Red LED Spectrum, Photopic Curve and Artificial Lighting.

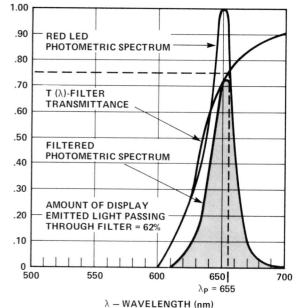

Figure 6.4.1-2   Effect of a Long Pass Wavelength Filter on a Standard Red Display.

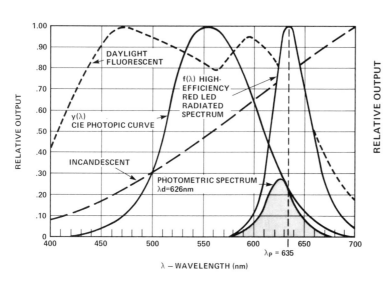

Figure 6.4.2-1   Relative Relationship Between the High-Efficiency Red LED Spectrum, Photopic Curve and Artificial Lighting.

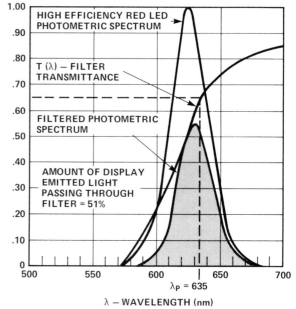

Figure 6.4.2-2   Effect of a Long Pass Wavelength Filter on a High-Efficiency Red LED Display.

green filters peak at 520 nm and drop off rapidly in the 550 nm to 570 nm range and are not recommended. The best possible filters for green LED displays are those which are yellow-green bandpass, peaking at 565 nm and dropping off rapidly between 575 nm and 590 nm. As shown in Figure 6.4.4-2, this filter passes wavelengths 550 to 570 while sharply reducing the longer wavelengths in the yellow

region. To effectively filter green LED displays in fluorescent light would require the use of a filter with a low transmittance value at the peak wavelength. This is due to the high concentration of green in the fluorescent spectrum. It is easier to filter green displays in bright incandescent light due to the low concentration of green in the incandescent spectrum, see Figure 6.4.4-1.

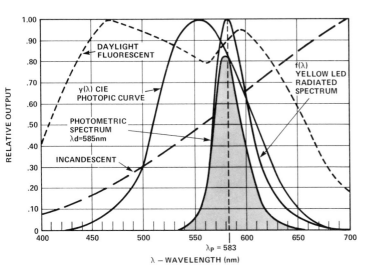

Figure 6.4.3-1 Relative Relationship Between the Yellow LED Spectrum, Photopic Curve and Artificial Lighting.

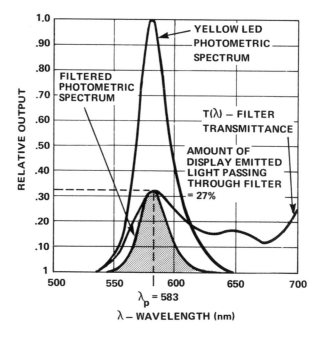

Figure 6.4.3-2 Effect of a Wavelength Filter on a Yellow LED Display.

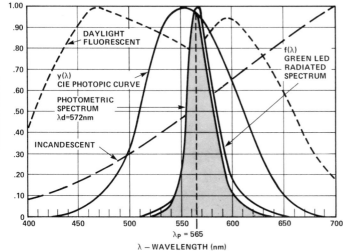

Figure 6.4.4-1 Relative Relationship Between the Green LED Spectrum, Photopic Curve and Artificial Lighting.

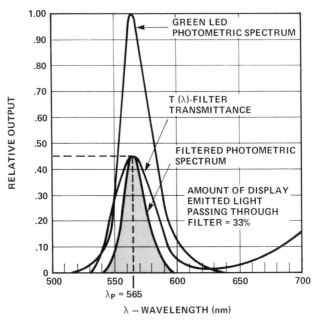

Figure 6.4.4-2 Effect of a Bandpass Wavelength Filter on a Green LED Display.

6.5 Reduction of Reflected Ambient Light as Provided by a Contrast Filter

Incident ambient light that is reflected back to an observer from the face of a filtered display travels twice through the contrast filter. The amount of light actually reflected from the face of the display is dependent upon the diffused reflectance, off the face of the display, as a function of wavelength, $R(\lambda)$.

Diffused reflectance is absorbtion dependent and is measured with the incident light beam normal to a surface. Those wavelengths which are not absorbed are reflected back as color. This is not the same as specular reflection. In specular reflection, the light is incident to a surface at some angle less than $90°$ where all wavelengths are essentially reflected equally. Figure 6.5-1 shows the diffused reflectance as a function of wavelength for stretched segment displays that have colored packages.

The amount of incident light reflected back to an observer from a filtered display as compared to an unfiltered display may be calculated from the ratio of the filtered reflected light to the unfiltered reflected light.

$$\frac{\text{Amount of Filtered}}{\text{Reflected Light}} = \frac{\int X(\lambda) \cdot R(\lambda) \cdot T^2(\lambda) \cdot Y(\lambda) \cdot d\lambda}{\int X(\lambda) \cdot R(\lambda) \cdot Y(\lambda) \cdot d\lambda}$$

where: $X(\lambda)$ = Spectral distribution of the incident ambient light.

$R(\lambda)$ = Relative diffused reflectance off the face of the display.

$T(\lambda)$ = Filter relative transmittance.

$Y(\lambda)$ = Photopic curve.

6.5.1 Effectiveness of a Wavelength Filter in an Ambient of Artificial Lighting

The contrast is very dependent upon the ambient lighting. If most of the spectral distribution of the ambient light is outside the radiated spectrum of the LED, it is very easy to reduce the reflected ambient light to a very low level while imposing only minimal attenuation on the emitted light from the display. Such is the case when using a red LED display in a fluorescent ambient or a green LED in an incandescent ambient. The opposite is also true. It is very difficult to effectively reduce the reflected incandescent light from the face of a red LED display, or to reduce

reflected fluourescent light from the face of a yellow or green LED display without significantly decreasing the emitted light from the LED.

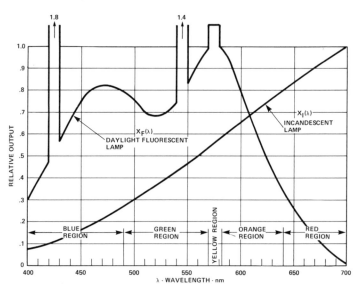

Figure 6.5.1-1   Spectral Distribution for a Daylight Fluorescent and an Incandescent Lamp used for Artificial Lighting.

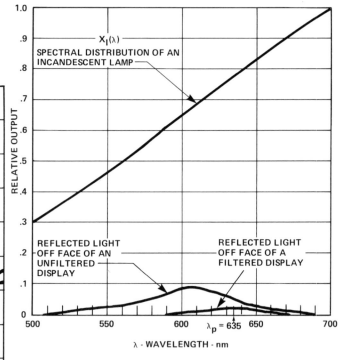

Figure 6.5.1-2   Reflected Incandescent Light Off an Unfiltered and Filtered High-Efficiency Red Display with a Colored Package, as Seen by an Observer. The Filtered Reflected Light is 17% of the Unfiltered Reflected Light.

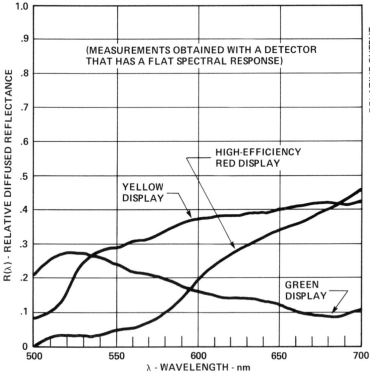

Figure 6.5-1 Diffused Reflectance for the Faces of Stretched Segment Displays that have Colored Packages.

Figure 6.5.1-1 reproduces the spectral distributions for fluorescent and incandescent lighting. Fluorescent lighting contains almost no red, yet contains a high level of yellow and long wavelength green. Incandescent, is just the

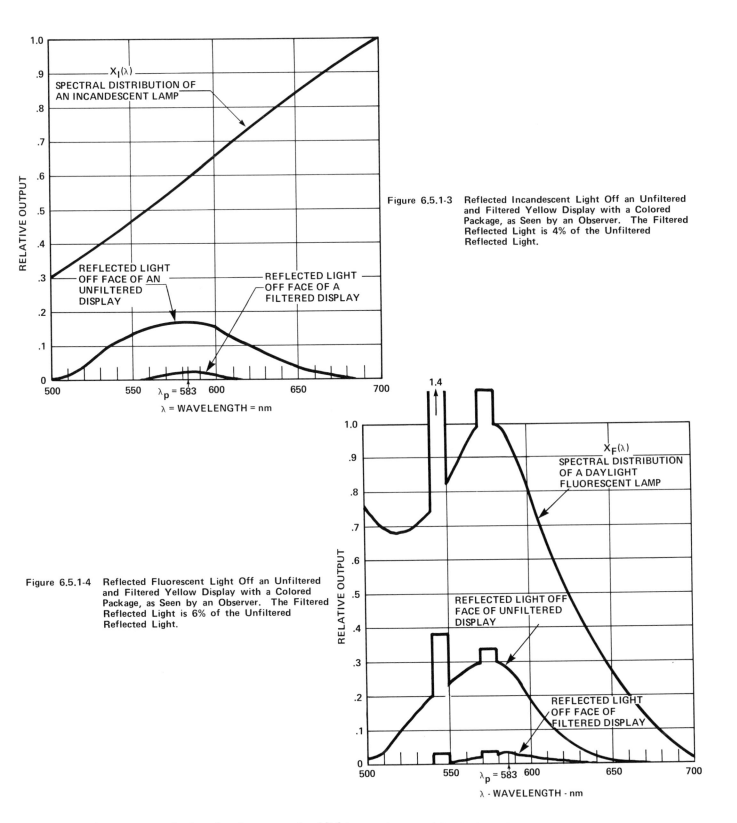

Figure 6.5.1-3 Reflected Incandescent Light Off an Unfiltered and Filtered Yellow Display with a Colored Package, as Seen by an Observer. The Filtered Reflected Light is 4% of the Unfiltered Reflected Light.

Figure 6.5.1-4 Reflected Fluorescent Light Off an Unfiltered and Filtered Yellow Display with a Colored Package, as Seen by an Observer. The Filtered Reflected Light is 6% of the Unfiltered Reflected Light.

opposite, being composed primarily of orange and red light. As can be seen from Figure 6.5.1-1 and the Figures in Section 6.4, the contrast filter may have to reduce the same wavelengths that are predominant in both the LED and ambient light in order to achieve adequate contrast. This is especially true for green LEDs under fluorescent lighting.

Figures 6.5.1-2 through 6.5.1-5 illustrate a comparative difference between the filtered and unfiltered diffused reflected light off the face of a stretched segment LED display as seen by the eye of an observer. The curve of the unfiltered diffused reflected light is obtained by the product of the lamp spectral distribution $[X(\lambda)]$, diffused

6.11

reflectance off the face of the display $[R(\lambda)]$ and the photopic curve $[y(\lambda)]$. Since the incident ambient passes through the filter twice, the curve of the filtered reflected light is the unfiltered curve multiplied by the square of the filter transmittance $[T^2(\lambda)]$. The filters used are the same as those used in Figures 6.4.2-2, 6.4.3-2 and 6.4.4-2. These curves illustrate that a wavelength filter substantially reduces the incident ambient light that is reflected off the face of a display. Even so, the peak of the reduced reflected ambient light is at or near the peak wavelength of the LED radiated spectrum.

6.5.2 Effectiveness of a Wavelength Filter in Daylight Ambients

The purpose of the wavelength filter is to reduce the luminous sterance of the background to a level that is very much less than the luminous sterance of the display's illuminated segments in order to achieve a high value of contrast. Referring to Equation 6.1-1 for contrast, the contrast goes to zero if the background luminance sterance equals the luminance sterance of the illuminated segments. This situation occurs when the ambient light level is sufficient to effectively mask the display's emitted light.

The spectral distribution of bright sunlight is nearly a flat curve across the complete color spectrum, as shown in Figure 6.5.2-1. Bright sunlight that is directly incident upon a display with a wavelength filter raises the background luminous sterance to a level that may actually exceed the luminous sterance produced by the emitted light from the illuminated segments. Therefore, the luminous sterance of the display segments, "ON" or "OFF", equals the luminous sterance of the background and the contrast goes to zero.

A similar situation occurs with overcast sky daylight that is directly incident onto the face of a wavelength filtered display. The situation is not as pronounced for standard red LED displays as for green LED displays, since the overcast sky filters out a considerable amount of the ambient red light, see Figure 6.5.2-1. It is obvious, then, that a wavelength filter by itself is not sufficient to achieve the necessary contrast in daylight ambient. Some shading or blocking out of the incident daylight is also required to achieve sufficient contrast between the display background and the illuminated segments.

It is almost impossible to achieve a minimum acceptable contrast if the incident daylight is parallel to the viewing axis (the incident daylight is perpendicular to the face of the display). If the incident daylight is not parallel to the viewing axis, other filter techniques may be employed to obtain acceptable contrast. These techniques include the use of louvered or cross hatch filters, incorporation of an eyebrow shade and recessing the display. Also, a circular polarizing filter may be employed with those displays that

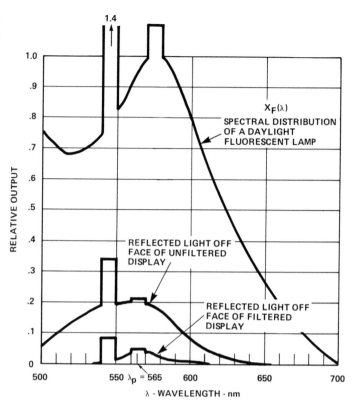

Figure 6.5.1-5   Reflected Fluorescent Light Off an Unfiltered and Filtered Green Display with a Colored Package, as Seen by an Observer. The Filtered Reflected Light is 10% of the Unfiltered Reflected Light.

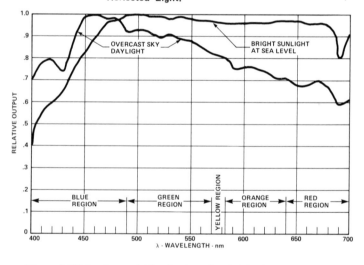

Figure 6.5.2-1   Spectral Distributions for Bright Sunlight and Overcast Sky Daylight.

have a face which is a specular reflecting surface. These techniques are discussed in more detail in later sections.

Indirect daylight that is incident to the face of a display is already reduced in level from direct daylight. Simple shading of a display, which is filtered by a louvered filter, will produce adequate contrast. It is the shading that reduces the indirect daylight to a low enough level to allow the filter to reduce the background luminous sterance to a level sufficient for adequate contrast.

6.6 Special Wavelength Filters and Filters in Combination

A designer is not limited to a single color wavelength filter to achieve the desired contrast and front panel appearance. Some unique wavelength filters and filter combinations have been successfully developed. One is the purple color filter for use with red LED displays, and another is the use of a neutral density filter in combination with a light amber filter to achieve a dark front panel for yellow LED displays.

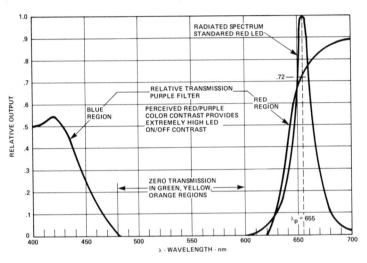

Figure 6.6.1-1   A Purple Color Wavelength Filter for Standard Red LED Displays.

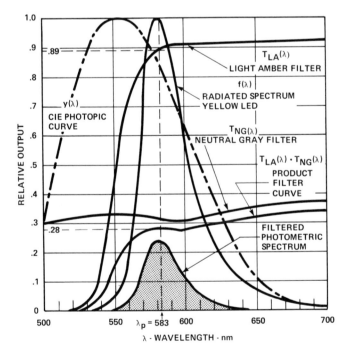

Figure 6.6.2-1   A Neutral Density Gray Filter in Combination with a Light Amber Filter for Use with Yellow Displays.

6.6.1   The Purple Contrast Filter for Red LED Displays

The wavelength filters that have been previously discussed have a distinct color that may be identified by a dominant wavelength. The contrast that they provide is essentially due to the high level luminous sterance of the display segments contrasted against the low level luminous sterance of a background of the same color. Another approach maintains the contrast ratio, but has as the background a different color than that of the LEDs in the display. This is easily accomplished by using a dark purple filter with standard red LED displays.

A most effective contrast filter is shown in Figure 6.6.1-1. The color, purple, is not defined by a dominant wavelength, see Figure 6.2-3. Purple is a mixture of red and blue light which is perceived by the eye as a distinct color from red. Psychologically, a purple contrast filter is more pleasing to many people than a red filter. The reason for this may be that when illuminated, the standard red display stands out so vividly against the purple background. Therefore, it is the color difference that enhances the contrast. This makes the purple contrast filter extremely effective in brightly lighted ambients.

6.6.2   Filters in Combination

A neutral density gray filter is often used in combination with other filters to provide a dark gray filter window as well as increased contrast in bright ambients. A typical example is given in Figure 6.6.2-1. The resulting filter is the product of the relative transmittance of the light amber, $T_{LA}(\lambda)$, and the relative transmittance of the neutral density gray, $T_{NG}(\lambda)$:

Product Filter = $[T_{LA}(\lambda) \cdot T_{NG}(\lambda)]$

The amount of light reaching the eye of a viewer is 24% of the unfiltered LED spectrum:

$$\begin{matrix}\text{Fraction of} \\ \text{Available Light} \\ \text{Through a} \\ \text{Combination Filter}\end{matrix} = \frac{\int f(\lambda) \cdot Y(\lambda) \cdot [T_{LA}(\lambda) \cdot T_{NG}(\lambda)] \cdot d\lambda}{\int f(\lambda) \cdot Y(\lambda) \cdot d\lambda}$$

The advantage is a dark gray front panel window with very low luminous sterance (zero transmission below 525 nm) that retains its appearance in bright ambients. The trade-off is a considerable reduction in the luminous sterance of the display which reduces the contrast ratio. This is somewhat offset by the distinct color difference between the illuminated yellow segments of the display and the dark gray background.

6.13

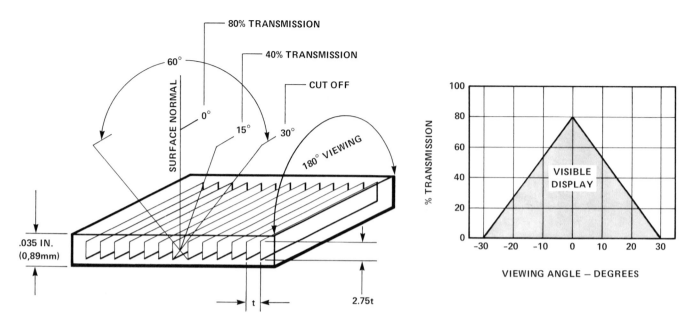

Figure 6.7-1 Construction Characteristics of a 0° Neutral Density
Louvered Filter.

6.7   Louvered Filters

Louvered filters are very effective in reducing the amount of bright artificial light or daylight reflected from the face of a display, without a substantial reduction in display emitted light. The construction of a louvered filter is diagrammed in Figure 6.7-1. Inside a plastic sheet are thin parallel louvers which may be oriented at a specific angle with respect to the surface normal. The zero degree louvered filter has the louvers perpendicular to the filter surface.

The operation of a louvered filter is similar to a venetian blind as shown in Figure 6.7-2. Light from the LED display passes between the parallel louvers to the observer. Off-axis ambient light is blocked by the louvers and therefore is not able to reach the face of the display to be reflected back to the observer. This results in a very high contrast ratio with minimal loss of display emitted light at the on-axis viewing angle. The trade-off is a restricted viewing angle. For example, the zero degree louvered filter shown in Figure 6.7-1 has a horizontal viewing angle of 180°; however, the vertical viewing included angle is 60°. The louver aspect

**AVAILABLE OPTIONS FOR LOUVERED FILTERS—
ANY COMBINATION IS POSSIBLE**

ASPECT RATIO AND VIEWING ANGLE	LOUVER ANGLE	LOUVER COLOR
2.75:1 = 60° 2.00:1 = 90° 3.50:1 = 48°	0° 18° 30° 45°	OPAQUE BLACK TRANSLUCENT GRAY TRANSPARENT BLACK

EXAMPLE: 2.75:1 — 18° — TRANSPARENT BLACK

Figure 6.7-2 The Operation of a Louvered Filter.

6.14

ratio (louver depth/distance between louvers) determines viewing angle. A list of louver option possibilities is given in Figure 6.7-2.

Some applications require a louver orientation other than zero degrees. For example, an 18 degree louvered filter may be used on the sloping top surface of a point of sale terminal. A second, is the use of a 45 degree louvered filter on overhead instrumentation to block out ambient light from ceiling mounted lighting fixtures.

Louvered filters are effective filters for enhancing the viewing of LED displays installed in equipment operating under daylight ambient conditions. In bright sunlight, the most effective filter is the crosshatch louvered filter. This is essentially two zero degree neutral density louvered filters oriented at 90 degrees to each other, as illustrated in Figure 6.7-3. Red, yellow and green digits may be mounted side by side in the same display. Using only the crosshatch filter, all digits will be clearly visible and easily read in bright sunlight as long as the sunlight is not parallel to the viewing axis. The trade-off is restricted vertical and horizontal viewing. The effective viewing cone is an included angle of 40° degrees (for a filter aspect ratio of 2.75:1).

Neutral density louvered filters are effective by themselves in most bright ambient lighting conditions without the aid of a secondary wavelength filter. However, colored louvered filters may be used for additional wavelength filtering at the expense of display emitted light.

A most effective filter which provides exceptional contrast in bright sunlight is the 45° louvered filter. The difference from standard cross hatch is that the louvers are at a 45° angle with respect to the edge of the filter material as shown in Figure 6.7-4. The louvers are transparent black, with a cross light transmission of 12% to 15%. This small amount of cross transmission virtually eliminates the double image due to ghosting which was a problem in earlier designs. The horizontal and vertical included viewing angles are increased to about 60° (for a louver aspect ratio of 2.75:1).

A combination that provides significant contrast improvement for standard red LED displays being used in bright sunlight is the 45° cross hatch with a purple tint to the filter plastic. The purple tint provides a color contrast for the illuminated display segments, so they may be recognized when the sunlight rays are somewhat parallel the viewing axis.

The combination of a 45° cross hatch filter and a fresnel lens in tinted plastic form a special magnifying filter for LED displays. Both the cross hatch and fresnel grooves are made extra fine so that they are virtually invisible to the eye. The result is a superior lens/filter combination of

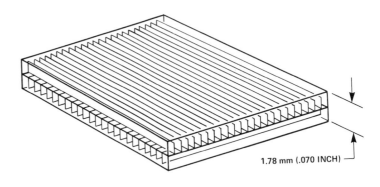

Figure 6.7-3 Conceptual Sketch of a Cross Hatch Filter. Two Louvered Filters Bonded Together in a 90° Orientation with Respect to Each Other.

minimum thickness that produces a magnified digit which is visible in bright daylight. Figure 6.7-5 illustrates a barrel lens magnifier in combination with a 45° cross hatch for use with unmagnified monolithic LED displays.

Louvered filters are available with either a "light matte" or "very light matte" anti-reflection surface which does not produce a fuzziness to the display appearance. Another option is a very hard scratch resistant surface which is incorporated directly into the plastic.

3M Company, Light Control Division, manufactures louvered filters for LED displays. Their product trade name is "Light Control Film", which is useable with all LED display and lamp products.

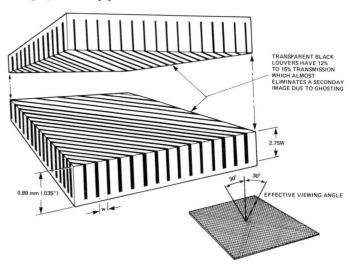

Figure 6.7-4 45° Cross Hatch Filter for Use with LED Displays in Bright Daylight Ambient.

6.8   Circular Polarizing Filters

Circular polarizing filters are effective when used with LED displays that have specular reflecting front surfaces. Specular reflecting surfaces reflect light without scattering. Displays that have polished glass or plastic facial surfaces belong to this category.

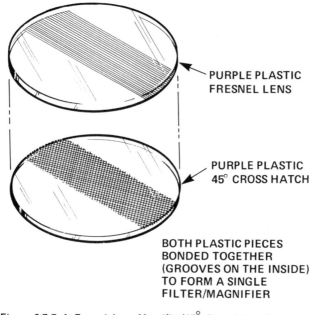

PURPLE PLASTIC
FRESNEL LENS

PURPLE PLASTIC
45° CROSS HATCH

BOTH PLASTIC PIECES
BONDED TOGETHER
(GROOVES ON THE INSIDE)
TO FORM A SINGLE
FILTER/MAGNIFIER

Figure 6.7-5 A Fresnel Lens Magnifier/45° Cross Hatch Purple Filter for Use with Unmagnified Standard Red Monolithic LED Displays.

Polaroid Corporation manufactures circular polarizing filters in the United States. In Europe, E. Käseman of West Germany produces high quality circular polarizers.

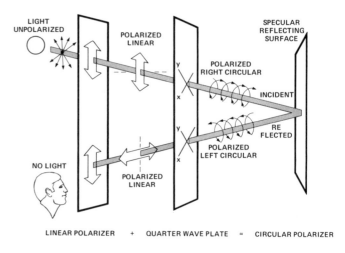

Figure 6.8-1 The Operation of a Circular Polarizer.

The operation of a circular polarizer may be described as follows. As shown in Figure 6.8-1, the filter consists of a laminate of linear polarizer and a quarter wave plate. The quarter wave plate has its optical axis parallel to the flat surface of the polarizer and is oriented at 45° to the linear polarization axis. Non-polarized light is first linearly polarized by the linear polarizer. The linearly polarized light has x and y components with respect to the quarter wave plate. As the light passes through the quarter wave plate, the x and y components emerge 90° out of phase with each other. The polarized light now has x and y forming a helical pattern with respect to the optical path, and is termed circular polarized light. As this circular polarized light is reflected by the specular reflecting surface, the circular polarization is reversed. When the light passes back through the quarter wave plate it becomes linearly polarized at 90° to the linear polarizer. Thus reflected ambient light is blocked. The advantage of a circular polarizer is that reflected ambient light is reduced by more than 95%. However, the trade-off is that display emitted light passing through the circular polarizer is reduced by approximately 65% at the peak wavelength. This then necessitates an increased drive current for the display, more than that required for a wavelength filter.

Circular polarizers are normally colored to obtain additional selected wavelength filtering. One Caution: outdoor applications will require the use of an ultraviolet, uv, filter in front of the circular polarizer. Prolonged exposure to ultraviolet light will destroy the filter's polarizing properties.

6.9 Anti-Reflection Filters, Mounting Bezels and Other Suggestions

Anti-reflection filters: A filtered display still may not be readable by an observer if glare is present on the filter surface. Glare can be reduced by the addition of an anti-reflection surface as part of the filter. Both sections of the display shown in Figure 6.9-1 are filtered. The left hand filter has an anti-reflection surface while the right hand filter does not.

An anti-reflection surface is a mat, or textured, finish or coating which diffuses incident light. The trade-off is that both incident ambient and display emitted light are diffused. It is therefore desirable to mount the filter as close to the display as possible to prevent the display image from appearing fuzzy.

Panelgraphic Chromafilters® come standard with an anti-reflection coating. SGL Homalite offers two grades of a molded anti-reflection surface. 3M Company and Polaroid also offer anti-reflection surface options. Optical coating companies will apply anti-reflection coating for specialized applications, though this is usually an expensive process.

Mounting bezels: It is wise to take into account the added appearance of a front panel that has the display set-off by a bezel. A bezel of black plastic, satin chrome or brushed aluminum, as examples, will accent the display and attract the eye of the observer. The best effect can be achieved by a custom bezel. Commercial black plastic bezels for digits up to 7.62mm (.3 inch) tall are available.

Other suggestions: When designing the mounting configuration of a display, consider recessing the display and filter 6.35mm (0.25 inch) to 12.7mm (0.5 inch) to add some shading effect. If a double sided printed circuit board is used, keep traces away from the normal viewing area or

Figure 6.9-1 The Effect of an Anti-Reflection Surface on an Optical Filter.

cover the top surface traces with a dark coating so they can not be seen. Mount the display panel in such a manner as to be easily removed if service should become necessary. If possible, mount current limiting resistors on a separate board to reduce the ambient temperature in the vicinity of the displays.

Filter Product	Type of LED Display	Ambient Lighting
Panelgraphic Chromafilter® with Anti-Reflection		
Ruby Red 60	Standard Red	Moderate
Dark Red 63		Bright
Purple 90		Bright
Scarlet Red 65	High-Efficiency Red	Moderate
Yellow 27	Yellow	Moderate to Bright
Green 48	Green	Moderate
SGL Homalite, Grade 100		
H100-1605	Standard Red	Moderate
H100-1804 (Purple)		
H100-1670	High-Efficiency Red	Moderate
H100-1726	Yellow	Dim
H100-1720		Moderate
H100-1440	Green	Dim
H100-1425		Moderate

Anti-Reflection
LR-72; 0.5 inch (12.70mm) Mounting Distance From Display
LR-92; Up to 3.0 inch (76.20mm) Mounting Distance From Display

Rohm & Haas		
Plexiglas 2423	Standard Red	Moderate
Oroglas 2444		
3M-Company — Louvered Filters		
Red 655	Standard Red	Bright
Violet		
Red 625	High-Efficiency Red	Bright
Amber 590	Yellow	Bright
Green 565	Green	Bright
Neutral Density	All	Moderate to Bright
Schott Optical Glass		
RG-645	Standard Red	Bright
RG-630		Moderate
RG-610	High-Efficiency Red	Moderate

TABLE 6.9-1 Specific Wavelength Filter Products

6.17

Manufacturer	Product

Panelgraphic Corporation
10 Henderson Drive
West Caldwell, New Jersey 07006
   Phone: (201) 227-1500
 	
Thorn/Panelgraphic
Great Cambridge Road
Enfield, Middlesex
ENGLAND
   Phone: 01-366-1291
	Chromafilter® — Wave length filters with anti-reflective coating; Red, Yellow, Green

SGL Homalite
11 Brookside Drive
Wilmington, Delaware 19804
   Phone: (302) 652-3686
 	
SGL Homalite
Comtronic GMBH
D8000 Munich 90
Theodolinden Str 4
GERMANY
   Phone: (089) 643 011
	Wavelength filters; two optional anti-reflective surfaces; three plastic grades; Red, Yellow, Green

3M - Company
Visual Products Division
3M Center, Bldg. 220-10W
St. Paul, Minnesota 55101
   Phone: (612) 733-0128
 	
3M Europe S.A.
53/54 Avenue Des Arts
1040 Bruxelles, Belgium
   Phone: 12-39-00
	3M — Brand Light control film; louvered filters

Rohm and Hass Independence Mall West Philadelphia, Penn. 19105 Phone: (215) 592-3000	Plexiglas; sheet and molding powder; wavelength filters
Roehm, GmbH Chemische Fabrik 6100 Darmstadt Kirschenallee WEST GERMANY Phone: (06161) 8061	Plexiglas; wavelength filters
Polaroid Corporation Polarizer Division 549 Technology Square Cambridge, Mass. 02139 Phone: (617) 864-6000	Circular polarizing
E. Käsemann GmbH D 8203 Oberaudorf WEST GERMANY Phone: (08033) 342	Circular Polarizing filters

Schott Optical Glass
Duryea, Pennsylvania 13642
   Phone: (717) 457-7485
 	
Physikalische Optik
Jenaer Glaswerk Schott Gen.
Hattenberg Stra. 10
Mainz, W. GERMANY
   Phone: (06131) 6061
	Glass filters

Norbex Division Griffith Plastics Corporation 1027 California Drive Burlingame, California 94010 Phone: (415) 344-7691	DIGIBEZEL®, Plastic bezels for LED displays
Industrial Electronic Engineers, Inc. 7720-40 Lemona Avenue Van Nuys, California 91405 Phone: (213) 787-0311	Plastic bezels for .30 inch (7,62mm) tall LED displays
Rochester Digital Displays, Inc. 120 North Main Street Fairport, New York 14450 Phone: (716) 223-6855	Complete mounting kits for H.P. 5082-7300, -7700 and -7600 displays.

TABLE 6.9-2 List of Filter and Bezel Product Manufacturers

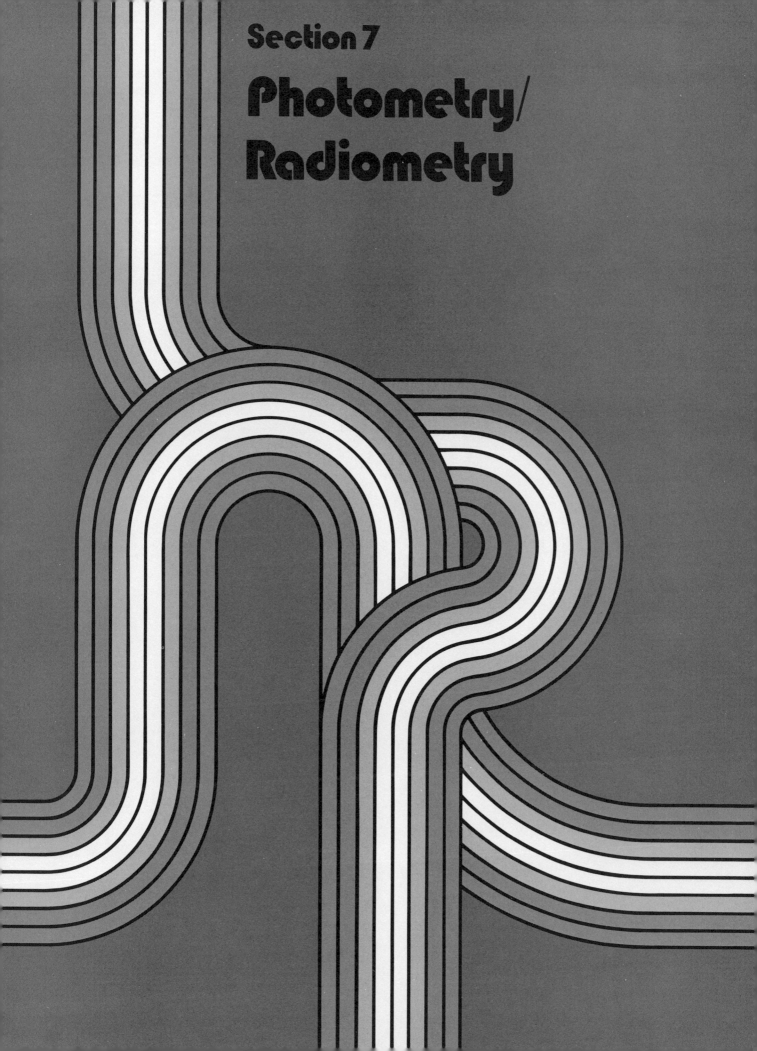

Section 7
Photometry/ Radiometry

7.0 PHOTOMETRY AND RADIOMETRY

7.1 Spectral Relationships

Photometry deals with flux (in lumens) at wavelengths that are visible, so the unit symbols have the subscript 'v' and the unit names have the prefix "luminous".

Radiometry deals with flux (in watts) at all wavelengths of radiant energy, so the unit synbols have the subscript 'e' and the unit names have the prefix "radiant".

Except for the difference in units of flux, radiometric and photometric units are identical in their geometrical concepts. Luminous flux is related to radiant flux by means of the "luminosity function", V_λ, also known as the "standard observer curve" or the "CIE curve". At the peak wavelength, 555 nm, of the luminosity function, the conversion factor is 680 lumens per watt. The luminosity function is given in Figure 7.1-1 on a log scale, so the accuracy is the same at all wavelengths. Values of V_λ taken from Figure 7.1-1 are accurate enough for all but the most exacting calculations.

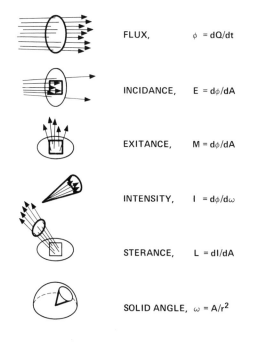

FLUX,	$\phi = dQ/dt$
INCIDANCE,	$E = d\phi/dA$
EXITANCE,	$M = d\phi/dA$
INTENSITY,	$I = d\phi/d\omega$
STERANCE,	$L = dI/dA$
SOLID ANGLE,	$\omega = A/r^2$

Figure 7.2-1 Generic Terms and Symbols and Their Geometrical Relationships.

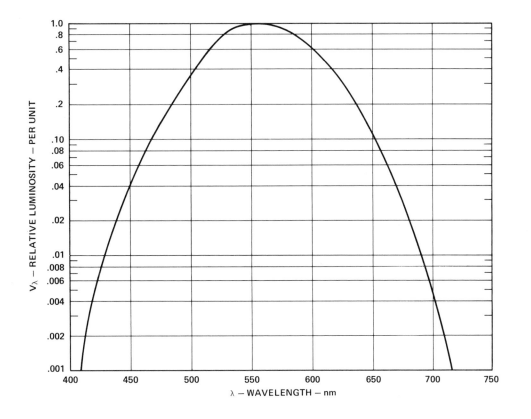

Figure 7.1-1 CIE Relative Luminosity Function. At Peak, 1 watt = 680 lumens.

For a spectrum of radiant flux, $(d\varphi_e/d\lambda)$, the total radiant flux is obtained by spectral integration:

$$\varphi_e(w) = \int_o^\infty (\frac{d\varphi_e}{d\lambda})\ d\lambda \qquad \text{(7.1-1)}$$

Luminous flux integration requires that each spectral element be weighted by the value of the luminosity function. It is therefore found as:

$$\varphi_v\ (\ell m) = 680\ (\frac{\ell m}{w}) \int_o^\infty (\frac{d\varphi_e}{d\lambda})\ V_\lambda\ d\lambda \qquad \text{(7.1-2)}$$

Conversion of photometric units to radiometric units is conveniently done by the luminous efficacy, η_v, which is just the ratio of the luminous flux to the radiant flux:

$$\eta_v\ (\frac{\ell m}{w}) = \frac{\varphi_v\ (\ell m)}{\varphi_e\ (w)} \qquad \text{(7.1-3)}$$

Luminous efficacy can be evaluated from any relative spectrum, i.e., one which is normalized, since the normalization factors cancel in the numerator and denominator of equation (7.1-3). Typical values of luminous efficacy for LEDs are:

	λ_{PEAK}(nm)	$\eta_v(\ell m/w)$
"STANDARD RED"	655	60
HIGH-EFFICIENCY RED	635	135
YELLOW	585	540
GREEN	565	640

7.2 Geometrical Relationships

Generically, there are only five units (and symbols) for radiant energy, as illustrated in Figure 7.2-1, and quantified in Table 7.2-1 and Table 7.2-2.

ϕ – FLUX, describes the rate at which energy is passing to, from, or through a surface or other geometrical entity.

E – INCIDANCE, describes the flux per unit area normally (perpendicularly) incident upon a surface.

M – EXITANCE, describes the flux per unit area leaving (diverging) from a source of finite area.

I – INTENSITY, describes the flux per unit solid angle radiating (diverging) from a source of finite area.

L – STERANCE, describes the intensity per unit area of a source.

ω – SOLID ANGLE; a solid angle, ω, with its apex at the center of a sphere of radius, r, subtends on the surface of that sphere an area, A, so that $\omega = A/r^2$ in steradians (sr).

Adding subscripts and prefixes quantifies these as radiometric or photometric units, e.g.:

I_e – RADIANT INTENSITY – watts per steradian, (W/sr)

I_v – LUMINOUS INTENSITY – lumens per steradian (lm/sr) or candelas (cd). cd = lm/sr

L_v – LUMINOUS STERANCE – candelas per square meter

As a practical matter, **flux** is used mainly in relating to the other terms. Few applications actually utilize all the flux available from a source. The same is true for **exitance**. The only situation in which "exitance" and "flux" have practical significance is with a receptor so tightly coupled to a source that virtually all the flux leaving the source enters the receptor, such as in "sandwich" type optoisolators.

Incidance has the same units as excitance but is of a vastly different nature. Exitance ignores the direction taken by the exiting flux. Incidance takes account of the flux component which is normal to a surface. If the flux direction at a surface is not normal, then the incidance is just the normal component of the angularly incident flux density. Incidance is most useful in describing photodetector properties.

Intensity is an extremely useful concept in both photometry and radiometry, and the candela unit of luminous intensity is the only universally utilized photometric unit. (Other photometric quantities are plagued by a profusion of English, metric, MKS, and CGS units of length.) Since flux passing through space is usually divergent, it is usually possible to define an equivalent point from which it diverges in terms of a solid angle, the flux therein, and hence the intensity of the equivalent point. As seen in Section 7.3, intensity is the most easily measured and most uniformly repeatable quantity. Therefore, although sterance is more fundamentally significant in many applications, the actual parameter to be measured for performance verification should be the intensity.

Sterance is most significant because of its constance in an optical (lens) system. This is discussed in much detail in the National Bureau of Standards publication NBS Technical Note 910-1 issued in March 1976. While magnification and image position varies, sterance does not.

Luminous sterance (or luminance) is the luminosity basis for visibility (i.e., distinguishing an object from its background). As such, luminous sterance is a useful measure for contrast and photographic exposure. However, luminous sterance is not the only basis for visibility -- color is another. For example, spots of brown gravy on a green tie are visible, even if the spots have the same luminous

GENERIC | RADIOMETRIC | PHOTOMETRIC

Term, Definition	Symbol	Defining Equation	New Term (Old Term)	Symbol	SI Units, Abbr.	New Term (Old Term)	Symbol	SI Units, Abbr.	Equivalent
FLUX, rate of flow of energy, Q, Q_e – radiant, Q_v – luminous, Q_q – photon [see note 1]	ϕ	$\dfrac{dQ}{dt}$	Radiant Flux (Radiant Power)	ϕ_e (P)	watts, W	Luminous Flux (Luminous Flux)	ϕ_v	lumens, lm	
INCIDANCE, flux per unit area on a reception surface	E	$\dfrac{d\phi}{dA}$	Radiant Incidance (Irradiance)	E_e (H)	watts per sq. meter, W/m^2	Luminous Incidance (Illuminance or Illumination)	E_v (E)	lux, lx [see note 3]	lumens per sq. meter, lm/m^2
EXITANCE, flux per unit area from an emitting surface	M	$\dfrac{d\phi}{dA}$	Radiant Exitance (Emittance)	M_e (W)	watts per sq. meter, W/m^2	Luminous Exitance	M_v	lumens per sq. meter, lm/m^2	[see note 2]
INTENSITY, flux per unit solid angle from a remote source	I	$\dfrac{d\phi}{d\omega}$	Radiant Intensity (Radiant Intensity)	I_e (J)	watts per steradian, W/sr	Luminous Intensity (Luminous Intensity)	I_v (I)	candelas, cd	lumens per steradian, lm/sr
STERANCE, flux per unit solid angle per unit area of emitting surface at angle θ with respect to surface normal.	L	$\dfrac{dI}{dA \cos\theta}$ $\dfrac{d^2\phi}{d\omega)dA \cos\theta)}$	Radiant Sterance (Radiance)	L_e (N)	watts per steradian per sq. meter, $W/sr/m^2$	(Luminous Sterance) (Luminance)	L_v (B)	candelas per sq. meter, cd/m^2 [see note 3]	lumens per steradian per sq. meter, $lm/sr/m^2$ nit, nt

Note 1. Quantametric terms use the prefix word "photon" and their symbols have a subscript, "q".

Note 2. Lux and other units of luminous incidance DO NOT APPLY to luminous exitance.

Note 3. Other units in Table 7.2-2.

TABLE 7.2-1 Terms, Definitions, Symbols, and Units for Energy Measurement

In general, confusion is averted by use of self-explanatory units, such as: lumens per square meter for luminous incidance; and, candelas per square meter for luminous sterance. At times, for brevity, where the risk of confusion is negligible, other units used are:

| Luminous Sterance | | Lambert L | Footlambert fL | Apostlib asb | *stilb sb | cd/ft² | **cd/m² | nit |
Unit, Abbr.	Equivalent							
Lambert, L	$1/\pi$ cd/cm² =	1	929	10,000	.3183	295.7	3183	3183
Footlambert, fL	$1/\pi$ cd/ft² =	.001076	1	10.76	.0003426	.3183	3.426	3.426
Apostilb, asb	$1/\pi$ cd/m² =	.0001	.0929	1	.00003183	.02957	.3183	.3183
*Stilb, sb	cd/cm² =	3.1416	2919	31,416	1	929	10,000	10,000
Candelas per sq. foot	cd/ft² =	.003382	3.1416	33.82	.001076	1	10.76	10.76
**Candelas/sq. meter	cd/m² =	.00031416	.2919	3.1416	.0001	.0929	1	1

| Luminous Incidance | | *Phot ph | *Footcandle fc | **,*Lux lx |
Unit, Abbr.	Equivalent			
*Phot, ph	lm/cm²	1	929	10,000
*Footcandle, fc	lm/ft²	.001076	1	10.76
**,*Lux, lx	lm/m²	.0001	.0929	1

*CIE Unit
**Recommended SI Unit
Many of the other units are in common usage, but efforts are being made to standardize on the SI units.

USE OF TABLE: In any row, the quantities are all equal; for example, in the table for luminous sterance, cd/m² = .00031416 L = .2919 fL = 3.1416 asb = .0001 sb = .0929 cd/ft² = 1 nt; in the table for luminous incidance, lx = .0001 ph = .0929 fc.

TABLE 7.2-2 Conversion Factors for Luminous Sterance and Luminous Incidance

DEVICE DESCRIPTION	APPARENT EMITTING AREA (mm^2)	STERANCE INTENSITY, L_v/I_v	
		cd/m^2 per mcd	fL per mcd
LEDs			
T-1 3/4 SIZE DIFFUSED	3.0	330	97
UNDIFFUSED	0.5	2,000	580
T-1 SIZE DIFFUSED	1.2	830	240
UNDIFFUSED	0.3	3,300	970
SUBMINIATURE	0.4	2,500	730
HERMETIC	2.0	500	150
RECTANGULAR	18.0	56	16
DOT MATRIX DISPLAYS	0.09	11,000	3,200
LARGE 7-SEGMENT DISPLAYS			
10.9 mm (.43 in.) TALL	4.4	230	66
7.6 mm (.3 in.) TALL	1.7	590	170
MONOLITHIC 7-SEGMENT			
LARGE CHIP (5082-74XX)	0.18	5,600	1,600
SMALL CHIP (5082-743X)	0.24	4,200	1,200
CALCULATOR DIGITS			
8, 9 CLUSTER (5082-7440 SERIES)	0.17	5,900	1,700
12, 14 CLUSTER (-7442/4/5/7)	0.15	6,700	1,900
8, 9 CLUSTER (5082-7240 SERIES)	0.18	5,600	1,600
5,15 CLUSTER (-7265/75/85/95)	0.27	3,700	1,100
CHIPS			
1.35 x 1.50 mm (-7811/21)	0.041	24,000	7,100
1.42 x 2.24 mm (-7832/42)	0.085	12,000	3,400
1.91 x 2.72 mm (-7851/61/52/62)	0.11	9,100	2,700
2.34 x 3.25 mm (-7871/81)	0.13	7,700	2,200
0.38 x 0.38 mm (-7890/93)	0.063	16,000	4,600

$$\frac{L_v\,(\text{cd/m}^2)}{I_v\,(\text{mcd})} = \frac{L_v\,(\text{cd/m}^2)}{I_v\,(\text{mcd})}\left(\frac{I_v\,(\text{mcd})}{10^3\,I_v\,(\text{cd})}\right) = \frac{10^{-3}}{A\,(\text{m}^2)} = \frac{10^3}{A\,(\text{mm}^2)} = \frac{10}{A\,(\text{cm}^2)}$$

$$\frac{L_v\,(\text{fL})}{I_v\,(\text{mcd})} = \frac{L_v\,(\text{fL})}{I_v\,(\text{mcd})}\left(\frac{\pi\,L_v\,(\text{cd/ft}^2)}{L_v\,(\text{fL})}\right)\left(\frac{I_v\,(\text{mcd})}{10^3\,I_v\,(\text{cd})}\right) = \frac{\pi}{10^3}\cdot\frac{1}{A\,(\text{ft}^2)}\left(\frac{92903\,A\,(\text{ft}^2)}{A\,(\text{mm}^2)}\right)$$

$$= \frac{\pi\cdot 92.9}{A\,(\text{mm}^2)} = \frac{291.9}{A\,(\text{mm}^2)} \qquad 1\,\text{cd/m}^2 = 0.2919\,\text{fL}$$

$$1\,\text{fL} = 3.426\,\text{cd/m}^2$$

TABLE 7.2-3 Intensity-to-Sterance Conversion ($\frac{L_v}{I_v} = \frac{1}{A}$)

sterance as the tie. On a brown tie of the same color as the gravy, a spot would be visible only if its luminous sterance is different from that of the tie by a factor of two or more.

While luminous sterance is the basis for visibility, luminous intensity is the popularly preferred performance parameter for LEDs and LED displays. The basic reasons are:

RELEVANCE — Most applications utilize the light radiated by all portions of the emitting area, but sterance is a measure of the intensity of only a small incremental portion. A large area device could have a higher intensity than a device of higher sterance but smaller area. Luminous intensity, being the product of sterance times area, ranks devices properly according to their effectiveness.

VIEWING ANGLE (RADIATION PATTERN) — Sterance of an LED remains nearly constant at all viewing angles but the projected light-emitting area does not. Luminous intensity, the product of sterance and area gives proper radiation pattern information.

EASE OF MEASUREMENT — Sterance measurement requires an optical system to image and field-stop the emitting surface; intensity measurement requires only a calibrated detector and a scale to find the LED-to-detector distance.

To understand the relationship between sterance and intensity, consider a uniform source whose area is one square meter and having an intensity of 1000 candelas. Its sterance is then 1000 cd/m^2. If this area were subdivided into 10^6 pieces, each having an area of one square millimeter, the intensity of each piece would be 10^{-3} candelas or one millicandela, but the sterance of each piece would still be 1000 cd/m^2.

Sterance can be an important consideration in product selection for a given application but is a poor parameter for specification. The figures in Table 7.2-3 give the sterance-to-intensity ratios for a number of product groups. They are simply the inverse of emitting area, with appropriate adjustment of units:

$$I_v \text{ (cd)} = L_v \text{ (cd/m}^2\text{)} \times A \text{ (m}^2\text{)} \qquad (7.2\text{-}1)$$

$$\frac{L_v}{I_v} \left(\frac{\text{cd/m}^2}{\text{mcd}} \right) = \frac{10^3}{A \text{ (mm}^2\text{)}} = \frac{10}{A \text{ (cm}^2\text{)}} \qquad (7.2\text{-}2)$$

7.3 Photometric and Radiometric Measurements

7.3.1 Spectral Effects

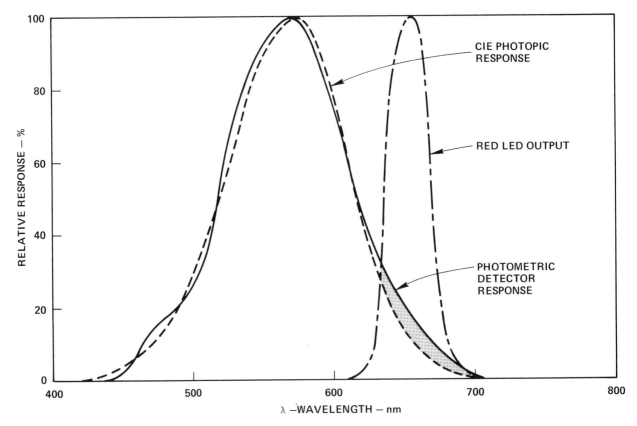

Figure 7.3.1-1   Spectral Response of a Photometer For Broad-Spectrum Radiation.

In radiometry, spectral effects are not much of a problem because radiometers can usually be made to have very nearly flat spectral response over the spectrum of interest. The flatest and broadest response is obtained from thermal-type radiometers (e.g. thermopile or thermocouple). Quantum-type (e.g. silicon diode) radiometers are much more sensitive than thermal-type, and have a higher speed of response; while the spectral flatness, is not as good, it is adequate for most LED measurements.

In photometric measurement of LEDs, it is important that the photometer be calibrated at the wavelength of the LED spectrum. This is necessary because most photometers do not have a spectral response that fits the luminosity function precisely at all wavelengths. To achieve the best overall fit, the response is adjusted (by colored filters) so that, over the spectrum, negative and positive deviations are balanced. This provides precise calibration for photometry of incandescent and other broad-spectrum sources, but for the narrow-spectrum LED sources, the entire spectrum might lie within a wavelength region where the photometer deviation is all positive or all negative, as seen in Figure 7.3.1-1.

In addition, for red LEDs, the spectral slope of the photometer must be very close to the $-.25$ dB per nm slope of the luminosity function. This is because LED spectra change with temperature and drive conditions, shifting as much as 10 nm. If a photometer has $-.20$ dB/nm (instead of the correct $-.25$ dB/nm), the incremental error would be 0.5 dB (+12%).

Suitable photometers are available from EG&G, Photo-Research Corp., Tektronix, and United Detector Technology.

7.3.2 Intensity Measurement

If the source to be measured has a plane of emission with respect to which intensity can be defined, then all that is necessary for intensity measurement is a photometer to read the incidance, E_v at a distance, d, from the source. The governing relationship is illustrated in Figure 7.3.2-1a. The flux, φ_v (lm) in the cone may be referred to the source as the product of the intensity, I_v (lm/sr) times the solid angle, ω (sr) of the cone, or it may be referred to the sensor as the product of the incidance, E_v (lm/m^2) times the area A (m^2). Since both relationships describe the same flux, they can be equated:

$$I_v \left(\frac{\ell m}{sr}\right) \times \omega \text{ (sr)} = E_v \left(\frac{\ell m}{m^2}\right) \times A \text{ (m}^2) \qquad (7.3.2\text{-}1)$$

If the distance, d, is large enough, then the value of ω can be expressed as:

$$\omega = A/d^2 \qquad (7.3.2\text{-}2)$$

Substitution in equation (7.3.2-1) gives the result:

$$I_v \left(\frac{\ell m}{sr}\right) = E_v \left(\frac{\ell m}{m^2}\right) \times d^2 \text{ (m}^2) \qquad (7.3.2\text{-}3)$$

The fact that the left side is in lumens per steradian while the right side is in lumens may seem incorrect, but note that the steradian is a dimensionless unit.

The relationship, $I_v = E_v d^2$, is valid only if d is large enough that the solid angle can be described as $\omega = A/d^2$. For the error in this assumption to be less than 1%, it is necessary that the distance, d, be at least ten times the diameter of the photodetector or ten times the diameter of the source, whichever is greater. This is called the TEN DIAMETERS RULE. Its purpose is to make the solid angle small enough that the flat photometer receptor can be regarded as a portion of a spherical surface.

To be absolutely correct, intensity should be measured with an incrementally small solid angle. The TEN DIAMETERS RULE defines a solid angle of 7.84 msr which is slightly larger than the 4.0 msr recommended by experienced observers.

When the distance, d, cannot be defined, intensity measurement can be made by placing the photometer at two different distances from the source, as in Figure 7.3.1-1 (b and c). With the "approximate" method, making θ small improves the accuracy, as far as equation (7.3.2-2) and (7.3.2-3) are concerned, but if θ is too small there may be other problems. One other problem is getting enough flux for an adequate signal-to-noise ratio. Another is the possible variation of the ratio φ/ω; undiffused LEDs, especially with very narrow radiation pattern, can cause the φ/ω ratio to vary substantially. The "precise" two-point method, reduces potential radiation pattern problems by keeping the solid angle nearly the same for both observations. Note in the expression for I_v by the "precise" method, that provision is made for $\varphi_2 \neq \varphi_1$; however, if optical bench positioning can be adjusted so that $\varphi_2 = \varphi_1$, the accuracy will be limited mainly by the accuracy obtained in measuring A_1 and A_2. These areas can be found as a ratio to the photometer area, A_0, as follows:

1. Place the photometer at large distance (>1m) from a small source (<10mm) of high intensity. (Wavelength and spectral effects are unimportant here.) Note the photometric reading as E_0.

2. Place A_1 over the photometer and note the reading as E_1. Then $A_1 = A_0(E_1/E_0)$.

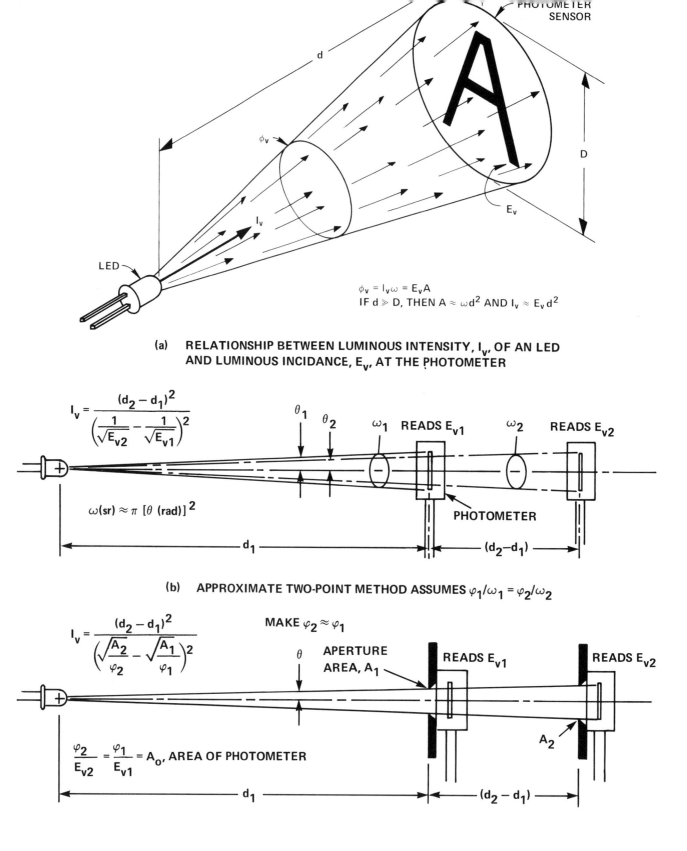

$$\phi_v = I_v\omega = E_v A$$
$$\text{IF } d \gg D, \text{ THEN } A \approx \omega d^2 \text{ AND } I_v \approx E_v d^2$$

(a)   RELATIONSHIP BETWEEN LUMINOUS INTENSITY, I_v, OF AN LED
AND LUMINOUS INCIDANCE, E_v, AT THE PHOTOMETER

$$I_v = \frac{(d_2 - d_1)^2}{\left(\frac{1}{\sqrt{E_{v2}}} - \frac{1}{\sqrt{E_{v1}}}\right)^2}$$

$$\omega(sr) \approx \pi \, [\theta \, (rad)]^2$$

(b)   APPROXIMATE TWO-POINT METHOD ASSUMES $\varphi_1/\omega_1 = \varphi_2/\omega_2$

$$I_v = \frac{(d_2 - d_1)^2}{\left(\sqrt{\frac{A_2}{\varphi_2}} - \sqrt{\frac{A_1}{\varphi_1}}\right)^2}$$

MAKE $\varphi_2 \approx \varphi_1$

$$\frac{\varphi_2}{E_{v2}} = \frac{\varphi_1}{E_{v1}} = A_o, \text{ AREA OF PHOTOMETER}$$

(c)   PRECISE TWO-POINT METHOD USES TWO APERTURES, A_1 AND A_2
SO THE SOLID ANGLE IS NEARLY THE SAME AT EACH POINT.
THIS REDUCES ERROR DUE TO RADIATION PATTERN PECULIARITIES.

Figure 7.3.2-1   Intensity Measurement — Optical Arrangements.

3.   Place A_2 over the photometer and note the reading as E_2. Thus $A_2 = A_0(E_2/E_0)$.

Several observations of A_1 and A_2 at different distances from the source will improve the accuracy.

7.3.3 Sterance Measurement

Sterance, being intensity per unit area, requires apparatus capable of resolving an incrementally small area. It is not actually necessary for this area to be precisely known because the measurement is usually done by comparing the unknown sterance to a sample of known sterance.

Figure 7.3.3-1 shows the arrangements that can be used for sterance measurement. Apparatus utilizing the beam splitter and aperture field stop is available commercially from Photo-Research Corp. and is called a brightness spot meter. Apparatus using the fiber-optic at the coaxial image plane is employed in the photometric microscope from Gamma Scientific Co. In both of these systems, a viewer can directly observe the precise portion of the LED for which the sterance is being measured. The area increment is the area of the fiber-optic core (or of the aperture field stop) divided by the square of the magnification of the objective lens. Spot diameters as small as .02 mm (.0008") can be resolved. The smaller the spot used, the larger will be the sterance variation observed across the surface of the source. This is the reason why sterance is a difficult specification to describe or use.

7.3.4 Flux Measurement

Total flux from a source is the integral of the flux radiated in all directions. If a source is placed at the center of a sphere, the total flux can be described as the integral of the incidance over the entire inside surface of the sphere. The completely general expression would require integration over both the azimuth angle (0 to 2π) and the polar angle (0 to π). However, most sources have radiation patterns that are very nearly a surface of revolution about the optical axis, so the integral is simplified as seen in Figure 7.3.4-1. That is, for a particular value of polar angle, θ, (also called the off-axis angle) the intensity is constant all the way around the polar axis. The differential flux can therefore be described as:

$$d\varphi = [I(\theta)] \times d\omega \qquad (7.3.4\text{-}1)$$

and the differential solid angle has the value:

$$d\omega = 2\pi \sin\theta \; d\theta \qquad (7.3.4\text{-}2)$$

Then the flux in a cone of half-angle θ is the integral:

$$\varphi(\theta) = \int_0^\theta I(\theta) \, 2\pi \sin\theta \; d\theta \qquad (7.3.4\text{-}3)$$

Total flux is that obtained for $\theta = \pi$:

$$\varphi = \int_0^\pi I(\theta) \, 2\pi \sin\theta \; d\theta \qquad (7.3.4\text{-}4)$$

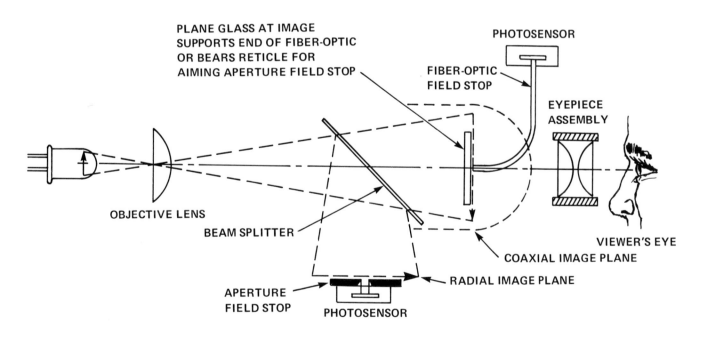

PLANE GLASS AT IMAGE SUPPORTS END OF FIBER-OPTIC OR BEARS RETICLE FOR AIMING APERTURE FIELD STOP

PHOTOSENSOR

FIBER-OPTIC FIELD STOP

EYEPIECE ASSEMBLY

OBJECTIVE LENS

BEAM SPLITTER

APERTURE FIELD STOP

PHOTOSENSOR

RADIAL IMAGE PLANE

COAXIAL IMAGE PLANE

VIEWER'S EYE

Figure 7.3.3-1   Sterance Measurement — Optical Arrangements.

AXIAL INTENSITY

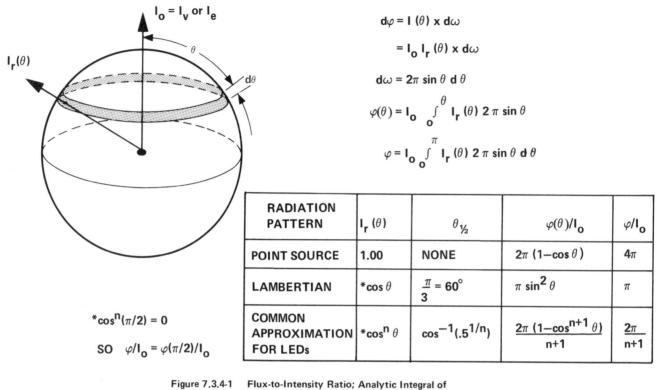

$$d\varphi = I(\theta) \times d\omega$$

$$= I_o\, I_r(\theta) \times d\omega$$

$$d\omega = 2\pi \sin\theta\, d\theta$$

$$\varphi(\theta) = I_o \int_o^\theta I_r(\theta)\, 2\pi \sin\theta$$

$$\varphi = I_o \int_o^\pi I_r(\theta)\, 2\pi \sin\theta\, d\theta$$

$*\cos^n(\pi/2) = 0$

SO $\varphi/I_o = \varphi(\pi/2)/I_o$

RADIATION PATTERN	$I_r(\theta)$	$\theta_{1/2}$	$\varphi(\theta)/I_o$	φ/I_o
POINT SOURCE	1.00	NONE	$2\pi(1-\cos\theta)$	4π
LAMBERTIAN	$*\cos\theta$	$\dfrac{\pi}{3} = 60°$	$\pi \sin^2\theta$	π
COMMON APPROXIMATION FOR LEDs	$*\cos^n\theta$	$\cos^{-1}(.5^{1/n})$	$\dfrac{2\pi(1-\cos^{n+1}\theta)}{n+1}$	$\dfrac{2\pi}{n+1}$

Figure 7.3.4-1   Flux-to-Intensity Ratio; Analytic Integral of Radiation Pattern.

*Exponent n is derived from the LED radiation pattern and usually is a valid approximation out to $\theta_{1/2}$, where $I(\theta) = .5\, I_o$. The value for n may be determined as follows: Assume $\cos^n(\theta_x) = y$, then $n = \mathrm{Log}(y)/\mathrm{Log}\cos\theta_x$.

$\varphi(\theta)$ and φ can be related to the axial intensity since the intensity at any angle, θ can be given as the product of the axial intensity, I_0, a constant, and the relative intensity, $I_r(\theta)$, for which $I_r(0) = 1.00$ as in customary radiation patterns. The axial intensity, of course, would be either I_v(cd) or I_e(w/sr).

$$\varphi(\theta) = I_o \int_o^\theta I_r(\theta)\, 2\pi \sin\theta\, d\theta \qquad (7.3.4\text{-}5)$$

Figure 7.3.4-1 gives the results of integrating radiation patterns with analytic functions. Also given is the value of $\theta_{1/2}$ which is the angle at which $I_r(\theta) = \frac{1}{2}$.

If the radiation pattern is not analytic but is a surface of revolution about the polar axis, then the integration is performed by converting equation 7.3.4-5 to a summation

$$\varphi(\theta) = I_o \sum I_r(\theta) \times [2\pi \sin\theta\, \Delta\theta] \qquad (7.3.4\text{-}6)$$

The term $[2\pi \sin\theta\, \Delta\theta]$ is called a zonal constant, C_Z, in which the incremental angle, $\Delta\theta$, is in radians. Of course, the magnitude of C_Z varies as $\sin\theta$, and depends on the size of the $\Delta\theta$ increment. A little experience will show that adequate accuracy (3%) is obtained by selecting $\Delta\theta$ equal to one fourth of the angle at which $I_r(\theta) = 0.5$ that is $\Delta\theta = (\theta_{1/2})/4$. Thus, if there are to be N equal parts over $180°$, choose $N = 180°(4/\theta_{1/2}) = 720°/\theta_{1/2}$ rounding off to the next larger integer that yields convenient values for $\delta = 180°/N$.

Figure 7.3.4-2 shows how δ and N are used in zonal integration. The customary method appears acceptable in Figure 7.3.4-2, but if δ had been $5°$, then it would have required using the values $2.5°$, $7.5°$, $12.5°$, etc. for $I_r(\theta)$ and $C_Z(\theta)$. An accuracy check, using the analytic $\cos^n\theta$ radiation pattern, shows that with the customary method $\varphi(\theta)$ is slightly above the analytic (correct) value, while the modified method yields $\varphi(\theta)$ slightly below the correct value. Both methods have adequate accuracy.

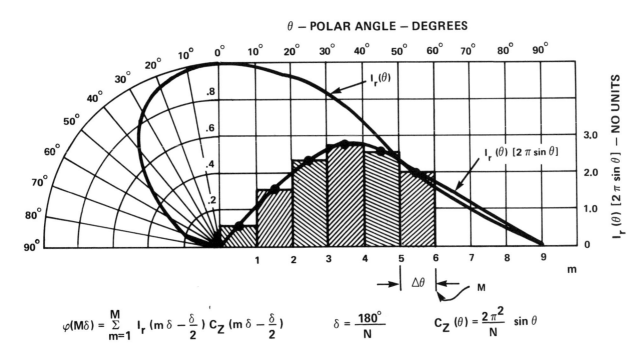

$$\varphi(M\delta) = \sum_{m=1}^{M} I_r \left(m\delta - \frac{\delta}{2}\right) C_Z \left(m\delta - \frac{\delta}{2}\right) \qquad \delta = \frac{180°}{N} \qquad C_Z(\theta) = \frac{2\pi^2}{N} \sin\theta$$

(a) CUSTOMARY METHOD — MAY REQUIRE AWKWARD VALUES FOR $(m\delta - \delta/2)$

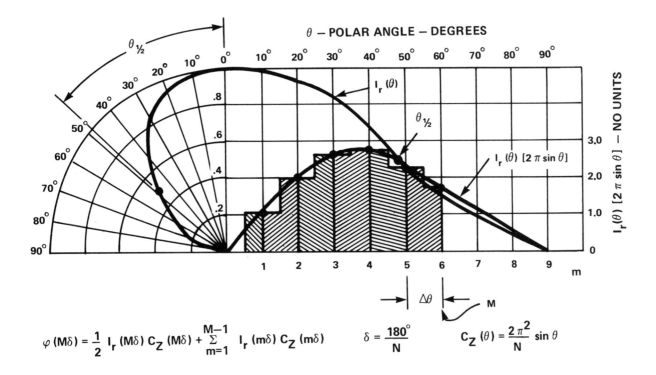

$$\varphi(M\delta) = \frac{1}{2} I_r (M\delta) C_Z (M\delta) + \sum_{m=1}^{M-1} I_r (m\delta) C_Z (m\delta) \qquad \delta = \frac{180°}{N} \qquad C_Z(\theta) = \frac{2\pi^2}{N} \sin\theta$$

(b) MODIFIED METHOD — CONVENIENT VALUES $(m\delta)$ FOR EVALUATING $I_r(m\delta)$ AND $C_Z(m\delta)$

Figure 7.3.4-2   Zonal-Constant Flux Integration; 3% Accuracy with $N \geqslant 720°/\theta_{1/2}$.

7.11

NOTES

Section 8

Reliability

8.0   RELIABILITY OF OPTOELECTRONIC DEVICES

LED optoelectronic devices are solid state semiconductor p-n junctions. They are fabricated and packaged utilizing many of the technologies which are used in the manufacture of silicon and germanium semiconductor products. It is reasonable to expect, therefore, that LED products will exhibit reliability aspects similar to those observed in these other semiconductor technologies. An LED offers many reliability advantages with respect to other available light sources. Long life-time, wide operating temperature range, and reliable operation in adverse environmental conditions are only a few of the important attributes of LEDs when compared to other light emitting technologies. These attributes can, however, only be realized when a device is properly designed, fabricated, packaged, and operated. Optoelectronic device data sheets define the limits within which devices should be operated in order to achieve the expected reliability performance. It is important that the optoelectronic system designer be aware of how variations in these limits will effect the long term performance of a device.

The reliability of an optoelectronic device can be considered to be dependent upon two separate variable factors. These are the reliability characteristics of the LED semiconductor chip and the reliability characteristics inherent to the package into which that chip is assembled. The variables are not necessarily independent because, for instance, package failure which is not in itself catastrophic can trigger eventual failure or degradation of the semiconductor device.

8.1   Reliability Aspects of the LED Semiconductor Chip

From a theoretical standpoint, an LED like any p-n junction should have a nearly infinite lifetime. This operating life would be limited only by the infinitely slow natural diffusion of the dopants. In practice, an LED device when operated under reasonable stress conditions is normally expected to exhibit an operating lifetime of about 100K hours. In some cases, the end of operating life is defined as the point at which a device light output has been reduced to 50% of the initial value. This could occur slowly over an extended period or catastrophically as a result of a short or an open. The value of 50% has been chosen as a ratio at which an observer would be expected to be capable of detecting a visible change in the light output. There are no known failure modes in LED semiconductors which should result in occurrence of catastrophic failures after extended periods of operation (such as notching in a tungsten filament incandescent lamp which will result eventually in a catastrophic failure of the filament).

Conversely, the slow reduction of light output with time is a well known (though not well understood) characteristic

of an LED device. This degradation of light output exhibits a direct correlation to the current density in the junction. Figures 8.1-1 and 8.1-2 depict the normalized light output vs. time for direct gap and indirect gap plastic package lamps when operated at various junction current densities. As can be seen, increased current density resulted in a more rapid degradation and greater total degradation for stress times up to about 5000 hours. The maximum operating levels for the devices of Figure 8.1-1 represent a stress level of 200% of maximum device rating and 350% of maximum device rating in Figure 8.1-2. Figure 8.1-3 represents long term operating life data for hermetic package lamps operated at the maximum allowable device stress level for stress times up to 40,000 hours. This data represents the potential for LED devices to be successfully used for periods in excess of 100k hours.

Close inspection of Figures 8.1-1 to 8.1-3 will reveal that the rate of light output degradation decreases as a function of increasing stress time. Figure 8.1-4 depicts the range of the expected rate of degradation (in % per 100 hours) vs. stress time. This plot represents a composite of several different device types from many different production lots when operated at 20 mA dc. It is important to note that with increasing stress time, even those device lots which start out with a very high rate of degradation converge quickly to a very low degradation rate. This implies that devices showing a high initial degradation will not necessarily continue to degrade to failure.

The degradation of LED light output demonstrates some dependency on junction temperature. As can be seen in Figures 8.1-1 and 8.1-3, however, this dependency is so small as to be inconclusive.

The data sheet maximum dc operating level for an LED is generally established somewhat below the level at which unacceptable degradation would be encountered within 10k hours of operation. To achieve the lowest possible LED degradation consistent with useful operating levels, it is best to select an operating point at 40% to 50% of the maximum data sheet rating.

8.2   Reliability Aspects of LED Packaging

The package into which a semiconductor device is assembled should provide environmental protection for the device as well as providing electrical contacts and a mounting technique. For an LED, the package is often also required to provide an optical system to enhance coupling of the emitted light to the detector. An ideally designed and utilized package would perform these functions very reliably. There are, however, many variables which can affect this performance. To evaluate a package, tests such as solderability, temperature cycling, thermal shock, moisture resistance, mechanical shock, acceleration and terminal

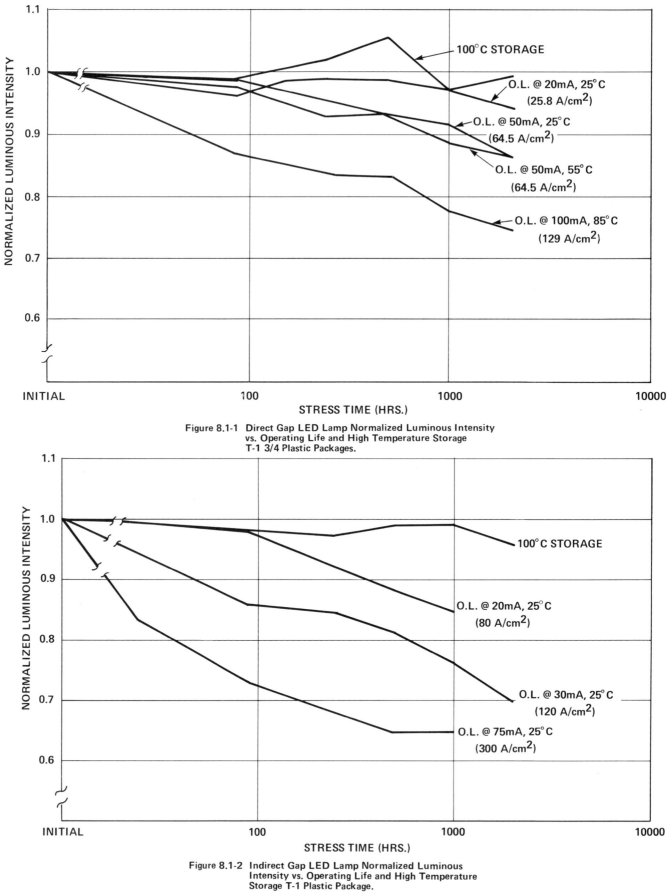

Figure 8.1-1 Direct Gap LED Lamp Normalized Luminous Intensity
vs. Operating Life and High Temperature Storage
T-1 3/4 Plastic Packages.

Figure 8.1-2 Indirect Gap LED Lamp Normalized Luminous
Intensity vs. Operating Life and High Temperature
Storage T-1 Plastic Package.

8.2

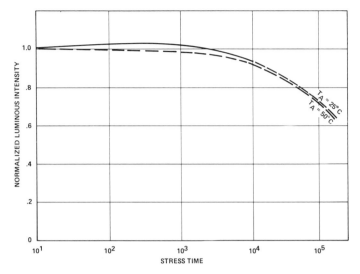

Figure 8.1-3 Direct Gap LED Lamp Normalized Luminous
Intensity vs. Operating Life TO-18 Hermetic Package.

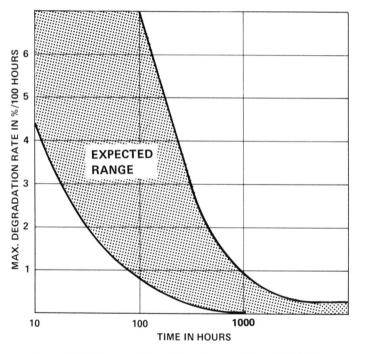

Figure 8.1-4 Expected Rate of Degradation vs. Time for T-1 3/4
Plastic Package.

strength are performed at various stress levels. These tests can be used to establish the worst case environmental stresses which should be applied to the package. Figure 8.2-1 is a copy of a typical plastic lamp reliability data sheet. Generally, such a data sheet will provide a reasonably good analysis of the performance a user can expect from a device. Deviation from this performance can, however, be caused by either variations in the manufacture of the package or by exceeding the defined environmental conditions in the end use product. For plastic packages, several environmental aspects are very important.

The package material is not impervious to the diffusion of water vapor, and long term exposure to high humidity will eventually subject the active elements to high humidity. The time required for moisture to saturate most plastic packages is in the order of 100 hours. High humidity can lead to failure from corrosion of the metallization or from increased surface leakage currents. The humdity associated with most normal applications does not harm plastic-packaged devices, but accelerated humidity testing should be considered a destructive test. Most units can be expected to withstand 10 days of non-operating environmental testing in high humidity and elevated temperature (e.g. 95% RH at 65°C).

Another package-related limitation of plastic products lies in the area of thermal fatigue life. Because the different materials (lead frame, plastic package, junction coating, die, bond wires, etc.) are in physical contact, and because coefficients of thermal expansion cannot be perfectly matched, temperature changes are accompanied by physical strain. Extended temperature cycling or thermal shock can lead to fatigue failure of the bond wires, or to die attach failures. It is not unusual to see 5% failures after 50 cycles temperature cycling -55°C to +125°C or thermal shock 0°C to +100°C.

Thermo-mechanical life could be a limiting factor for plastic products in markets where reliability is required through many temperature excursions, such as automotive applications. A sizeable effort of product improvement is currently underway to develop the materials that will realize the reliability objectives of these new markets.

There are theoretical limitations to the maximum temperature at which the device materials are stable and compatible. The gold wires bonded to aluminum metallization form intermetallic compounds at a rate which depends very strongly on temperature. At ordinary temperatures the reaction is so slow that it is not a significant contributor to failure rate. At +150°C bond life may be a few months. At +200°C, bond life may be a few days. Trace impurities and bonding conditions can significantly affect bond life. Stress from surroundings package materials can break a bond weakened by intermetallics.

The thermal expansion coefficients of epoxy package materials are relatively constant over normal service temperature, but at some high temperature the expansion coefficient increases significantly. This temperature is called the "glass transition temperature", or Tg. Tg of most of the presently utilized epoxy casting compounds is specified at +120°C, but it varies by several degrees, depending on batch history and assembly conditions. Also, the glass transition effect occurs over a range of several degrees, so that any exposure over a +115°C is potentially harmful. The epoxy and dye materials, being organic, are also subject to chemical degradation if overheated.

If proper care is not taken in assembly, plastic packages are potentially susceptible to the occurrence of thermal intermittents. The solid encapsulation around the bond wires keeps adjacent parts of a broken wire together, so that a defective bond, or a bond which failed for some reason, may not be a consistent failure. A unit with this phenomenon will often be good at room temperature and open when hot. A continuity test monitored over temperature ranging from room ambient to the maximum storage temperature is quite effective at detecting thermal intermittents. A high temperature function test is less effective, but still quite good.

Failure resulting from improper soldering operations have occurred frequently in LED applications. This is generally the result of soldering at too high a temperature and/or for too long a time period. As pointed out above, excess temperature above Tg, can cause substantial changes in the mechanical properties of the epoxies commonly available for LED products. Since most devices are relatively small, there is very little thermal capacitance available to distribute heat conducted into the package from the leads. Exposure of the epoxy to extreme excess temperature conditions can result in softening of the epoxy around the leads where mechanical stress could cause loosening of the die attach or lead bond posts or the formation of voids in the epoxy in the vicinity of the leads. Either condition may result in catastrophic failure of the device from broken wires or wire bond separation.

From a manufacturing standpoint, there has historically been a high degree of parametic variability in the epoxy materials used for LED encapsulation. Incoming inspection criteria such as gell time, hardness, clarty, and viscosity can be utilized by the LED manufacturer to improve product uniformity.

OED SOLID STATE VISIBLE LIGHT EMITTER

The following cumulative test results have been obtained from testing performed at OED Division, in accordance with the latest revisions of Military Semiconductor Specifications MIL-S-19500, MIL-STD-202 and MIL-STD-750. Because this device has a non-hermetic, cast epoxy enclosure intended for commercial and industrial markets, we do not recommend its use in applications requiring military high reliability performance.

TEST	MIL-STD-750 REFERENCE	TEST CONDITIONS	UNITS TESTED	FAILED
Physical Dimensions	2066	Device profile at 20X	50	0
Solderability	2026	SN 60, Pb 40, solder at 230°C	50	0
Temperature Cycling	1051.1	5 cycles from -65°C to +100°C, .5 hrs. at extremes, 5 min. transfer	50	0
Thermal Shock	1056.1	5 cycles from 0°C to +100°C, 3 sec. transfer	50	0
Moisture Resistance	1021.1	10 days, 90-98% RH, -10 to +65°C, non operating	50	0
Shock	2016.1	5 blows each X_1, Y_1, Y_2, 1500 G, 0.5 msec. pulse		0
Vibration Fatigue	2046	32 ±8 hours each X, Y, Z, 96 hr. total, 60 Hz, 20 G min.	50	0
Vibration Variable Frequency	2056	4, 4 minute cycles each X, Y, Z, at 20 G min., 100 to 200 Hz	50	0
Constant Acceleration	2006	1 minute each X_1, Y_1, Y_2, at 20,000 G.	50	0
Thermal Strength	2036.3	Condition A, 1 lb. for 10 sec.	50	0
Lead Fatigue	2036.3	Condition F, Method A, 8 oz., 10 sec.	50	0

Figure 8.2-1 Typical Mechanical Reliability Data Sheet for a
T-1 3/4 Plastic Lamp.

Section 9
Mechanical

9.0 MECHANICAL HANDLING CONSIDERATIONS FOR LED DEVICES

The first stage in designing a circuit utilizing an optoelectronic device is selecting the proper device for the application. The second step is to establish the electrical operating conditions and design the circuit. The third step is to install the optoelectronic device into the physical assembly, be it a printed circuit board, front panel mounting or some other mounting arrangement. The mounting considerations are primarily mechanical in nature, requiring attention to such items as the similarity of LED packages, the bending of leads, silver plated lead frames, soldering and post solder cleaning, socket mounting and heat sinking if required. Reliable operation of the LED device is more positively assured when all of these mechanical considerations have been given careful attention.

9.1 Similarity in LED Packages

Most plastic encapsulated LED devices are assembled using the lead frame technology. The exceptions are some stretched segment display devices which are assembled on substrates. Independent of the lead frame or package design, a lead frame device has the LED die attached directly to one lead and wire bonded across to another lead, as shown in Figure 9.1-1.

The primary thermal path to the LED is the cathode lead. Any mechanical and thermal stress applied to the leads is transmitted directly to the LED, die attach and wire bonds. The plastic encapsulant forms the device package and is the only supporting element for the lead frame. Therefore, the integrity of the encapsulation must be maintained to insure reliable operation of the LED device throughout its expected operating life.

Devices that do not use an encapsulating epoxy are most likely assembled on a ceramic substrate. Thick film metallization is usually applied to the face of the substrate. The LED is die attached to one metallization pad and wire bonded across to another pad, as shown in Figure 9.1-2. The circuit metallization is the primary thermal path from the LED to the leads, and the substrate acts as a secondary thermal path to the external ambient. The substrate isolates the LEDs and wire bonds from mechanically applied stresses; the amount of isolation dependent upon the type of substrate and package configuration. Devices of this construction are mechanically more rugged than a plastic encapsulated device and are more tolerant of temperature extremes.

One other type of device is assembled on a TO header, such as the high-reliability LED lamp which is packaged on a

TO-18 header. The devices will withstand considerable mechanical and temperature stress without any effect upon performance.

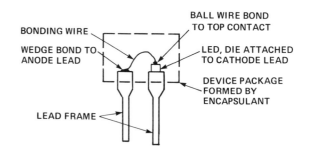
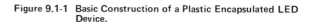

Figure 9.1-1 Basic Construction of a Plastic Encapsulated LED Device.

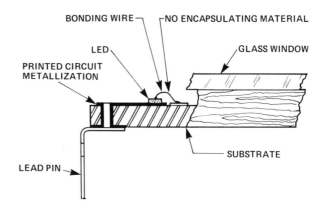

Figure 9.1-2 Basic Construction of an Unencapsulated, Substrate LED Device.

9.2 The Bending of Leads

In many LED lamp applications, it is necessary to bend the leads in order to mount the lamp at some angle other than 90° to the surface of a PC board. The leads of a lamp may be easily bent without introducing mechanical stresses inside the plastic package. The proper procedure for bending leads is illustrated in Figure 9.2-1. Bend the leads prior to soldering. Firmly grip the leads at the base of the lamp package with a pair of needle nose pliers. The pliers form a mechanical ground to absorb the stresses when bending. Bend the leads, one at a time, to the angle desired.

9.3 The Silver Plated Lead Frame

Since the price of gold has increased several times during the past few years, the cost of a gold plated lead frame has increased substantially, necessitating the search for an alternative. The impact of this cost increase has been industry wide. Many plating material alternatives were examined, and silver plating offered most of the desired properties of gold, while remaining price competitive when compared to other materials.

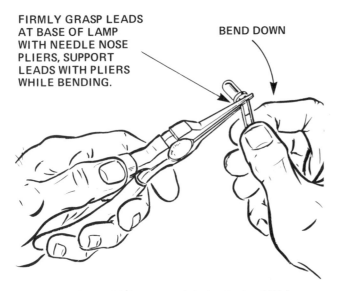

FIRMLY GRASP LEADS AT BASE OF LAMP WITH NEEDLE NOSE PLIERS, SUPPORT LEADS WITH PLIERS WHILE BENDING.

BEND DOWN

Figure 9.2-1 Correct Method to Bend the Leads of an LED Lamp.

By using silver plating, no additional manufacturing process steps are required. Silver has excellent electrical conductivity. LED die attach and wire bonding to a silver lead frame is accomplished with the same reliability as with a gold lead frame. Also, soldering to a silver lead frame provides a reliable electrical and mechanical solder joint. Soldering silver plated lead frame LED devices into a printed circuit board is no more complicated than soldering LED devices with gold plated lead frames.

9.3.1 The Silver Plating

The silver plating process is performed as follows: the lead frame base metal is cleaned and then plated with a copper strike, nominally 50 microinches (0.00127 mm) thick. Then a 150 microinch (0.00381 mm) thick plating of silver is added. A "brightener" is usually added to the silver plating bath to insure an optimum surface texture to the silver plating. The term "brightener" comes from the medium bright surface reflectance of the silver plate. Figure 9.3.1-1 illustrates the metallographic cross-section of the silver plating system as it would appear with a 1200X magnification.

150 μinch MINIMUM SILVER PLATING

50 to 80 μinch COPPER STRIKE

LEAD FRAME BASE METAL

Figure 9.3-1 Metallographic Cross-Section Through a Silver Plated Lead Frame, 1200X Magnification.

Since silver is porous with respect to oxygen, the copper strike acts as an oxygen barrier for the lead frame base metal. Thus, oxide compounds of the base metal are prevented from forming underneath the silver plating. Copper is miscible and readily diffuses into silver to form a solution that has a low eutectic point. This inter-diffusion between the copper strike and the silver overplate improves the solderability of the overall plating system. If basic soldering time and temperature limits are not exceeded, a lead frame base metal-copper-silver-solder metallurgical system will be obtained.

9.3.2 The Effect of Tarnish

Silver resists attack by most dry and moist atmospheres, such as carbon monoxide or high temperature steam. Halogen gases do attack silver, however, once the initial film layer is formed the process does not continue.

Silver reacts chemically with sulfur to form tarnish, silver sulfide (Ag_2S). The build-up of tarnish is the primary reason for poor solderability. However, the density of the tarnish and the kind of solder flux used actually determine the solderability. As the density of the tarnish increases, the more active the flux must be to penetrate and remove the tarnish layer.

9.3.3 Storage and Handling

The best technique for insuring good solderability of a silver plated lead frame device is to prevent the formation of tarnish. This is easily accomplished by preventing the leads from being exposed to sulfur and sulfur compounds. The two primary sources of sulfur are free air and most paper products, such as paper sacks and cardboard containers. The best defense against the formation of tarnish is to keep silver lead frame devices in protective packaging until just prior to the soldering operation. One way to accomplish this is to store the LED devices unwrapped in their original packaging. For example, Hewlett-Packard ships its seven segment display products in plastic tubes which are sealed air tight in polyethylene. It is best to leave the polyethylene intact during storage and open just prior to soldering.

Listed below are a few suggestions for storing silver lead frame devices.

1. Store the devices in the original wrapping unopened until just prior to soldering.

2. If only a portion of the devices from a single tube are to be used, tightly re-wrap the plastic tube containing the unused devices in the original or a new polyethylene sheet to keep out free air.

3. Loose devices may be stored in zip-lock or tightly sealed polyethylene bags.

4. For long term storage of parts, place one or two petroleum naphthalene mothballs inside the plastic package containing the devices. The evaporating naphthalene creates a vapor pressure inside the plastic package which keeps out free air.

5. Any silver lead frame device may be wrapped in "Silver Saver" paper for positive protection against the formation of tarnish. "Silver Saver" is manufactured by:

   The Orchard Corporation
   1154 Reco Avenue
   St. Louis, Missouri 63126
   314/822-3880

6. To reduce shelf storage time, it is worthwhile to use inventory control to insure that the devices first received will be the first devices to be used.

   One caution: The adhesives used on pressure sensitive tapes such as cellophane, electrical and masking tape can soak through silver protecting papers and may leave an adhesive film on the leads. This film reduces solderability and should be removed with Freon T-P35, Freon T-E35, or equivalent, prior to soldering.

   Petroleum naphthalene in the form of mothballs or chips may be obtained from the Frank Curran Company, 8101 South Lemont Road, Downers Grove, Illinois 60515, 312/969-2200.

9.4 Solders, Fluxes, and Surface Conditioners

Optoelectronic devices are usually soldered into a printed circuit board along with other components. It is necessary to achieve reliable soldering in order to obtain reliable circuit operation. A basic understanding of the types of solder, fluxing agents and surface conditioners that are typically used in the electronic industry will aid the process engineer in establishing a reliable soldering process.

9.4.1 Solders

The solder most widely used for wave soldering electronic components into printed circuit boards is Sn60 (60% tin and 40% lead) per Federal Standard QQ-S-571. Two alternatives are the eutectic composition Sn63 and the solder SN62 which contains 2% silver. Table 9.4.1-1 lists the composition, maximum acceptable contaminant levels and temperatures for these solders.

As the device leads pass through the solder wave of a flow solder process, the tin in the solder scavenges silver from the silver plating and forms one of two silver-tin intermetallics (Ag_6Sn or Ag_3Sn). This silver in the molten solder should not be considered a contaminant. As the silver content increases, the rate of scavenging decreases and the probability of obtaining the desired base metal-copper-silver-solder metallurgical system is improved. The result is that the silver content in solder, which reaches a maximum of 2½% in Sn60 at 230°C, aids in producing reliable solder joints on silver plated lead frames. Periodic replenishing of the solder as it becomes contaminated, with fresh solder, helps assure reliable soldering on a continual basis.

For hand solder operations, a high quality resin core wire solder is recommended. The solder may be obtained in wire diameters from .254mm (.010 inch) to 3.175mm (.125 inch) as common sizes. The core is typically an RMA flux in the amount of 2.2% or 3.3% by weight.

9.4.2 Fluxes

Solder flux classification per Federal Standard QQ-S-571, listed in order of increasing strength, are as follows:

Type R: Non-Activated Rosin Flux
Type RMA: Mildly Activated Rosin Flux
Type RA: Activated Rosin Flux
Type AC: Organic Acid Flux, Water Soluable

The Type R flux is a pure water white (WW) rosin without any additives. Rosin is a complex natural product obtained from the gum of live trees. It is mixed with a suitable solvent to form a homogeneous solution. Fluxes are graded as to the percentage by weight of rosin solids in the solution. The flux and its residue are non-corrosive and non-conductive.

The RMA flux is a homogeneous mixture of WW rosin in a blended alcohol vehicle into which a small amount of activating agent has been added. The flux and its residue are non-corrosive and non-conductive.

The RA flux is the same as RMA flux except that a greater amount of activating agents have been added. The flux and its residue are non-corrosive, and the residue is non-conductive only if all the solvent is volatilized and the residue remains dry.

An AC flux is considered full active and has a greater fluxing ability than a rosin flux. An AC flux may contain acids, organic or inorganic chlorides and therefore is corrosive. Due to their organic nature, the AC flux residues

Compo- sition	Tin	Lead	Silver	Anti- mony	Bis- muth max	Copper, max	Iron, max	Zinc, max	Alumi- num, max	Arse- nic, max	Cad- mium, max	Total of all others, max	Approximate melting range	
													Solidus	Liquidus
	%	%	%	%	%	%	%	%	%	%	%	%	°C	°C
Sn60	59.5 to 61.5	Remainder	----	0.20 to 0.50	0.25	0.08	0.02	0.005	0.005	0.03	----	0.08	183	191
SN62	61.5 to 62.5	Remainder	1.75 to 2.25	0.20 to 0.50	0.25	0.08	0.02	0.005	0.005	0.03	----	0.08	179	179
Sn63	62.5 to 63.5	Remainder	----	0.20 to 0.50	0.25	0.08	0.02	0.005	0.005	0.03	----	0.08	183	183

—————— MAXIMUM CONTAMINANT LEVELS ——————

TABLE 9.4.1-1 Composition, Maximum Contaminant Levels and Temperatures for Three Solders Commonly Used in Wave Soldering Operations.

decompose at soldering temperatures, thereby eliminating a large amount of the corrosive residue. All residues must be removed to prevent the formation of conductive ionic paths and corrosion caused by residual chloride salts.

Suggested applications of these flux types with respect to various tarnish levels are as follows:

Silver plated lead frames that are clean, contaminant and tarnish free may be soldered using a Type R flux such as Alpha 100.

Minor Tarnish: Since some minor tarnish or other contaminant may be present on the leads, a type RMA flux such as Alpha 611 or 611 Foam, Kester 197 or equivalent is recommended. Minor tarnish may be identified by reduced reflectance of the ordinarily medium bright surface of the silver plating. Type RMA fluxes which meet MIL-F-14256 are used in the construction of telephone communication, military and aerospace equipment.

Mild Tarnish: For a mild tarnish, a type RA flux such as Alpha 711-35, Alpha 809 foam, Kester 1544, Kester 1585 or equivalent should be used. A mild tarnish may be identified by a light yellow tint to the surface of the silver plating.

Moderate Tarnish: A type AC water soluable flux such as Alpha 830, Alpha 842, Kester 1429 or 1429 foam, Lonco 3355 or equivalent will give acceptable results on surface conditions up to a moderate tarnish. A moderate tarnish may be identified by a light yellow-tan color on the surface of the silver plating.

9.4.3 Surface Conditioners

If a more severe tarnish is present, such as a heavy tarnish identified by a dark tan to black color, a cleaner/surface conditioner must be used. Some possible cleaner/surface conditioners are Alpha 140, Alpha 174, Kester 5560 and Lonco TL-1. The immersion time for each cleaner/surface conditioner will be just a few seconds and each is used at room temperature. For example, Alpha 140 will remove severe tarnish almost upon contact; therefore, the immersion time need not exceed 2 seconds. These cleaner/surface conditioners are acidic formulations. Therefore, immediately, thoroughly wash all devices which have been cleaned with a cleaner/surface conditioner in cold water. A hot water wash will cause undue etching of the surface of the silver plating. A post rinse in deionized water is advisable.

CAUTION: These cleaner/surface conditioners may etch exposed glass surfaces and may have a detrimental effect upon the glass filled encapsulating epoxies used in optoelectronic devices. Complete immersion of an optoelectronic device into a surface conditioner solution is NOT recommended. For best results, immerse only the tarnished leads.

Three major suppliers of soldering supplies are:

Alpha Metals, Inc.
56 G Water Street
Jersey City, New Jersey 07304
302/434-6778

Kester Solder Company
4201 G Wrightwood Avenue
Chicago, Illinois 60639
312/235-1600

London Chemical Company (Lonco)
240 G Foster
Bensenville, Illinois 60106
312/287-9477

9.5 The Soldering Process

Before the actual soldering begins, the printed circuit boards and components to be soldered should be free of dirt, oil, grease, finger prints and other contaminants. Fluorinated cleaners such as Genesolve DI-15 or DE-15 may be used to preclean both the printed circuit boards and LED devices. Operators may wear cotton gloves to prevent finger prints when loading components into the printed circuit boards.

If the silver lead frames have acquired an unacceptable layer of tarnish, remove this tarnish layer with a cleaner/surface conditioner just prior to soldering. Since a cleaner/surface conditioner does slightly etch the surface of the silver plating, the silver leads are now more susceptible to tarnish formation. Therefore, use a cleaner/surface conditioner only on those silver lead frame devices which will be soldered within a four hour time period.

9.5.1 Wave Soldering

The temperature of a solder wave should be at least 38°C (100°F) above the melting temperature (solidus) of the solder. For Sn60 solder, this 221°C (430°F) minimum. Most wave solder operations maintain a solder wave temperature between 230°C (446°F) and 260°C (500°F). At 230°C, Sn60 solder dissolves silver at the rate of 60 microinches per second, and at 260°C it dissolves silver at the rate of 80 microinches per second. Therefore, with an initial silver plating thickness of 150 microinches, a dwell time of less than 2 seconds in a 230°C solder wave will provide the desired lead base metal-copper-silver-solder

Figure 9.5.1-1   The Desired Metallurgical System After a Solder Operation.

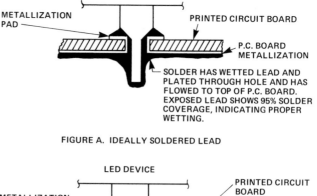

FIGURE A. IDEALLY SOLDERED LEAD

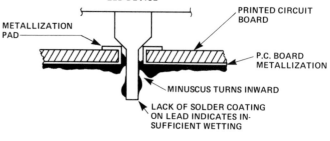

FIGURE B. UNDESIRABLY SOLDERED LEAD

Figure 9.5.4-1   Comparison Between an Ideally Soldered Lead and an Undesirably Soldered Lead.

metallurgical system, as illustrated in Figure 9.5.1-1.   The copper strike is intact. At 4 seconds dwell time in a 230°C solder wave, all of the silver has been dissolved and the solder must adhere to the copper strike; at 5 seconds, it is possible to have completely dissolved the copper and now the solder must adhere to the lead base metal. For plastic LED devices, the 230°C solder temperature is preferable to 260°C.

9.5.2   Hand Soldering

Contrary to popular belief, hand soldering can be more injurious to an LED device than wave soldering, the reason being that it is difficult to control dwell time and temperature. To effectively hand solder, it is best that the operator realize that "the hotter and longer the better" is not true. It is best that a 15 watt iron (25 watt maximum) be used. Maintain the iron in contact with the lead for only that time which is sufficient to allow the solder to flow evenly around the joint. Heat sinking the leads may be worthwhile, though this should not be necessary if sufficient care is observed. Always keep the tip of the iron clean and well tinned.

9.5.3   Cutting the Leads

The best rule to follow is to cut the leads of an LED device after soldering, not before. This accomplishes two things. First, the additional lead length provides heat sinking during the soldering operation. Secondly, the soldered joint provides a mechanical ground which prevents the mechanical stresses due to cutting from being transmitted into the device package. If it is necessary to cut the leads prior to soldering, support the device by the leads with a pair of long nose pliers during the cutting operation as is illustrated in Figure 9.2-1.

9.5.4   Printed Circuit Board

Printed circuit boards, either single sided, double sided or multi-layer, may be manufactured with plated through holes with a metal trace pad surrounding the hole on both sides of the printed circuit board. The plated through hole is desirable to provide a sufficient surface for the solder to wet, and thereby be pulled up by capillary attraction along

the lead through the hole to the top of the printed circuit board. This provides the best possible solder connection between the printed circuit board and the leads of the LED device. Figure 9.5.4-1A illustrates an ideally soldered lead. The amount of solder which has flowed to the top of the printed circuit board is not critical. A sound electrical and mechanical joint is formed. Figure 9.5.4-1B illustrates a soldered lead which is undesirable.

9.6   Post Solder Cleaning

Unlike most semiconductor devices, LED devices for visual applications are limited in the kinds of materials that may be used for encapsulation or hermetic sealing. In the case of encapsulating materials, the optical properties required limit the selection to only a few highly specialized epoxies. Exposure to a solvent used in a cleaning operation must not alter these optical properties in any way. For this and other reasons, only certain cleaning solvents may be used to post solder clean LED devices.

9.6.1   Type of Cleaners

It is important to remove both the rosin and ionic residues after soldering to insure reliable operation of the complete circuit. This is best accomplished by using an azeotrope of fluorocarbon and alcohol. The fluorocarbon is used to dissolve the residual rosin and the ionic contaminants are removed by the alcohol.

The type of fluorocarbon is very important. Tests have demonstrated that the only fluorocarbon that is compatible

CLEANING AGENT	COMPOSITION	CLEANING OPERATION
WATER		60°C (140°F) Wash to remove an AC flux residue.
ETHANOL, ISOPROPANOL	Alcohol	General cleaning after hand soldering.
FREON TF GENESOLV D ARKLONE P	100% Fluorocarbon (F113)	General cleaning agent for removing grease, oils, etc.
FREON TE ARKLONE A	≈4% Ethanol	Improved general cleaning agent.
GENESOLV DE-15 DI-15 BLACO-TRON DE-15 DI-15	≈15% Ethanol ≈15% Isopropanol ≈15% Ethanol ≈15% Isopropanol	Vapor cleaning at boiling — up to 2 minutes in the vapors — best for post solder cleaning.
ARKLONE K	≈25% Isopropanol	Vapor cleaning at boiling — up to 2 minutes in vapors. Best for post solder cleaning.
FREON T-E 35 T-P 35	≈35% Ethanol ≈35% Isopropanol	Room temperature post solder cleaning.

TABLE 9.6.1-1 Table of Suggested Post Solder Cleaning Agents that are Compatible with LED Devices.

with plastic LED devices is trichloro-tri-fluoroethane (F113), sold under tradenames as Freon, Genesolv D and Arklone. Some suggested cleaning products are listed in Table 9.6.1-1.

Cleaning solvent mixtures based on the fluorocarbon tetrachloro-di-fluoroethane (F112) are not recommended for cleaning plastic LED parts. Also, such cleaning agents from the keystone family (acetone, methyl ethyl ketone, etc.) and from the chlorinated hydrocarbon family (methylene chloride, trichloroethylene, carbon tetrachloride, etc.) are not recommended for cleaning LED parts. All of these various solvents attack or dissolve the encapsulating epoxies used to form the packages of plastic LED devices.

9.6.2. Bulk Cleaning Processes

Post cleaning of soldered assemblies when a type RMA or Type RA flux has been used may be accomplished via a vapor cleaning process in a degreasing tank, using an azeotrope of fluorocarbon and 15% to 25% alcohol as the cleaning agent. A recommended method is a 15 second suspension in vapors, a 15 to 30 second spray wash in liquid cleaner, and finally a one minute suspension in the vapors. When a water soluable Type AC flux such as Alpha 830 or Kester 1429/1429F is used, the following post cleaning process is suggested: thoroughly wash with water, neutralize using Alpha 2441 or Kester 5761 foaming, then thoroughly wash with water and air dry.

Four major suppliers of solvents are:

Allied Chemical Corp. (Genesolv)®
Speciality Chemicals Division
P.O. Box 1087R
Morristown, New Jersey 07960
201/455-5083

E.I. DuPonte de Nemours & Company
Freon Products Division
Wilmington, Delaware 19898
302/774-8341

Baron-Blakeslee (Blaco-Tron)®
1620 S. Laramie Avenue
Chicago, Illinois 60650
312/656-7300

Imperial Chemical Industries, Ltd. (Arklone)
Imperial Chemical House, Millbank
London SW1P3JF, England

9.6.3 Special Cleaning Instructions for Monolithic PC Board Displays

The monolithic printed circuit board display devices use a lens made from a special plastic chosen for its superior optical properties. This plastic does not lend itself to vapor cleaning processes. Also, the lens does not environmentally protect the LED chips. Therefore, the following special cleaning procedures are suggested.

For cleaning after a solder operation, the following process is recommended: wash display in clean liquid Freon T-P 35 or Freon T-E 35 solvent for a time period up to 2 minutes maximum. Air dry for a sufficient length of time to allow solvent to evaporate from beneath display lens. Maintain solvent temperature below 30°C (86°F). Methanol, isopropanol, or ethanol may be used for hand cleaning at room temperature. Water may be used for hand cleaning if it is not permitted to collect under display lens.

Solvent vapor cleaning at elevated temperatures is not recommended as such processes will damage display lens. Ketones, esters, aromatic and chlorinated hydrocarbon solvents will also damage display lens. Alcohol base active rosin flux mixtures should be prevented from coming in contact with display lens.

9.7   Socket Mounting

LED devices may be socket mounted in the same fashion as other semiconductor devices. The selection of a suitable socket is generally based on a quality vs. cost trade-off, with cost usually the primary factor.

When selecting a socket to mount silver lead frame devices, it is the performance of the socket that becomes the important factor. The socket contacts must have sufficient force and contact area to form an air tight mechanical seal at the lead/contact interface to prevent the formation of tarnish at this interface. Tarnish (silver sulfide) is a semiconductor, thus would introduce an undesirable resistance if allowed to form in the contact area. Figure 9.7-1 illustrates the desired condition. A good quality

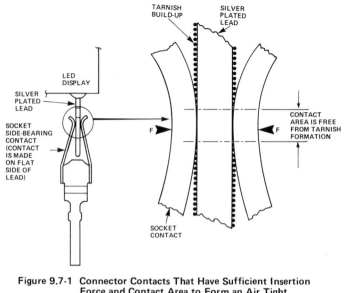

Figure 9.7-1   Connector Contacts That Have Sufficient Insertion Force and Contact Area to Form an Air Tight Mechanical Seal with the Silver Plated Lead to Prevent Tarnish Build Up in the Contact Area.

socket will require considerable insertion force and provide good wiping action upon insertion to scrape away any initial tarnish. The socket should be capable of securely holding the device without relative movement of the device lead with respect to the socket contact in the presence of mechanical shock or vibration. Sockets with side-bearing contacts that grip the flat sides of a lead are preferred for the mounting of DIP LED devices that have silver plated leads. Sockets that have edge-bearing contacts may be used to mount devices that have square silver plated leads, such as plastic LED lamps, providing the insertion force is sufficient to provide the required air tight mechanical seal along the edge of the lead.

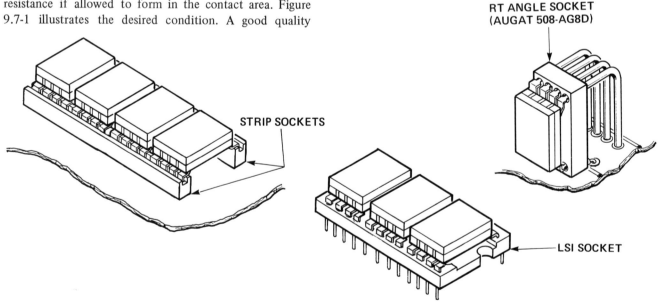

Figure 9.7.1-1   Optional Socket Mounting Configurations for the HP 5082-7300 Series of OBIC Displays.

MANUFACTURER	PART NUMBER	PINS	APPLICATIONS
AUGAT	508-AG8D	8	(1) HP 5082-7300; RT Angle Mtg
33 Perry Avenue	514-AG21D	14	90°
Attleboro, Mass. 02703	514-AG25D		45° (1) Stretched 7-Segment
617/222-2203	514-AG26D		60°
	314-AG5D-2R	14	(1) Stretched 7-Segment
	324-AG6D	24	(3)
	336-AG6D	36	(4) HP 5082-7300
	340-AG6D	40	(5)
	314-AG39D Low Profile	14	(1) Stretched 7-Segment
	324-AG39D Low Profile	24	(3) HP 5082-7300
	340-AG39D Low Profile	40	(5)
	325-AG1D	25	Strip Socket
BURNDY	HBRB2S-1	2	(1) LED Lamp; T-1 or T-1 3/4
Richards Avenue	DILB14P-108	14	(1) Stretched 7-Segment
Norwalk, Conn. 06856	DILB24P-108	24	(3) HP 5082-7300
203/838-4444	DILB40P-108	40	(5) HP 5082-7300
CAMBION	703-3777-04-12	14	(1) Stretched 7-Segment
445 Concord Avenue	703-5151-01-04-16		
Cambridge, Mass. 02138	703-5153-07-04-16	24	(3) HP 5082-7300
617/491-5400			
CIRCUIT ASSEMBLY	CA-14S-10SD	14	(1) Stretched 7-Segment
3169 Redhill Avenue	CA-24S-10SD	24	(3) HP 5082-7300
Costa Mesa, Calif. 92626	CA-40S-10SD	40	(5)
714/540-5490	CA-25STL-10SD	25	Strip Socket
	CA-S36 SP100	36	Stripline Plug (for pc board Monolithic displays)
ROBINSON-NUGENT	IC-143-S1	14	(1) Stretched 7-Segment
800 E. 8th Street	IC-163-S1	16	(2)
New Albany, Indiana 47150	IC-246-S1	24	(3) HP 5082-7300 Displays
812/945-0211	IC-406-S2	40	(5)
	ICN-143-S3 Low Profile	14	(1) Stretched 7-Segment
	ICN-406-S1	40	(1) HP 5082-7300 Display
	SB-25	25	Strip Socket
JERMYN	A1237	14	(1) Stretched 7-Segment
712 Montgomery Street	A1252AM Low Profile		
San Francisco, CA 94111	A23-2023 Low Profile	24	(3) HP 5082-7300
415/362-7431	A23-2030	40	(5)
AMP	583640 Series	12	(1)
449 Eisenhower Blvd.		24	(2) HP HDSP-2000
Harrisburg, Penn. 17105	Low Profile	36	(3)
717/564-0100	583773 Series	3 to 22	Strip Connectors

TABLE 9.7-1 A List of Sockets for Use With LED Displays

9.7.1 Special Socket Assemblies for LED Displays

LED display devices may be mounted in DIP sockets rather than being directly soldered into a PC board. Stretched seven segment displays may be mounted in standard 14 pin dip sockets. OBIC displays may require the use of strip sockets, LSI sockets or special sockets as illustrated in Figure 9.7.1-1. A small alphanumeric OBIC display may be mounted in standard DIP sockets that have been machined down in length to accept the 12 pin devices so as to form an end stacked display string.

Front panel mounting kits are available for LED displays. Rochester Digital Displays, Inc., 120 North Main Street, Fairport, New York 14450, 716/223-6855, offer two such kits. One kit accepts the HP 5082-7300 series display and the other accepts stretched seven segment displays.

PC board monolithic display products may be mounted using any one of several techniques. The most straight forward is the use of standard PC board edge connectors. A more cost effective approach can be implemented through the use of standard (or custom) stamped or etched metal mounting clips such as those available from Burndy, Richards Avenue, Norwalk, Connecticut 06850, 203/838-4444, (series LED-B) or J.A.V. Manufacturing, Inc., 20 Lucon Drive, Deer Park, New York 11729, (series 1255). Circuit Assembly Corporation, 3169 Red Hill Avenue, Costa Mesa, California 92626, 714/540-5490 and Burndy each manufacture low cost connectors especially designed for mounting the display board at a given angle as illustrated in Figure 9.7.1-2. A third approach would be to use a series of etched clips which are first soldered to the PC mother board. The display board is then pressed into place, with each clip being inserted into one of the plated through holes at the edge of display board.

Front panel mounting hardware for use with T-1 3/4 plastic LED lamps is manufactured by the Eldema Division of the Genisco Technology Corporation, 18435 Susana Road, Compton, California 90221, 213/537-4750. Two of their devices are illustrated in Figure 9.7.1-3.

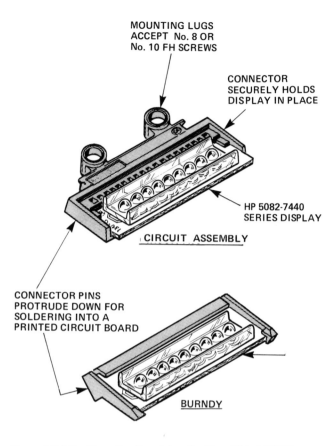

Figure 9.7.1-2   Mounting Connectors for Monolithic LED PC Board Display Devices.

9.8   Heat Sinking

Most LED devices may be operated in elevated ambient temperatures without heat sinking by utilizing input power derating. Devices which usually fall into this category are those devices that do not contain on-board integrated circuits. Heat sinking may be required for those on-board integrated circuit (OBIC) devices that have considerable power dissipation occuring within the integrated circuit.

The specific devices which may require heat sinking are the OBIC LED numeric and alphanumeric displays. An OBIC alphanumeric display, for example, combines a significant amount of integrated circuit logic and LED display capability in a very small package. As such, on board power dissipation is relatively high and the thermal design of the mounting assembly becomes an important consideration. The display is designed to operate over a wide temperature range with full power dissipation. This capability may be utilized only if the mounting assembly has a thermal resistance to ambient of $35°C/W$ per device, or less.

The primary thermal path for power dissipation is through the device leads. The thermal resistance junction-to-lead is typically 15 to $25°C/W$, depending upon the package configuration. For a plastic device, the maximum LED junction temperature is $110°C$. A hermetic device will tolerate a much higher IC junction temperature, but the maximum LED junction temperature should not exceed $125°C$. Reliable operation is obtained when the junction temperatures are maintained below these maximum values.

The thermal resistance to ambient of $35°C/W$ for the display mounting structure assumes that the mounting surface for the display is an isothermal plane. If only one OBIC display is operated on this isothermal plane at a power level of 1.7 watts maximum, as an example, the temperature rise above ambient would be $42.5°C$:

$$T_{RISE} = (35°C/W)(1.7W) = 42.5°C$$

If a second display is placed on this same thermal plane, with no increase in thermal dissipation capability, the temperature rise would double to $85°C$, reaching catastrophic levels very quickly. Hence, for each OBIC device added to the display string, there must be an appropriate increment of heat dissipating capability added to the mounting plane. For short display strings (4 to 6 OBIC devices), the $35°C/W$ per device may be achieved by utilzing a printed circuit board design which maximizes the amount of metal surface area remaining on the board. For longer display strings, or if sufficient metal surface area cannot be achieved, an external heat sink must be employed to obtain the $35°C/W$ per device case-to-ambient thermal resistance.

In practice, heat sink design involves the optimization of techniques used to dissipate heat through the device leads. The heat transfer is from the device leads to the PC board metallization to the heat sink and is dissipated in the surrounding ambient. A heat sink of approximately 52 square centimeters (8 square inches) per OBIC device will

Q086 LED SOCKET ASSEMBLY

Q084 LED MOUNTING ASSEMBLY

Figure 9.7.1-3   Front Panel Mounting Hardware for T-1 3/4 LED Lamps.

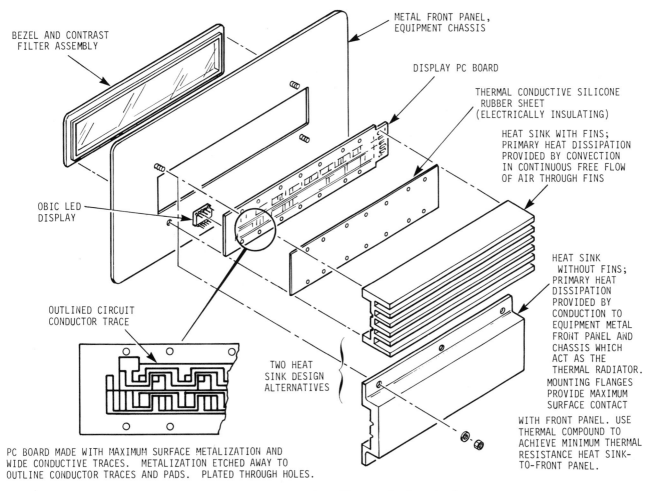

Figure 9.8.1-1   Heat Sink and Panel Mounting Configuration for an OBIC Alphanumeric LED Display.

typically permit operation at a power dissipation of 1 watt/device in an ambient temperature of 70°C.

As an alternative to soldering, OBIC displays may be mounted in DIP or stripline sockets which are themselves soldered into a PC board with maximum surface area metallization. These sockets will allow enough space between the PC board and the OBIC displays to permit a heat sink bar to be inserted to conduct heat to an external sink. These sockets add a thermal resistance of about 2°C/W between the device leads and the PC board.

9.8.1   OBIC Display Assembly with On-Board Heat Sink

Heat sinking an OBIC display is most effectively accomplished by constructing a direct thermal path between the device leads and the heat sink. A direct thermal path may be achieved by implementing the design concept illustrated in Figure 9.8.1-1. The printed circuit board is designed with a maximum amount of metallization remaining on the board. Thin lines are etched in the

metallization to outline the circuit conductors. The PC board metallization is used as part of the thermal path between the device leads and the heat sink.

The heat sink is designed to fit flush against the PC board, with grooves cut into it to provide clearance for the device leads. A sheet of thin thermally conductive silicone rubber is inserted between the heat sink and the printed circuit board. This silicone rubber completes the thermal path and at the same time, electrically insulates the heat sink from the PC board metallization.

Heat transfer from the heat sink to the ambient air may be accomplished in one of two ways. One method is to mechanically fasten the heat sink to the metal chassis or front panel. This extends the display heat sink by utilizing the complete chassis as a thermal radiator. If this is not possible, then fins may be made as part of the display heat sink to increase its surface area. Heat is now transferred to the local ambient air and carried off by convection. This process is assisted by insuring a continuous free flow of air passing through the fins of the heat sink.

Thermal conductive silicone rubber is sold under the trade name "Sil-Pad", by the Bergquist Company, Inc., 4350 West 78th Street, Minneapolis, Minnesota 55435. The sheet is a nominal 0.305mm (.012 inch) thick, has a thermal resistance of approximately .08°C/W per square centimeter and a Shore A hardness of 76. Anticipated useful life at a temperature of 140°C is 300 thousand hours, in a non condensing atmosphere.

The heat sink may be made from a length of aluminum extrusion or purchased from any one of a number of heat sink manufacturers. In either case, the heat sink will be a custom part and should be designed so as to (1) have maximum surface area contact with the PC board, (2) have as much surface contact area with the supporting chassis or (3) have the maximum possible surface area if the local ambient air must carry off the heat by convection.

9.8.2 A Display Assembly with Heat Pipe

An emerging technology for the efficiency transfer of heat from one point to another is the heat pipe. When it is not possible to incorporate an on-board heat sink, these devices are a cost effective means of efficiently transferring the heat from the display PC board to a remote heat sink or to the equipment chassis. The heat pipe by itself does not function as the heat sink. The input to output thermal resistance of a 305mm (12 inch) long heat pipe is typically .10°C/W when transporting 10 watts. The contact thermal resistance at either end of the heat pipe can be maintained to about .062°C/W per square centimeter (.4°C/W per square inch).

9.8.3 List of Manufacturers

The following list of manufacturers are only suggestions, as those not listed may well be equally qualified:

Heat Sinks:

Aham
968 W. Foothill Blvd.
Azusa, California 91702

Aham
2 Gill Street, Bldg. 5
Woburn, Massachusetts 01810

International Electronic Reserach Corp. (IERC)
135 W. Magnolia Blvd.
Burbank, California 91502

Industrial Heat Sink Corporation
5338 Alhambra Avenue
Los Angeles, California 90032

Heat Pipes:

Noren Products, Inc.
846 Blandford Blvd.
Redwood City, California 94062

Tecknit
129 Dermody Street
Cranford, New Jersey 07016

Jermyn
712 Montgomery Street
San Francisco, California 94111

NOTES